NEUROMETHODS

Series Editor
Wolfgang Walz
University of Saskatchewan
Saskatoon, SK, Canada

For further volumes:
http://www.springer.com/series/7657

Animal Models
of Neurodevelopmental Disorders

Edited by

Jerome Y. Yager

Division of Pediatric Neurology, Department of Pediatrics,
University of Alberta, Edmonton, AB, Canada

Editor
Jerome Y. Yager
Division of Pediatric Neurology
Department of Pediatrics
University of Alberta
Edmonton, AB, Canada

ISSN 0893-2336　　　　　　　　ISSN 1940-6045　(electronic)
Neuromethods
ISBN 978-1-4939-4065-3　　　　ISBN 978-1-4939-2709-8　(eBook)
DOI 10.1007/978-1-4939-2709-8

Series Preface

Experimental life sciences have two basic foundations: concepts and tools. The *Neuromethods* series focuses on the tools and techniques unique to the investigation of the nervous system and excitable cells. It will not, however, shortchange the concept side of things as care has been taken to integrate these tools within the context of the concepts and questions under investigation. In this way, the series is unique in that it not only collects protocols but also includes theoretical background information and critiques which led to the methods and their development. Thus it gives the reader a better understanding of the origin of the techniques and their potential future development. The *Neuromethods* publishing program strikes a balance between recent and exciting developments like those concerning new animal models of disease, imaging, in vivo methods, and more established techniques, including, for example, immunocytochemistry and electrophysiological technologies. New trainees in neurosciences still need a sound footing in these older methods in order to apply a critical approach to their results.

Under the guidance of its founders, Alan Boulton and Glen Baker, the *Neuromethods* series has been a success since its first volume published through Humana Press in 1985. The series continues to flourish through many changes over the years. It is now published under the umbrella of Springer Protocols. While methods involving brain research have changed a lot since the series started, the publishing environment and technology have changed even more radically. Neuromethods has the distinct layout and style of the Springer Protocols program, designed specifically for readability and ease of reference in a laboratory setting.

The careful application of methods is potentially the most important step in the process of scientific inquiry. In the past, new methodologies led the way in developing new disciplines in the biological and medical sciences. For example, Physiology emerged out of Anatomy in the nineteenth century by harnessing new methods based on the newly discovered phenomenon of electricity. Nowadays, the relationships between disciplines and methods are more complex. Methods are now widely shared between disciplines and research areas. New developments in electronic publishing make it possible for scientists that encounter new methods to quickly find sources of information electronically. The design of individual volumes and chapters in this series takes this new access technology into account. Springer Protocols makes it possible to download single protocols separately. In addition, Springer makes its print-on-demand technology available globally. A print copy can therefore be acquired quickly and for a competitive price anywhere in the world.

Wolfgang Walz

Preface

The definition of the Developmental Disabilities has evolved over the last number of years, and thankfully as a result, encompasses a broad range of individuals with both cognitive and motor impairment. One recent definition put forward is that of a "disability manifested before a person reaches maturity and is attributable to mental retardation or related conditions and includes cerebral palsy, epilepsy, autism, or other neurological condition, when such conditions result in the impairment of general intellectual functioning or adaptive behavior." Of course, many jurisdictions and governmental ministries have definitions which vary and are at times exclusive to some disabilities, which the health and scientific field would include. Nonetheless, as it relates to the research community, we recognize this terminology as encompassing a very wide range of diseases that span a broad spectrum of brain maturation and include numerous etiologies, at times resulting in similar and overlapping phenotypes.

Because of this, identifying animal models which reflect and are inclusive to this host of diseases is a daunting challenge. As a goal to success, at least as this area of investigation relates to health, the animal model must be able to provide a reflection of the human condition, both short and long term, the ability to investigate mechanisms of injury and outcome, biomarkers of disease, and a portal by which therapeutic interventions for the long-term benefit of the individual may be tested. Herein lies a particular challenge, as many of the disabilities discussed in this text relate to injuries evolving during pregnancy, have both genetic and environmental components influencing their outcome, and in the human population, often require years of follow-up to determine effectiveness. The lack of translatability of many animal models, from a therapeutic perspective, has significantly challenged researchers to seek novel in vivo models of disease that better reflect the human condition and will allow for outcomes in therapy that will transfer to the human.

As diseases become better understood, the healthcare community and the applied basic sciences have identified the need to become person- and certainly disease-specific. Undoubtedly in the "adult" world, there have been numerous potential therapies in animal research that have not been successfully translated to the human, as determined by clinical trials. One of the explanations, and probably just one, is that our animal models seek to identify specific variables, which cannot then be broadly applied. Hence, both the clinical and applied animal research communities are becoming more specific in their models, and so they should be for their application to the human phenotype. One of the more successful translations has been the use of hypothermia as a rescue therapy for perinatal asphyxia. Hypothermia has been successful in several models of hypoxic-ischemic brain injury, across several species, and in clinical trials in humans. This is partially due to the fact that it was recognized as a specific therapy for specific circumstances, that being term and near-term acute neonatal encephalopathy caused by hypoxia-ischemia.

While it was not the intent of this manual to be all inclusive, we have accumulated a host of authors well recognized in their fields of expertise, with all contributors involved in the investigation of the causes, outcomes, treatment, and prevention of the developmental disabilities. We have further attempted to provide a spectrum of models that is reflective of the various species that can be utilized in experimentation on these disorders across a broad range of the disabilities. This text is meant to introduce and entice those interested in better understanding and treating the developmental disorders, to animal models that best reflect that spectrum of disability in the human. Certainly, as a clinician-scientist who has worked in the field for the last two decades, the area of investigation has grown substantially. But what is just as clear is that no one animal model will suffice to provide us with all of the answers. And as we move increasingly to patient-specific forms of therapy, the need for animal models that encompass the specifics of the underlying mechanisms, reflect the phenotypic expression, and take into account environmental and genetic influences, both ante nataly and post nataly, will be required.

My deepest thanks go to Dr. Robert C. Vannucci, who introduced me to the world of perinatal brain injury and the developmental disabilities, to the role of animal models in the science of investigation, and for allowing me the opportunity to learn from one of the pioneers of modern perinatal neurology. I also wish to thank each of the authors in this text for their patience and their tremendous contributions to the field of brain injury in the newborn, its understanding, and our ability to move forward in diminishing the burden of the developmental disabilities. And, of course to the patients I see every day with cerebral palsy and other developmental disabilities, from whom I learn far more than I could ever contribute.

Edmonton, AB, Canada   *Jerome Y. Yager*

Contents

Contributors

MARIE-JULIE ALLARD • *Pediatric Neurology Laboratory, Départements de Pédiatrie, Université de Sherbrooke, Sherbrooke, QC, Canada*

ALAURA ANDROSCHUK • *Division of Pediatric Neuroscience, Department of Pediatrics, University of Alberta, Edmonton, AB, Canada*

KRISTIAN AQUILINA • *School of Clinical Sciences, University of Bristol, Bristol, UK; Department of Pediatric Neurosurgery, Great Ormond Street Hospital, London, UK*

EDWARD A. ARMSTRONG • *Department of Pediatrics, University of Alberta, Edmonton, AB, Canada*

STEPHEN A. BACK • *Neuroscience Section, Pape' Family Pediatric Research Institute, Oregon Health and Science University, Portland, OR, USA*

JESSICA BAKER • *Department of Pediatrics, University of Chicago, Chicago, IL, USA*

PRAVEEN BALLABH • *Department of Pediatrics, Anatomy & Cell Biology, New York Medical College, Maria Fareri Children's Hospital, Westchester Medical Center, Valhalla, NY, USA; Regional Neonatal Center, Maria Fareri Children's Hospital, Westchester Medical Center, Valhalla, NY, USA*

LAURA BENNET • *Department of Physiology, University of Auckland, Auckland, New Zealand*

JULIE BERGERON • *Pediatric Neurology Laboratory, Départements de Pédiatrie, Université de Sherbrooke, Sherbrooke, QC, Canada*

DAVID L. BIGAM • *Department of Surgery, University of Alberta, Edmonton, AB, Canada*

MARC DEL BIGIO • *Department of Pathology, College of Medicine, Faculty of Health Sciences, University of Manitoba, Winnipeg, MB, Canada*

TAMARA BODNAR • *Department of Cellular and Physiological Sciences, University of British Columbia, Vancouver, BC, Canada*

FRANCOIS V. BOLDUC • *Division of Pediatric Neuroscience, Department of Pediatrics, University of Alberta, Edmonton, AB, Canada; Neuroscience and Mental Health Institute, University of Alberta, Edmonton, AB, Canada*

LINDSEA C. BOOTH • *Department of Physiology, University of Auckland, Auckland, New Zealand*

MARIE-ELSA BROCHU • *Pediatric Neurology Laboratory, Départements de Pédiatrie, Université de Sherbrooke, Sherbrooke, QC, Canada*

ELAVAZHAGAN CHAKKARAPANI • *School of Clinical Sciences, St Michael's Hospital, Level D, University of Bristol, Bristol, UK*

PO-YIN CHEUNG • *Department of Surgery, University of Alberta, Edmonton, AB, Canada; Department of Pediatrics, University of Alberta, Edmonton, AB, Canada; Department of Pharmacology, University of Alberta, Edmonton, AB, Canada*

MATHILDE CHEVIN • *Pediatric Neurology Laboratory, Départements de Pédiatrie, Université de Sherbrooke, Sherbrooke, QC, Canada*

WENDY COMEAU • *Department of Cellular and Physiological Sciences, University of British Columbia, Vancouver, BC, Canada*

JENNIFER CORRIGAN • *Department of Pediatrics, University of Alberta, Edmonton, AB, Canada*

JOANNE O. DAVIDSON • *Department of Physiology, University of Auckland, Auckland, New Zealand*

MATTHEW DERRICK • *Department of Pediatrics, NorthShore University Health System, Evanston, IL, USA; Department of Pediatrics, University of Chicago, Chicago, IL, USA*

PAUL P. DRURY • *Department of Physiology, University of Auckland, Auckland, New Zealand*

STUART FAULKNER • *Department of Genetics and Development, Toronto Western Hospital, Toronto, ON, Canada; Department of Neurosurgery, University of Toronto, Toronto, ON, Canada*

MICHAEL FEHLINGS • *Department of Genetics and Development, Toronto Western Hospital, Toronto, ON, Canada; Department of Neurosurgery, University of Toronto, Toronto, ON, Canada*

RICHDEEP S. GILL • *Department of Surgery, University of Alberta, Edmonton, AB, Canada*

DJORDJE GRBIC • *Pediatric Neurology Laboratory, Départements de Pédiatrie, Université de Sherbrooke, Sherbrooke, QC, Canada*

PIERRE GRESSENS • *Inserm, U1141, Paris, France; Univ Paris Diderot, Sorbonne Paris Cité, UMRS 1141, Paris, France; PremUP, Paris, France; Centre for the Developing Brain, Department of Division of Imaging Sciences and Biomedical Engineering, King's College London, King's Health Partners, St. Thomas' Hospital, London, UK*

CLÉMENCE GUIRAUT • *Pediatric Neurology Laboratory, Départements de Pédiatrie, Université de Sherbrooke, Sherbrooke, QC, Canada*

ALISTAIR JAN GUNN • *Department of Physiology, Faculty of Medical and Health Sciences, University of Auckland, Auckland, New Zealand*

A. ROGER HOHIMER • *Department of Obstetrics and Gynecology, Oregon Health & Science University, Portland, OR, USA; Maternal and Fetal Medicine, Oregon Health & Science University, Portland, OR, USA*

JANANI KASSIRI • *Division of Pediatric Neurology, University of Alberta Hospital, Edmonton, AB, Canada*

BRYAN KOLB • *Department of Neuroscience, University of Lethbridge, Lethbridge, AB, Canada*

VIVIAN LAM • *Department of Cellular and Physiological Sciences, University of British Columbia, Vancouver, BC, Canada*

KEHUAN LUO • *Department of Pediatrics, NorthShore University Health System, Evanston, IL, USA*

BRAD MATUSHEWSKI • *Physiology and Pharmacology and Pediatrics, Departments of Obstetrics and Gynaecology, Schulich School of Medicine and Dentistry, University of Western Ontario, London, ON, Canada*

ANTOINETTE NGUYEN • *Department of Pediatrics, University of Alberta, Edmonton, AB, Canada*

BRYAN S. RICHARDSON • *Physiology and Pharmacology and Pediatrics, Departments of Obstetrics and Gynaecology, Schulich School of Medicine and Dentistry, University of Western Ontario, London, ON, Canada; Department of Obstetrics and Gynaecology, London Health Sciences Centre - Victoria Hospital, London, ON, Canada*

ART RIDDLE • *Departments of Pediatrics BRB-353A, Oregon Health & Science University, Portland, OR, USA; Oregon Health & Science University, Portland, OR, USA*

CRYSTAL A. RUFF • *Department of Genetics and Development, Toronto Western Hospital, Toronto, ON, Canada; Department of Neurosurgery, University of Toronto, Toronto, ON, Canada*

DEBORAH SAUCIER • *Provost and VP Academic, University of Ontario Institute of Technology, Oshawa, ON, Canada*

ALEXANDRE SAVARD • *Pediatric Neurology Laboratory, Départements de Pédiatrie, Université de Sherbrooke, Sherbrooke, QC, Canada*

LESLIE SCHWENDIMANN • *Inserm, U1141, Paris, France; Univ Paris Diderot, Sorbonne Paris Cité, UMRS 1141, Paris, France; PremUP, Paris, France*

GUILLAUME SÉBIRE • *Pediatric Neurology Laboratory, Départements de Pédiatrie, Université de Sherbrooke, Sherbrooke, QC, Canada; Child Neurology Division, Department of Pediatrics, McGill University, Montreal, QC, Canada*

KATARZYNA STEPIEN • *Department of Medical Genetics, University of British Columbia, Vancouver, BC, Canada*

SIDHARTHA TAN • *Department of Pediatrics, NorthShore University Health System, Evanston, IL, USA; Department of Pediatrics, University of Chicago, Chicago, IL, USA*

MARIANNE THORESEN • *School of Clinical Sciences, University of Bristol, Bristol, UK; Division of Physiology at Institute of Basic Medical Sciences, University of Oslo, Oslo, Norway; School of Clinical Sciences, St Michaels Hospital, Child Health, Level D, University of Bristol, Bristol, UK*

LUIGI TITOMANLIO • *Inserm, U1141, Paris, France; Univ Paris Diderot, Sorbonne Paris Cité, UMRS 1141, Paris, France; PremUP, Paris, France; AP-HP, Pediatric Emergency Department, Hôpital Robert Debré, Paris, France*

KRISTINA UBAN • *Developmental Cognitive Neuroimaging Laboratory, Children's Hospital Los Angeles, Los Angeles, CA, USA*

GUIDO WASSINK • *Department of Physiology, University of Auckland, Auckland, New Zealand*

JOANNE WEINBERG • *Department of Cellular and Physiological Sciences, University of British Columbia, Vancouver, Canada*

JEROME Y. YAGER • *Division of Pediatric Neurology, Department of Pediatrics, University of Alberta, Edmonton, AB, Canada*

Chapter 1

Unilateral Common Carotid Artery Ligation as a Model of Perinatal Asphyxia: The Original Rice–Vannucci Model

Antoinette Nguyen, Edward A. Armstrong, and Jerome Y. Yager

Abstract

Hypoxic-ischemic encephalopathy (HIE) is a detrimental event leading to unfavorable neurological outcomes in the newborn, the clinical phenotype of which is typically referred to as cerebral palsy. The high incidence of HIE results in a need for animal models that can replicate this human experience in order to determine the pathophysiology of injury and develop therapeutic interventions. One of the first models to be developed was, the now commonly referred to as the Rice–Vannucci model, after the student and principle investigator who first developed and described the model. Now, perhaps the best characterized and certainly the most commonly utilized model to reflect perinatal hypoxic-ischemic injury, the "Rice–Vannucci" model is the cornerstone to investigating neonatal brain injury and hypoxic-ischemic encephalopathy. This chapter describes the methodology for utilizing this model, attempt to recognize aspects of the model which have since evolved since its inception, and identify areas of caution when undertaking its use for hypoxic-ischemic encephalopathy.

Key words Hypoxia-ischemia, Neonatal, Asphyxia, Brain injury, Stroke

1 Introduction

The incidence of neonatal encephalopathy is 2–4/1,000 term births [1–4]. The outcomes of neonatal encephalopathy include intellectual disability, seizures, learning disabilities and cerebral palsy. This acquired brain injury can arise from single or multiple risk factors, including hypoxic-ischemic (HI) injury to the developing brain, most often resulting from a final common pathway of compromised blood flow to the fetus or new born infant. Neurological outcomes of HIE depend on the maturation stage and regional distribution of blood vessels, and temporal onset of injury [2, 5]. Currently, the only therapeutic intervention available to improve outcome following an HI brain injury, is hypothermia, a form of rescue therapy effective in mild to moderate forms of injury [6, 7]. Experimental evidence has shown that reducing body temperature by 2–4 °C can reduce brain injury and improve

Jerome Y. Yager (ed.), *Animal Models of Neurodevelopmental Disorders*, Neuromethods, vol. 104, DOI 10.1007/978-1-4939-2709-8_1, © Springer Science+Business Media New York 2015

1

neurological outcomes [6, 8–11]. Clinical trials have shown effectiveness in the human condition of term perinatal asphyxia, and in many developed countries, post-ischemic hypothermia has become standard of care [12–14].

Several animal models have been utilized to replicate HI injury [15–18]; however, the most prominent model of neonatal HIE is the Rice–Vannucci model. The Rice–Vannucci model is adapted from Levine's adult rat model of anoxic-ischemia [19], which involves ligation of the right common carotid artery followed by exposure to nitrogen/nitrous oxide. However, application of the Levine method cannot be directly applied to newborn animal models because the newborn brain is significantly different from an adult [20, 21]. The Vannucci model aims to reflect neonatal HI injury, involving unilateral ligation of a common carotid artery to produce ischemia. Ligation of the common carotid artery alone does not produce injury because the Circle of Willis is able to compensate for the loss of blood due to the anastomosis between the left and right hemispheres. However, brain damage is established once the ligation is paired with hypoxia. In this regard, ligation alone results in an increased blood flow to the contralateral hemisphere. The introduction of hypoxia results in an overall reduction of blood pressure, and therefore an absolute reduction of blood flow to the hemisphere ipsilateral to the carotid artery ligation [22]. The ischemia is therefore accomplished by hypoxia from exposure to 8 % oxygen, producing injury to the ipsilateral hemisphere while leaving the contralateral hemisphere intact (Fig. 1) [22]. This model has been translated to several different species; however, the percentage and timing of hypoxia must be carefully adjusted since mouse studies have shown varying degrees of brain injury. In addition, different strains of mice and rats used have shown different degrees of vulnerability to HI. Sex differences should be noted as well as studies have identified different mechanisms of cell death occurring in males and females [15, 23].

The mechanisms of injury occurring in the Vannucci model have been widely characterized. An observable injury to the naked eye is edema in the hemisphere ipsilateral to the ligation, which arises from the movement of water from the extracellular to the intracellular environment, and depending on the severity of injury, generally maximized at 48 h post-ischemia [24] Metabolically, and coincident with the ischemic injury, serum and brain concentrations rapidly fall, alongside the depletion of ATP. Anerobic metabolism increases the production of lactate, resulting in metabolic acidosis. Hypocapnia counteracts the metabolic acidosis, caused by hyperventilation of the spontaneously breathing rat pup during the hypoxic and maintains pH at physiological levels. The loss of blood flow and consequently oxygen to the developing tissue leads to a cascade of cytotoxic events. Energy failure leads to the disruption of the Na^+/ATPase pump, causing an increase of glutamate release

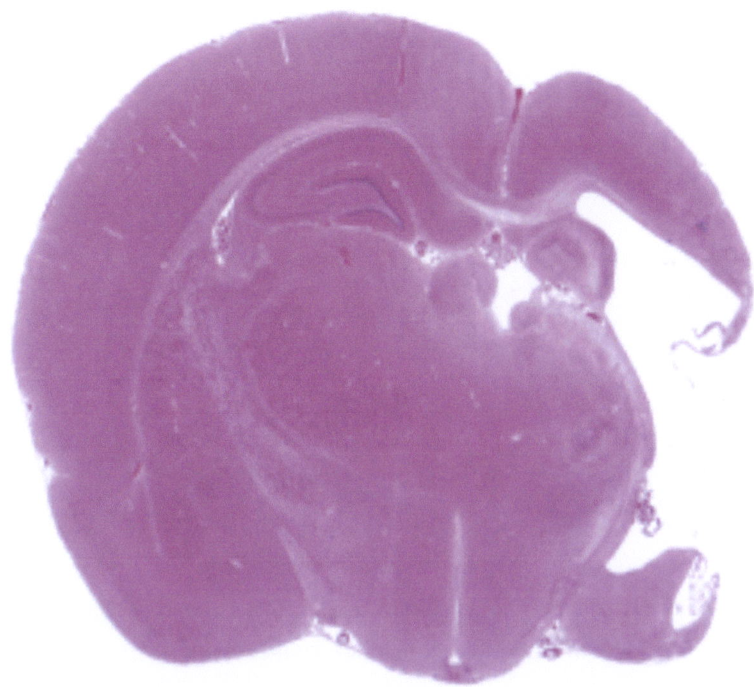

Fig. 1 Coronal H and E stained section of rat pup brain following exposure to hypoxia-ischemia in 8 % oxygen for 3 h following right common carotid artery ligation. The animal was sacrificed at 21 days post insult. Note the large porencephalic cyst on the ipsilateral hemisphere, sparing of the contralateral hemisphere. Reminiscent of a perinatal brain injury or stroke

into the cytoplasm, activating the glutamate receptor NMDA, thus promoting an influx of Ca^{2+}. This increase in Ca^{2+} results in mitochondrial dysfunction, thereby facilitating the production of free radicals. The increase in free radicals causes DNA and protein damage, as well as lipid peroxidation [25–32]. The first cascade of events results in necrotic cell death, followed by a reperfusion stage, and finally a second wave of events ending with apoptotic cell death. Depending on the amount of time the rat pups are exposed to hypoxia following ischemic ligations, mild-to-severe forms of injury can be obtained. Injuries include minimal damage to maximal infarct affecting the entire hemisphere. Generally, the damage produced follows the middle cerebral artery distribution, where the first affected region is the deeper structures of the grey matter and cerebral cortex. Specifically, the subcortical and periventricular white matter, hippocampus, striatum, and neocortical layers 3, 5, and 6 have been documented to be vulnerable to neonatal HI brain injury [33].

2   Materials

- Anesthetic machine
- Anesthesia of choice, i.e., isoflurane
- Carbon filter
- Nitrogen and oxygen tanks
- Lamp or heat pad
- Incubator (set at 34.5 °C)
- Water bath (set at 37 °C)
- 500 ml glass jars
- Fine Iris Scissors
- Halsey Needle Holder
- Standard serrated pattern forceps,
- Finer Graefe forceps
- Coarser Eye Dressing forceps
- 5-0 black braided silk

3   Methods

The Rice–Vannucci model consists simply of permanently ligating one of the common carotid arteries, exposing the pup to 8 % oxygen for variable periods of time, and allowing the pup to recover with their Dam for variable periods of time, depending on the experimental paradigm. There are, however, a number of nuances which require experience with the model. My laboratory (Yager) has studied this model for 25 years.

1. A variety of species of rats can be utilized. The original work was done in Wistar rats (Fig. 2). However, as albinos they are not as well suited for behavioral studies. Our laboratory now uses Long-Evans rats, as they are more reliable for behavioral studies.

2. Rat pups, born on E23 (Term) are kept and nourished with their Dams till day of surgery on Postnatal Day 7. We record the day of birth as PD1. In our laboratory we cull the litters to ten pups. We only use pups between 12 and 16 g on the day of experimentation.

3. Collect postnatal day 7 pups from the dam and place them in a box in the incubator set at 34.5 °C. Weigh and sex each of the pups, and record the data. If the study is long term, you may need to "mark" the pups for later identification.

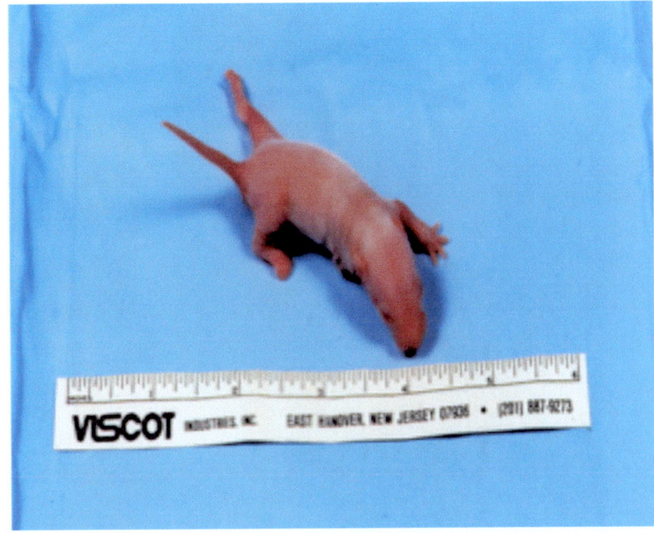

Fig. 2 Wistar rat pup at 7 days of age. Usual weight is between 12 and 15 g

4. The rectal temperature of the rat pup under these circumstances or with the Dam is 36.5 ± 0.5 °C [34, 35].

5. Anesthetize the pup with a nose cone (often simply constructed from the end of a 10 ml syringe), delivering 4 % isoflurane (anesthetic). Place the pup in a supine position for the duration of the surgery. Keep the rat pup on the heating pad or underneath a heat lamp to maintain the pup's temperature (Fig. 3).

6. While the pup is being anesthetized, cut an 8 cm length of 5-0 silk and fold it in half. Set it aside.

7. Once the pup is anesthetized, indicated by an absence of pain reflex tested by a tail or toe pinch, you can begin surgery. At this point, turn the anesthetic down to approximately 1 %, or the animals will die during anesthesia.

8. With a pair of sharp nosed scissors, make a 1 cm ventral midline neck incision just slightly above the collarbones. Lower incisions may damage the carotid bodies leading to breathing problems. At a higher incision, the carotid artery runs behind the trachea making to difficult to retrieve. Some ethics groups will require local anesthetic to be injected, in addition to the inhaled anesthesia.

9. Once the incision is made, adipose and thymus tissue appears. Gently move and hold the tissue to the side by releasing and closing the coarser forceps.

10. Once the trachea is reached, move laterally, to the right or left, depending on which carotid artery you wish to ligate. Immediately next to the trachea are the sternohyoid muscle

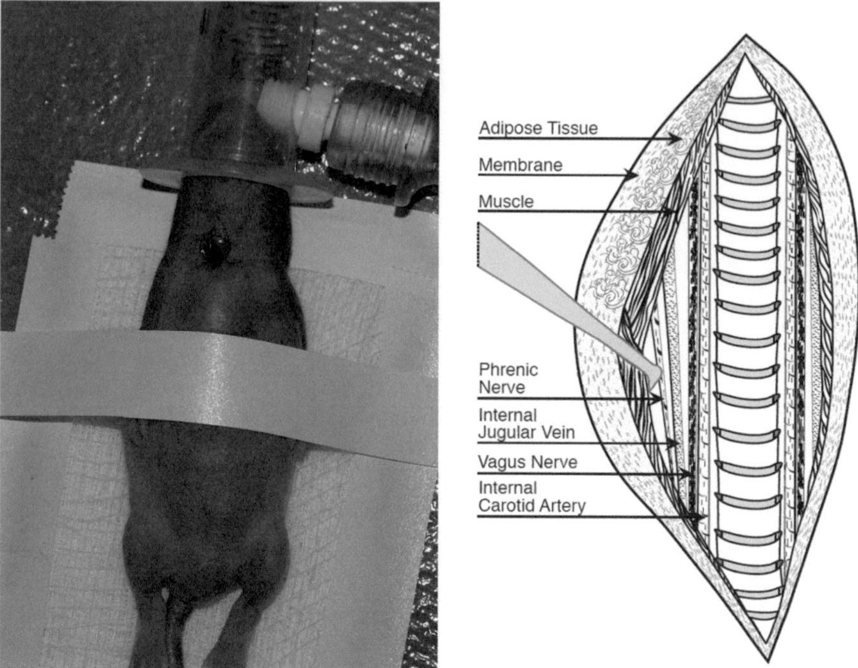

Fig. 3 On the *left* is the rat pup under general anesthesia. The front paws are gently taped down for better exposure. There is a heat lamp overhead. The midline neck incision has been made. A cartoon of the anatomy is placed on the *right* hand side

and a membrane. Puncture the membrane and the common carotid artery should be visible running along the trachea. If you do not see it immediately, you may have moved it aside or have it on your holding forceps. The artery will be found right next to the trachea running alongside the vagus nerve. Damage to the vagal nerve can affect control of the diaphragm causing, breathing problems, and in the ability of the pup to open the ipsilateral eye (Figs. 3 and 4).

11. Hold the muscle away from the artery with the coarser forceps and scoop the carotid artery with the finer forceps (Fig. 6). Hook the carotid with the finer forceps and gently pull it up so it lies directly above the adipose tissue. Be careful not to pull it too far from the body or it will break.

12. Once the artery is isolated, use the tip of the curved forceps and gently pull the previously folded silk underneath the carotid artery to the other side of the neck. Pull slowly as the tension from the silk may shear the carotid artery.

13. Cut the silk at the fold. Pull half of the silk gently up towards the face. Pull the bottom half of the silk gently down towards the heart. Be sure that there is enough distance between the two sutures for you to comfortably cut.

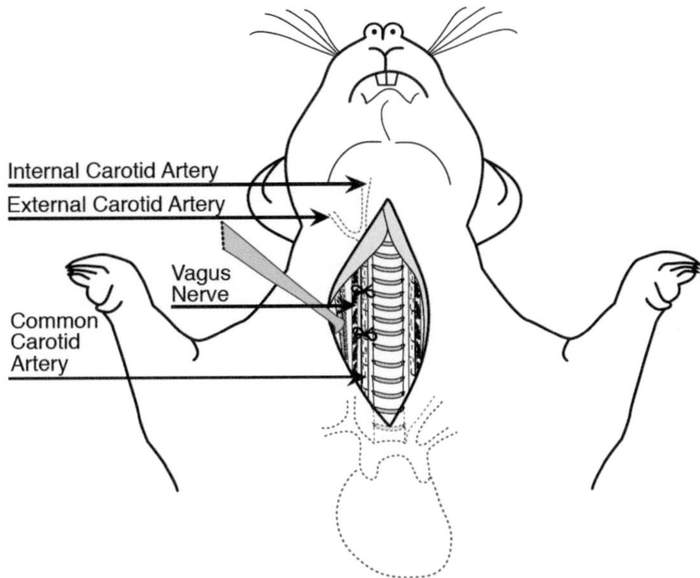

Fig. 4 A cartoon of the rat pup with anatomy of carotid artery, vagus nerve, and trachea exposed

14. Once enough distance is established, perform two single knots on each suture.

15. Ligate the artery to ensure complete cessation of blood flow from the carotid artery. There should be no blood. Trim the excess silk. Some labs will cut the artery between the sutures. Either is effective.

16. With the Halsey needle driver, close the incision with 5-0 silk, and the appropriate amount of sutures using one double knot followed by two single knots for each suture. Remove excess silk.

17. The surgery should be completed between 5 and 7 min.

18. Place the rat pup back into the incubator and continue with the next pup, repeating steps 1–13. Once all pups have undergone the carotid ligation, remove them from the incubator and place them back with the dam to recuperate.

19. Leave the pups with the dam for 2 h to allow full removal of the anesthetic from the pup's physiological system because prolonged anesthesia exposure has been shown to be neuroprotective.

20. While the pups are with the dam, be sure that the water bath is maintained at 36.5–37 °C.

21. After the recovery period, place the rat pups (1–2 pups) into 500 ml glass jars in the water bath. Original studies had placed 3–4 pups per jar. However, our experience has shown that with

Fig. 5 Pictures of the water bath with four glass jars, each containing one rat pup. Eight percent oxygen is blown threw the inlet portal via plastic tubing and outlet portals release the gas. Measured O_2 in the chambers is 8 %. The water bath is maintained at 37 ± 0.5 °C. Water bath height is about mid-way up the jars

an increased number of pups, the humidity in the jar increases and so does the temperature. Hence, it increases the likelihood of variable outcomes. One pup per jar is ideal (Fig. 5).

22. Once the rat pups are placed in the jar, they are subjected to a positive flow of 8 % oxygen balanced nitrogen. This is accomplished by placing inlet/outlet portals in the jars. The inlet portal receives premixed 8 % oxygen. Be sure that oxygen is analyzed to be at 8 % O_2 since minute changes (as little as 0.5 %) have been shown to affect the severity of brain damage. In this regard, our laboratory always uses premixed gas containers.

23. Flow of oxygen through the jars should be barely perceptible.

24. In rat pups 7 days of age, hypoxia for 90 min will cause mild to moderate injury, while 3 h will cause severe injury by gross pathology. The degree of injury in each laboratory will vary due to the environmental conditions. Alterations, even as little as using plastic jars compared to glass jars will make a difference.

25. After hypoxia, pups are returned to their dam. Brain injury and metabolic changes are seen throughout the course of recovery, and even during the HI event. Timing of sacrifice will depend on the experimental paradigm.

4   Notes

The Rice–Vannucci rodent model is by far the most commonly utilized model in laboratories exploring the mechanisms and therapeutic interventions regarding perinatal asphyxia. It has been the most broadly characterized, and has the advantage of a model that can survive for extended periods of time. The latter is of increasing importance, as guidelines for preclinical data are requiring, not only pathological confirmation of effectiveness, but functional/ behavioral robustness as well. Other laboratories have modified this technique, and utilized variations of the model in different species, including mice and guinea pigs [36–38], all of which are effective. Unfortunately space does not allow for elaboration in these models. However, what follows are NOTES which will help to understand the nuances of the model and help prevent and understand issues that face all researchers undertaking in vivo methodology in the immature animal.

1. To prevent litter bias, culling the pups to ten following delivery will allow equal maternal attention, thereby facilitating the same rate of growth and development.

2. Perinatal brain injury is highly dependent on the age of the newborn, with premature infants being more susceptible to white matter injury and term infants being more vulnerable to grey cortical and subcortical injury. Whereas all experimentation used to occur in the PD7 day old, investigators are now using rat pups which are more sensitive to the appropriate age of the experimental paradigm. In this regard, PD 3–5 is more in keeping with the premature human [39], PD 7—late prematurity, and PD 10–12 a term infant. Remember the rat pup is a non-precocial animal [17, 40, 41].

3. Depending on the age of the pup, the brain will be more or less sensitive to a similar degree of HI. Towfighi et al. identified a change in the pattern of injury in the cerebral cortex and white matter with the age of the pup [42, 43].

4. In spite of the fact that rat pups are obviously of an age prior to the influence of sex hormones, there is definitely a gender related difference in outcomes, particularly as it relates to response to therapies. Though different laboratories have found somewhat different outcomes, this variable should always be kept in mind [20, 44–46].

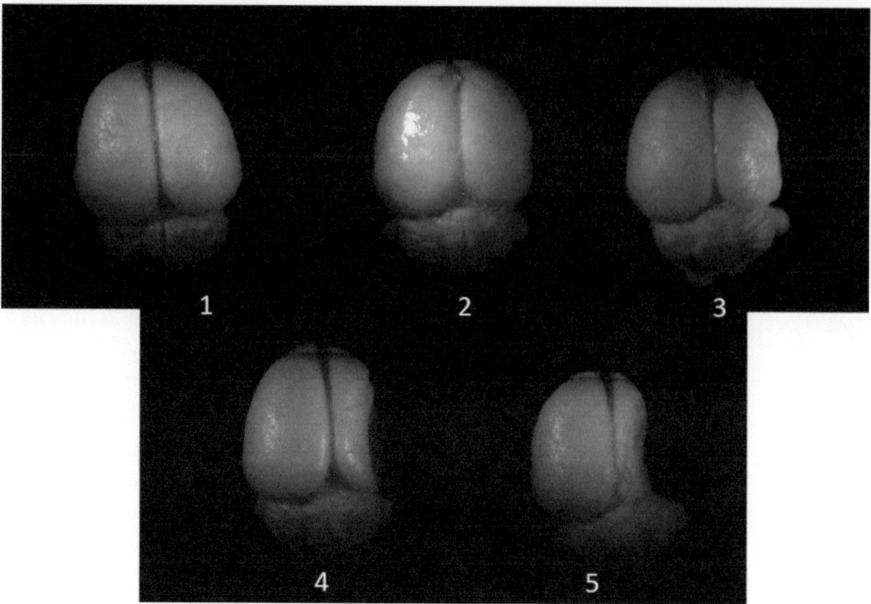

Fig. 6 Variability of injury to rat brains following hypoxia-ischemia. The above photograph indicates the broad range of variability that can be seen, even within a single litter. Normal
(*1*) Normal (no damage)
(*2*) Mild <25 % injury
(*3*) Moderate <50 % injury
(*4*) Moderate <75 % injury
(*5*) Severe >75 % injury

5. Additional variabilities that affect the extent of brain injury include: body temperature, length of exposure to anesthetic, timing of hypoxia exposure from the surgical procedure, duration of hypoxic exposure, litter size, and environmental humidity.

6. In spite of the fact that all pups in a specific litter will be exposed to the same insult, variability of damage will occur. It is unclear what is responsible for this, though many theories exist. Our laboratory has done significant work to determine this factor, but have found that neither sex, temperature, and parentage, nor humidity is responsible for the variability, though they do influence the extent of injury (Fig. 6).

7. Prolonged anesthetic exposure to either xenon or isoflurane will provide neuroprotection for the brain, so exposure should be minimized.

8. Rat pups do not have a mature thermoregulatory system and thus, their body temperature should be maintained constant throughout the surgical procedure, and throughout the experimental paradigm. As is well known, cooling has neuroprotective properties, whereas above normal temperatures will be detrimental [47].

5   Conclusion

The Rice–Vannucci model has certainly stood the test of time with respect to its value as a preclinical model of perinatal brain injury. It is clearly the most well-characterized and defined rodent model and offers the advantage of being able to determine outcomes across the age spectrum, a valuable commodity when investigating therapeutic interventions. Whereas modifications to this model have been adapted by some laboratories, a more important aspect is recognizing some of the details of the model regarding sex, age, and environmental influences and most importantly, of course is the need to maintain consistency and randomization between litters.

References

1. Badawi N, Kurinczuk JJ, Keogh JM, Alessandri LM, O'Sullivan F, Burton PR et al (1998) Intrapartum risk factors for newborn encephalopathy: the Western Australian case-control study. BMJ 317(7172):1554–1558

2. Badawi N, Kurinczuk JJ, Keogh JM, Alessandri LM, O'Sullivan F, Burton PR et al (1998) Antepartum risk factors for newborn encephalopathy: the Western Australian case-control study. BMJ 317(7172):1549–1553

3. Glass HC, Ferriero DM (2007) Treatment of hypoxic-ischemic encephalopathy in newborns. Curr Treat Options Neurol 9(6):414–423

4. Vannucci RC, Vannucci SJ (2005) Perinatal hypoxic-ischemic brain damage: evolution of an animal model. Dev Neurosci 27(2–4):81–86

5. Towfighi J, Zec N, Yager J, Housman C, Vannucci RC (1995) Temporal evolution of neuropathologic changes in an immature rat model of cerebral hypoxia: a light microscopic study. Acta Neuropathol 90(4):375–386

6. Thoresen M, Satas S, Loberg EM, Whitelaw A, Acolet D, Lindgren C et al (2001) Twenty-four hours of mild hypothermia in unsedated newborn pigs starting after a severe global hypoxic-ischemic insult is not neuroprotective. Pediatr Res 50(3):405–411

7. Thoresen M (1999) Cooling the asphyxiated brain - ready for clinical trials? Eur J Pediatr 158(Suppl 1):S5–S8

8. Thoresen M, Satas S, Puka-Sundvall M, Whitelaw A, Hallstrom A, Loberg EM et al (1997) Post-hypoxic hypothermia reduces cerebrocortical release of NO and excitotoxins. Neuroreport 8(15):3359–3362

9. Srinivasakumar P, Zempel J, Wallendorf M, Lawrence R, Inder T, Mathur A (2013) Therapeutic hypothermia in neonatal hypoxic ischemic encephalopathy: electrographic seizures and magnetic resonance imaging evidence of injury. J Pediatr 163(2):465–470

10. Roka A, Azzopardi D (2010) Therapeutic hypothermia for neonatal hypoxic ischaemic encephalopathy. Early Hum Dev 86(6):361–367

11. Bona E, Hagberg H, Loberg EM, Bagenholm R, Thoresen M (1998) Protective effects of moderate hypothermia after neonatal hypoxia-ischemia: short- and long-term outcome. Pediatr Res 43(6):738–745

12. Jacobs SE, Berg M, Hunt R, Tarnow-Mordi WO, Inder TE, Davis PG (2013) Cooling for newborns with hypoxic ischaemic encephalopathy. Cochrane Database Syst Rev 1: CD003311

13. Shankaran S, Laptook A, Wright LL, Ehrenkranz RA, Donovan EF, Fanaroff AA et al (2002) Whole-body hypothermia for neonatal encephalopathy: animal observations as a basis for a randomized, controlled pilot study in term infants. Pediatrics 110(2 Pt 1):377–385

14. Shankaran S, Laptook AR, Ehrenkranz RA, Tyson JE, McDonald SA, Donovan EF et al (2005) Whole-body hypothermia for neonates with hypoxic-ischemic encephalopathy. N Engl J Med 353(15):1574–1584

15. Northington FJ (2006) Brief update on animal models of hypoxic-ischemic encephalopathy and neonatal stroke. ILAR J 47(1):32–38

16. Semple BD, Blomgren K, Gimlin K, Ferriero DM, Noble-Haeusslein LJ (2013) Brain development in rodents and humans: identifying benchmarks of maturation and vulnerability to injury across species. Prog Neurobiol 106–107:1–16

17. Hagberg H, Ichord R, Palmer C, Yager JY, Vannucci SJ (2002) Animal models of developmental brain injury: relevance to human disease. A summary of the panel discussion from the Third Hershey Conference on Developmental Cerebral Blood Flow and Metabolism. Dev Neurosci 24(5):364–366

18. Hagberg H, Bona E, Gilland E, Puka-Sundvall M (1997) Hypoxia-ischaemia model in the 7-day-old rat: possibilities and shortcomings. Acta Paediatr Suppl 422:85–88

19. Levine S (1960) Anoxic-ischemic encephalopathy in rats. Am J Pathol 36:1–17

20. Yager JY, Wright S, Armstrong EA, Jahraus CM, Saucier DM (2006) The influence of aging on recovery following ischemic brain damage. Behav Brain Res 173(2):171–180

21. Yager JY, Shuaib A, Thornhill J (1996) The effect of age on susceptibility to brain damage in a model of global hemispheric hypoxia-ischemia. Brain Res Dev Brain Res 93(1–2): 143–154

22. Rice JE 3rd, Vannucci RC, Brierley JB (1981) The influence of immaturity on hypoxic-ischemic brain damage in the rat. Ann Neurol 9(2):131–141

23. Hill CA, Fitch RH (2012) Sex differences in mechanisms and outcome of neonatal hypoxia-ischemia in rodent models: implications for sex-specific neuroprotection in clinical neonatal practice. Neurol Res Int 2012:867531, Pubmed Central PMCID: 3306914

24. Vannucci RC, Christensen MA, Yager JY (1993) Nature, time-course, and extent of cerebral edema in perinatal hypoxic-ischemic brain damage. Pediatr Neurol 9(1):29–34

25. Vannucci RC, Lyons DT, Vasta F (1988) Regional cerebral blood flow during hypoxia-ischemia in immature rats. Stroke 19(2): 245–250

26. Vannucci RC, Brucklacher RM, Vannucci SJ (2005) Glycolysis and perinatal hypoxic-ischemic brain damage. Dev Neurosci 27(2–4): 185–190

27. Vannucci RC, Brucklacher RM, Vannucci SJ (2001) Intracellular calcium accumulation during the evolution of hypoxic-ischemic brain damage in the immature rat. Brain Res Dev Brain Res 126(1):117–120

28. Vannucci RC, Brucklacher RM, Vannucci SJ (1999) CSF glutamate during hypoxia-ischemia in the immature rat. Brain Res Dev Brain Res 118(1–2):147–151

29. Vannucci RC, Brucklacher RM (1994) Cerebral mitochondrial redox states during metabolic stress in the immature rat. Brain Res 653(1–2):141–147

30. Vannucci RC (1993) Mechanisms of perinatal hypoxic-ischemic brain damage. Semin Perinatol 17(5):330–337

31. Vannucci RC (1993) Experimental models of perinatal hypoxic-ischemic brain damage. APMIS Suppl 40:89–95

32. Vannucci RC (1990) Experimental biology of cerebral hypoxia-ischemia: relation to perinatal brain damage. Pediatr Res 27(4 Pt 1):317–326

33. Towfighi J, Yager JY, Housman C, Vannucci RC (1991) Neuropathology of remote hypoxic-ischemic damage in the immature rat. Acta Neuropathol 81(5):578–587

34. Yager J, Towfighi J, Vannucci RC (1993) Influence of mild hypothermia on hypoxic-ischemic brain damage in the immature rat. Pediatr Res 34(4):525–529

35. Yager JY, Asselin J (1996) Effect of mild hypothermia on cerebral energy metabolism during the evolution of hypoxic-ischemic brain damage in the immature rat. Stroke 27(5):919–925, discussion 26

36. Comi AM, Johnston MV, Wilson MA (2005) Immature mouse unilateral carotid ligation model of stroke. J Child Neurol 20(12): 980–983

37. Sheldon RA, Sedik C, Ferriero DM (1998) Strain-related brain injury in neonatal mice subjected to hypoxia-ischemia. Brain Res 810(1–2):114–122

38. Sheldon RA, Chuai J, Ferriero DM (1996) A rat model for hypoxic-ischemic brain damage in very premature infants. Biol Neonate 69(5):327–341

39. Back SA, Luo NL, Borenstein NS, Levine JM, Volpe JJ, Kinney HC (2001) Late oligodendrocyte progenitors coincide with the developmental window of vulnerability for human perinatal white matter injury. J Neurosci 21(4):1302–1312

40. Dobbing J, Sands J (1979) Comparative aspects of the brain growth spurt. Early Hum Dev 3(1):79–83

41. Yager JY (2004) Animal models of hypoxic-ischemic brain damage in the newborn. Semin Pediatr Neurol 11(1):31–46

42. Towfighi J, Mauger D, Vannucci RC, Vannucci SJ (1997) Influence of age on the cerebral lesions in an immature rat model of cerebral hypoxia-ischemia: a light microscopic study. Brain Res Dev Brain Res 100(2):149–160

43. Towfighi J, Mauger D (1998) Temporal evolution of neuronal changes in cerebral hypoxia-ischemia in developing rats: a quantitative light microscopic study. Brain Res Dev Brain Res 109(2):169–177

44. Hurn PD, Vannucci SJ, Hagberg H (2005) Adult or perinatal brain injury: does sex matter? Stroke 36(2):193–195

45. Johnston MV, Hagberg H (2007) Sex and the pathogenesis of cerebral palsy. Dev Med Child Neurol 49(1):74–78

46. Yager JY, Wright S, Armstrong EA, Jahraus CM, Saucier DM (2005) A new model for determining the influence of age and sex on functional recovery following hypoxic-ischemic brain damage. Dev Neurosci 27(2–4): 112–120

47. Yager JY, Asselin J (1999) The effect of pre hypoxic-ischemic (HI) hypo and hyperthermia on brain damage in the immature rat. Brain Res Dev Brain Res 117(2):139–143

Chapter 2

Bilateral Uterine Artery Ligation (BUAL): Placental Insufficiency Causing Fetal Growth Restriction and Cerebral Palsy

Jennifer Corrigan, Edward A. Armstrong, Stuart Faulkner, Crystal A. Ruff, Michael Fehlings, and Jerome Y. Yager

Abstract

Placental insufficiency is the leading cause of intrauterine growth restriction in the western world. The fetus, when exposed to a compromised environment, is vulnerable to a number of disorders later in life, as a consequence of the reduction in oxygen and nutrition during gestation and the resulting fetal growth restriction. These conditions include neurological disabilities such as cerebral palsy (CP), intellectual disability, epilepsy, and mental health issues in childhood (Autism and ADHD) and in later life (schizophrenia). Certainly, fetal growth restriction as a result of placental insufficiency has been strongly associated with adult onset disease including cardiovascular disease, diabetes, and stroke. Current therapeutic interventions are limited, and there are no standard therapies utilized which examine prevention of these disorders, in the face of a growth restricted fetus. In this regard, models that impact the intrauterine environment are vital to the development of preclinical phenotypes that mimic human conditions. In this case, our laboratory along with several others has developed a model of placental insufficiency with fetal growth restriction and a cerebral palsy phenotype as the outcome. The result is a model that reflects mild to moderate cerebral palsy and expresses the pathologic, radiologic, and functional deficits that closely reflect that of the human, and that persist into adulthood. In addition, our group and has shown that the inherent fetal growth restriction characteristics of this model exhibits evidence of later onset adult disease. The latter findings reflect the overlap of fetal growth restriction as a risk factor for cerebral palsy. This chapter provides an informative background and the methodologies required for the development and duplication of this model of placental insufficiency in the rodent.

Key words Placental insufficiency, Perinatal brain damage, Bilateral uterine artery ligation, Fetal growth restriction, Cerebral palsy

---

1 Introduction

Cerebral palsy continues to occur with an incidence of 2–3/1,000 live births in term newborns, but its incidence is increasing, given the higher rate of premature births. In this latter population, the incidence rises tenfold, to approximately 20/1,000 births [1–3],

Jerome Y. Yager (ed.), *Animal Models of Neurodevelopmental Disorders*, Neuromethods, vol. 104, DOI 10.1007/978-1-4939-2709-8_2, © Springer Science+Business Media New York 2015

and in developing countries, the incidence rises yet another ten times. Placental insufficiency, resulting in hypoxia and ischemia to the developing fetus, has a role to play in both the antepartum and intrapartum causes of CP in over 60 % of cases [4, 5]. In addition to leading to CP, chronic placental insufficiency in the third trimester, leads to asymmetrical fetal growth restriction. The latter is a condition where nutritional deprivation to the fetus results in birth weights below the 3rd percentile, compared to controls, in association with "relative" sparing of brain growth. Growth restriction, in this manner, is also recognized as a significant risk factor for increasing the incidence of CP [5–7].

Brain injury resulting from placental insufficiency and fetal growth restriction (FGR) leads to significant adverse neurological outcomes [8, 9]. Early in infancy, the most common phenotypic outcome is cerebral palsy. Other adverse outcomes, however, both neurologic and systemic, continue to emerge throughout life. Conditions that have been identified as occurring with increased frequency as comorbidities of cerebral palsy include seizures and epilepsy, as well as behavioral issues such as attention deficit hyperactivity disorder, and learning disabilities. Given that the majority of children who develop CP are in the mild to moderate category [10, 11], later onset comorbidities are now being recognized, as the children get older. Particularly as it relates to fetal growth restriction, a significant risk factor for CP, schizophrenia, and other mental health issues is increasingly found [12, 13]. Furthermore, fetal growth restriction (FGR) has been associated with a predisposition to adult onset disease including cardiovascular disease (hypertension and myocardial infarction, diabetes, and stroke) compared to control populations of normal birth weight [14–16].

Despite advances in maternal care and prenatal screening, therapeutic strategies to protect the developing fetal brain remain limited. Current interventions most often include "rescue therapy" that targets newborns in distress. These include mild to moderate hypothermia [17], as well as rehabilitation (physiotherapy, constraint induced therapy, occupational therapy) providing long-term improvement. Regenerative medicine, in the form of stem cell therapy, is currently being developed, but has only recently reached the clinical trial phase for some neurologic paradigms. However, even if all these therapies were to work, given that they approach the problem "after" the injury has occurred, in the postnatal window, they will only benefit between 10 and 20 % of patients.

Studies in the past number of years, have clearly shown that the timing of injury to the newborn begins during gestation in the majority of cases (80–90 %) [18–20]. It may alternatively begin antepartum, with the fetus experiencing a "second-hit" at the time of delivery [5]. In this regard, our laboratories recognized that a preclinical animal model was needed that would reflect the antepartum nature of most injuries, and take into account some of the risk factors that were associated with CP during gestation. This would provide the advantage of reflecting the "human condition" more accurately, allow for the development of "preventive" therapies of CP, and utilize a more appropriate model for later regenerative and rescue therapies. It is for these reasons that a model of uterine artery ligation was developed.

Others had utilized the model of intrauterine artery ligation in the past, and our laboratories have elaborated on this model, and proven its effectiveness as a model of permanent disability, with the phenotype of moderate CP. The latter is particularly important, given the need to utilize models that can reflect not only pathologic alterations consistent with CP and its comorbidity phenotype, but its functional deficits as well. In this regard, the rodent model is ideal, due to our ability to follow the animal long-term into the human equivalent of late childhood and adulthood.

Wigglesworth JS [21] was the first to utilize a unilateral uterine artery ligation model. In this model, a single horn is occluded, leaving the other freely open. The occlusion ensured a reduction of blood flow, and hence oxygen and nutritional support, to the uterine horn supporting approximately half of the fetuses carried by the dam. The fetuses most proximate to the occlusion experienced the most significant ischemia. Hence, these animals became growth restricted, with their weights being between 15 and 60 % below normal compared to controls. The fetuses on the ipsilateral side of the occlusion, but increasingly distal to it, receive a significant degree of collateral blood supply from the ovarian and uterine arteries. Consequently, the weight of these pups is preserved, and they are difficult or impossible to distinguish from pups originating on the contralateral side to the occlusion. Olivier et al. [22], found that unilateral uterine artery ligation on gestational day 17 resulted in white matter damage to the cingulum, accompanied by microgliosis and astrogliosis. In addition, the number of positive O_4 cells, indicative of pre-oligodendrocytes, was decreased, presumably resulting in the reduced axonal myelination into adulthood. The results of the newborn brain damage afforded by unilateral uterine artery ligation reveals brain injury similar to those seen in patients with cerebral palsy. This suggests that placental insufficiency, with its inherent risk of hypoxia and reduction in blood flow to the fetus (ischemia), leading to fetal growth restriction, provides the substrate for a model of cerebral palsy and its comorbidities.

However, as pointed out earlier, determining which rat pups are exposed to a reduced blood flow, and which are control, in the unilateral model, confounds ones ability to interrogate the effects of the occlusion, particularly since the alterations seen are subtle. For this reason, our laboratory has modified a bilateral uterine artery ligation (BUAL) model, developed by Lane et al. [23, 24].

In this regard, both uterine horns and their respective fetuses are experiencing placental insufficiency during pregnancy. In our experience, surviving rat pups are all growth restricted, although there is variation between those most proximal to the ligation compared to those most distal, providing the ability to distinguish between moderate and severely growth restricted animals.

2 Findings

The BUAL model described below, consistently results in fetal growth restriction, of both a moderate and severe nature. We define severe growth restriction as greater than 2 SD below the mean (as is the definition in humans) and moderate growth restriction as being below the 10 percentile. Mortality in this model, as determined by the mean number of pups born following BUAL, compared to controls, who experience sham surgery without ligation, is approximately 50 %. In the short term, these rat pups show significant abnormalities in their developmental reflexes. Later in adulthood, we have documented persistent abnormalities of gross motor control (as determined by gait analysis and foot faults), fine motor control (as determined by single-pellet reaching), as well as evidence of comorbidities including anxiety and attentional abnormalities. Pathologically, we see white matter disturbances early on, and which can be documented neuro-radiologically by MRI (Fig. 5a), as a biomarker of injury.

Hence, the advantages of this model are severalfold:

1. The model phenotypically (behavior) mimics many of the early and late developmental disabilities (both motor and cognitive) that are hallmarks of cerebral palsy.

2. We have documented the persistence of these abnormalities into adulthood, allowing this model to be ideal for the preclinical determination of effective therapies, and the determination of mechanisms involved in pathophysiology.

3. The model incorporates risk factors known to be involved in cerebral palsy, including:

    (a) Prematurity—The rat pup at birth is equivalent to a very premature infant.

    (b) Fetal Growth Restriction—A phenotypic marker of placental insufficiency and known to increase the risk of cerebral palsy.

4. We have further shown that we can develop biomarkers of injury, in the form of behavior and radiologic parameters, that will allow for the assessment of therapies during the evolution of recovery.

3 Materials

1. Sterile surgical pack: 1 surgical drape, 1 leg clamp, 1 scalpel handle, 1 stainless steel feather surgical blade, 1 needle driver, toothed forceps, blunt nosed forceps, 1×1″ sterile gauze (Fig. 1).

2. Non-sterile surgical supplies: 1×1″ gauze, povidone–iodine 10 % topical antiseptic prep solution, electric shaver, 1 mL syringe and needle, 10 mL syringe and needle, normal saline (0.9 % NaCl)

3. Suture materials: 1 sterile 5-0 silk with reverse cutting needle, 1 sterile 4-0 coated vicryl* (polyglactin 910) absorbable suture with tapered needle.

4. Anesthesia supplies: rodent sized anesthesia box, rodent size anesthesia nose cup, oxygen, isoflurane, bupivacaine hydrochloride, heating pad

5. Other: sterile gown, sterile gloves, mask, 4 % chlorhexidine gluconate surgical scrub brush with nail cleaner

6. Two people are required, one to perform the surgery and one to assist, in order to maintain sterility during surgery.

7. Bilateral Uterine Artery Ligation (BUAL) is performed in the middle of the third trimester, requiring timed-pregnant naive animals.

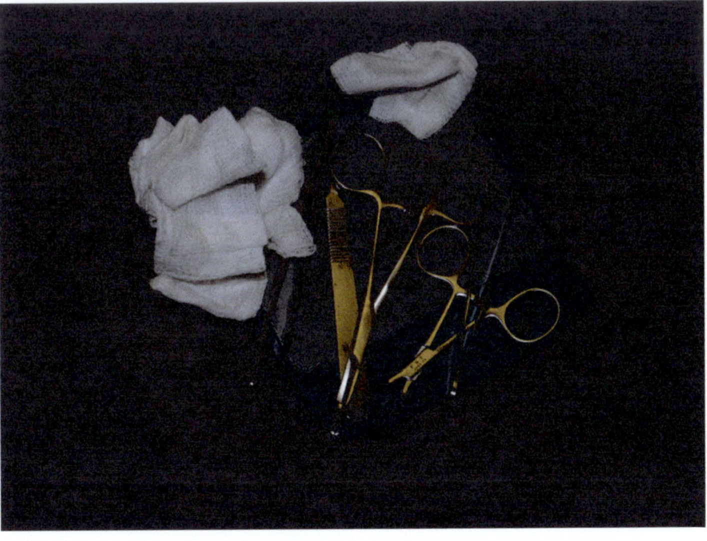

Fig. 1 Sterile pack for performing BUAL

4   Methods

1. BUAL surgery is performed in Long-Evans rat dams on gestational day 20 of their 23-day gestation. Our results indicated that (in our hands) BUAL at gestational age 20 days provides the most appropriate results with both moderate and severe IUGR, and an acceptable 50 % mortality rate. Earlier ligation of the uterine arteries resulted in a marked increase in mortality, while later ligation resulted in the inability to produce significant IUGR. It would be wise to do some preliminary work in one's own laboratory to set those parameters.

2. Prior to surgery, prepare and autoclave a surgical pack (Fig. 1).

3. The procedure for BUAL is best accomplished with two people. One person scrubs in for surgery, while the other person prepares the dam for surgery.

4. Prepare by connecting the anesthesia box to a mixture of 4 % isoflurane and 2,000 cm^3/min of oxygen, before placing the rat dam inside, for rapid induction.

5. Once the sealed anesthesia box is filled with gas, quickly place the dam inside and wait approximately 5 min or until she appears to be asleep for 1 min. Be sure to observe the dam is adequately ventilating while under anesthetic. Respiration should appear slow, deep and rhythmic.

6. During this time, the assistant (non-sterile person) should prep the surgical supplies. Draw up 10 mL of saline with 10 mL syringe. Draw up 0.2 mL of bupivacaine hydrochloride with the 1 mL syringe.

7. Immediately prior to transferring the dam from the anesthesia box to the anesthetic nose cone, unhook the gas hose from the box and attach it to the nose cone to begin filling it with the anesthetic gas mixture. At this time, turn down the isoflurane to between 2.0 and 2.5 % and the Oxygen to 500 cm^3/min. This level of anesthetic should be used as maintenance for the duration of the surgery.

8. Place the nose of the dam in the nose cone on top of a covered heating pad. The dam should be in the supine position. Ensure that body and legs are square with the nose cone and table. Pinch a hind toe of the dam. A leg jerk pain response indicates that the dam is not sufficiently anesthetized to begin surgery. Adjust anesthetic level is necessary.

9. Begin shaving the abdominal hair, in the opposite direction of hair growth. A large enough area must be shaven to allow for sufficient margins around the skin incision in order to prevent contamination with hair. The caudal border should be roughly 1 cm above to the diaphragm and the rostral border should be

1 cm above the urethra. Take care not to injure the nipples of the dam when shaving, as they may affect subsequent feeding of offspring.

10. Clean the freshly shaven area with soap and water, using gauze in a circular motion moving outwards. Discard the used gauze. Repeat with clean gauze until all the hair has been cleaned off. Pour iodine onto the cleaned area (approximately 3–4 drop or until the area is sufficiently covered). Wipe off excess iodine in the method described above.

11. The sterile person will now open the sterile pack, remove the drape and place it over the rat. The opening in the drape should line up with the clean, bare abdomen. Secure the drape with a leg clamp. The clamp should be secured to the superficial skin of the upper thigh of the hind limb.

12. Using the scalpel blade, make a longitudinal incision down the midline of the abdomen, cutting through the skin only. Depending on the size of dam, the incision should be approximately 5–7 cm. For ease of surgery, use a longer incision (Fig. 2).

13. Visualize the midline connective tissue of the muscle layer. Using the toothed forceps lift the muscle layer up, and make an incision with the scalpel blade. The cutting edge of the blade should be facing upwards. Pass forceps through the created opening in the midline fascia, and lift muscle up and away from the protruding uterus. Make a longitudinal incision through the muscle that 0.5 cm short of the skin incision on

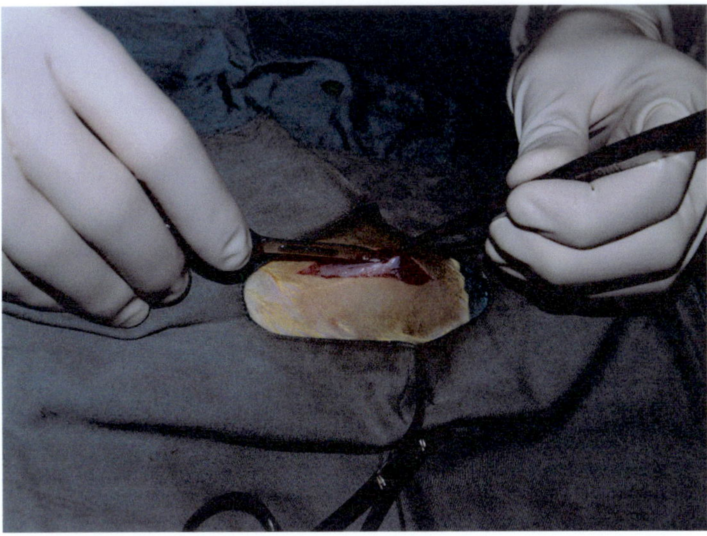

Fig. 2 After making the midline incision through skin, use the toothed forceps to pick up the muscle and make a vertical incision

either side. This will prevent the muscle layer from sliding under the skin, allowing for proper closer.

14. *N.B.* It is crucial that the incision does not deviate from the midline fascia. Incisions made off-center and through the muscle will result in increased bleeding, poorer healing and an increased propensity for the rat to chew at their sutures.

15. Once through all the muscle layers, the uterine horns and pups can be visualized. Gently push on the abdominal flanks of the dam, in order to slowly externalize the uterine horns. Do NOT yank on the uterine horns in order to pull them out, as this will stretch and cause damage to the uterus. Have the assistant generously spray the uterine horns with saline, to help the slide out. The uterine horns need to be kept moist with saline for the entire duration outside the abdomen (Fig. 3).

16. Once the uterine horns have been externalized, locate the uterine bifurcation. This is generally the most caudal point. The uterine artery bifurcates into the left and right uterine artery at the level of the uterine horn bifurcation.

17. Isolate the base right uterine artery using a 4-0 vicryl suture with a tapered needle. An individual branch off of the uterine artery supplies each pup, thus the uterine artery must be ligated proximal to all of its branches, in order to ensure all pups receive ischemia. Secure the suture needle with a pair of needles. Starting superiorly on the left side of the artery, drive the needle inferiorly through the surrounding fascia, around the artery and up superiorly on the right side of the artery. The needle should be placed as close to the artery as possible, without damaging it. Tying off the suture will ligate the uterine artery. In order to do so, use one double knot, followed by two single knots. Pull the double knot tightly until the vicryl begins to slip through the forceps. The single knots can be pulled tightly, as they will not increase the tightness of the double knot. Ensure that all knots are square to prevent them from coming undone. Repeat on the left uterine artery (Fig. 4).

18. *N.B.* To ensure consistency, the suture should be instrument tied using the needle drivers and a blunt nosed forceps.

19. Use the toothed forceps to lift one side of the muscle and gently replace the uterine horn. Repeat on the other side. Try to arrange the horns so that the muscle layer lies as flat as possible and the edges are close to adhering.

20. Close the muscle layer with 4-0 vicryl absorbable sutures using a single interrupted pattern. Sutures should be placed approximately 0.5 cm apart. Bring the muscle layers together so that the edges are just touching and slightly everted. Secure with a double knot, making sure not to over tighten. Follow with two tight single knots.

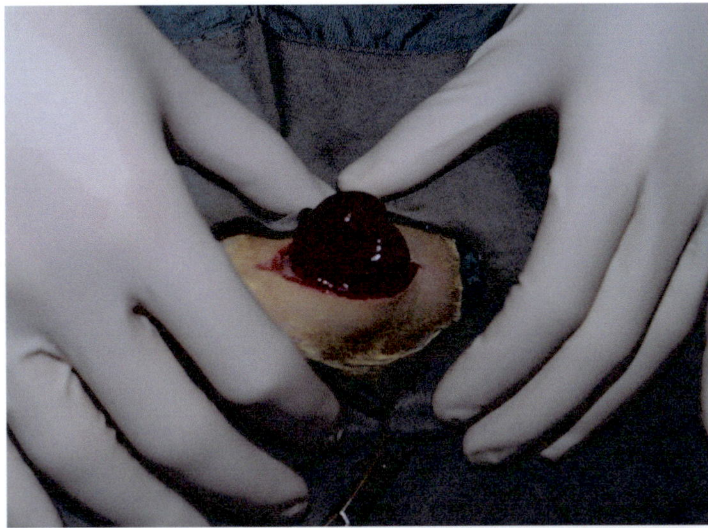

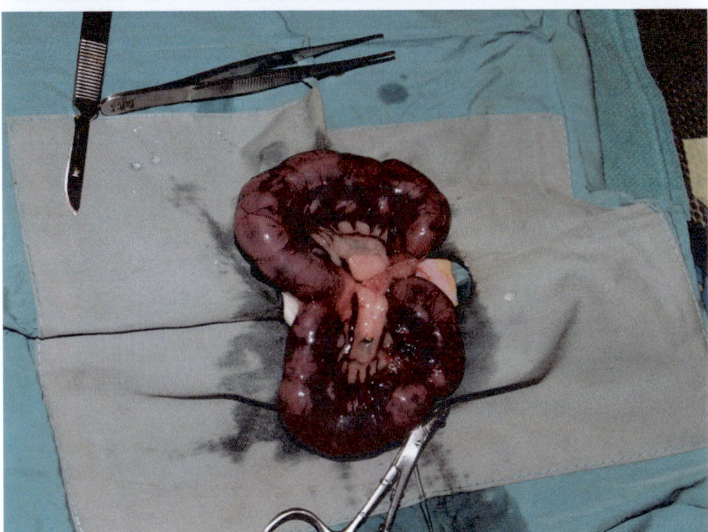

Fig. 3 Externalizing the uterine horns by applying pressure to the lateral walls of the abdomen and gently pushing the horns along with the fetuses outside of the abdomen. Keep the horns moist by irrigating them with saline

21. Apply 0.5–1.0 mL of bupivacaine hydrochloride to muscle incision.

22. Close the skin with 5-0 silk sutures using a single interrupted pattern. Sutures should be placed approximately 0.25–0.5 cm apart. Bring the edges of the skin together so that just touching and slightly everted. Secure with a double knot, making sure not to over tighten. Follow with two tight single knots. The incision and sutures should lie flat.

23. Clean any blood residue off the abdomen using soap and warm water. Return dam to a clean cage with paper towel on the bottom.

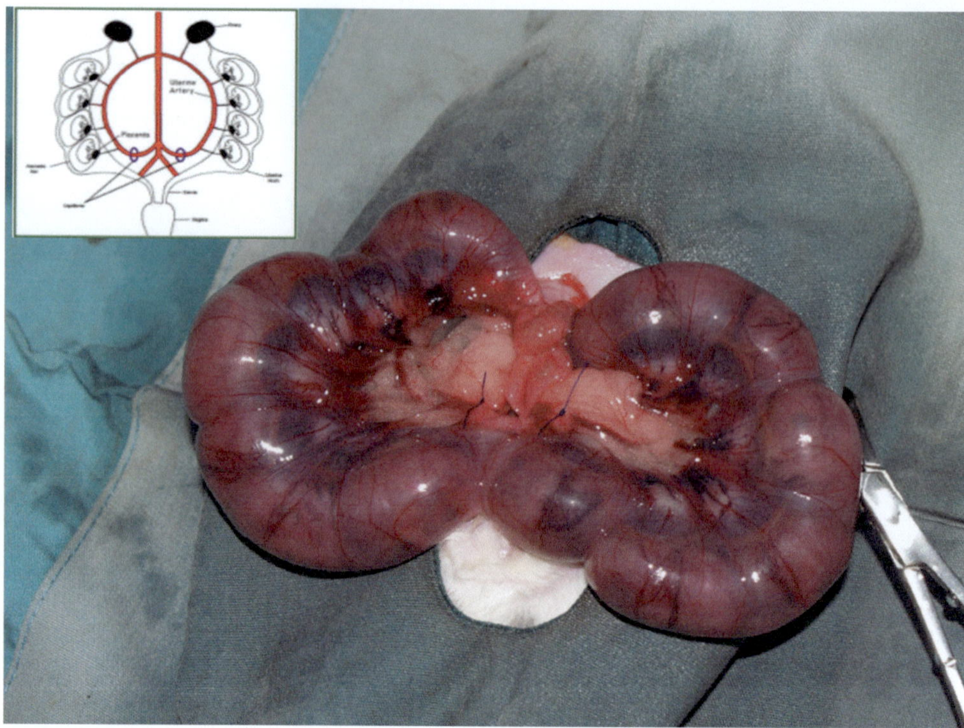

Fig. 4 *Upper left hand corner* is a cartoon depiction of the uterine horns, and the location of ligation proximal to the uterus in the rodent. Color photograph of the uterine horns ligated in our rodent model

24. Carefully monitor the dam for the next 24 h. Ensure that she does not attempt to remove any of sutures. The incision should be inspected daily for the next 7 days.

5   Results

After performing a BUAL procedure, rat pups are generally born on their expected date of delivery (E23). The dams will give birth vaginally, which is again an advantage of the model, as this is typically what occurs in the human condition. In some laboratories, the pups are removed by cesarean section and cross-fostered. However, we believe this introduces significant confounders to the experimental paradigm. The BUAL pups are significantly smaller than their controls, both in numbers per litter, and in birth weights. This difference in weight continues till about 21 postnatal days, after which the pups begin to normalize compared to controls. Moreover, measurements of their head size and body weights have shown us that this procedure produces an asymmetrical growth restriction.

Confirmatory findings in this model [25], reveal:

1. A decrease in MBP staining (Fig. 5a–c), and
2. Thinning of the corpus callosum on MRI imaging.
3. A reduction in CA_1 and CA_3 cell numbers.
4. Abnormalities of developmental reflexes.
5. When followed to adulthood, we find a persistence of:
    (a) Abnormal gait analysis.
    (b) Abnormal tapered beam and foot fault testing.
    (c) Abnormal single-pellet reaching.
    (d) Abnormalities in open-field and elevated plus maze.
    (e) Abnormalities in water maze testing
    (f) Reduction in oligodendroglial cell count.

6 Notes

1. Before beginning the surgery, it is absolutely critical that the area of incision is clear of any abdominal hair, and thoroughly cleaned with iodine, since the risk of infection is high. It is important that the dam remains healthy, as we do not use surrogates in our paradigm.

2. Once the uterine horn is exposed, be sure to have saline ready and continuously pour the saline over the uterus to prevent it from drying. We find that the procedure is more efficiently done with two people, a sterile surgeon and a non-sterile assistant.

3. Ligate the uterine arteries at the point immediately distal to the bifurcation, in order to get all of the artery feeders. The uterine artery should be ligated using an instrument tie, as described above, in order to decrease intra-operator variability. Despite this, variability remains an issue, and thus consistency of results may be improved by having the same surgeon perform all operations.

4. Successful ligation can be confirmed visually by observing the decreased perfusion to the corresponding uterine horn. The arteries distal to the ligation will turn dark after several minutes as an indication of the cessation of blood flow.

5. Saline should be applied generously to each uterine horn before replacing them into the abdomen. This will aid in both easing the process of replacing the horns into the abdomen, and reduce any insensible water losses due to exposure.

6. As outlined in the methods, the muscle should be sutured using a single interrupted, not continuous technique. Each suture should be secured by a double knot and then two single

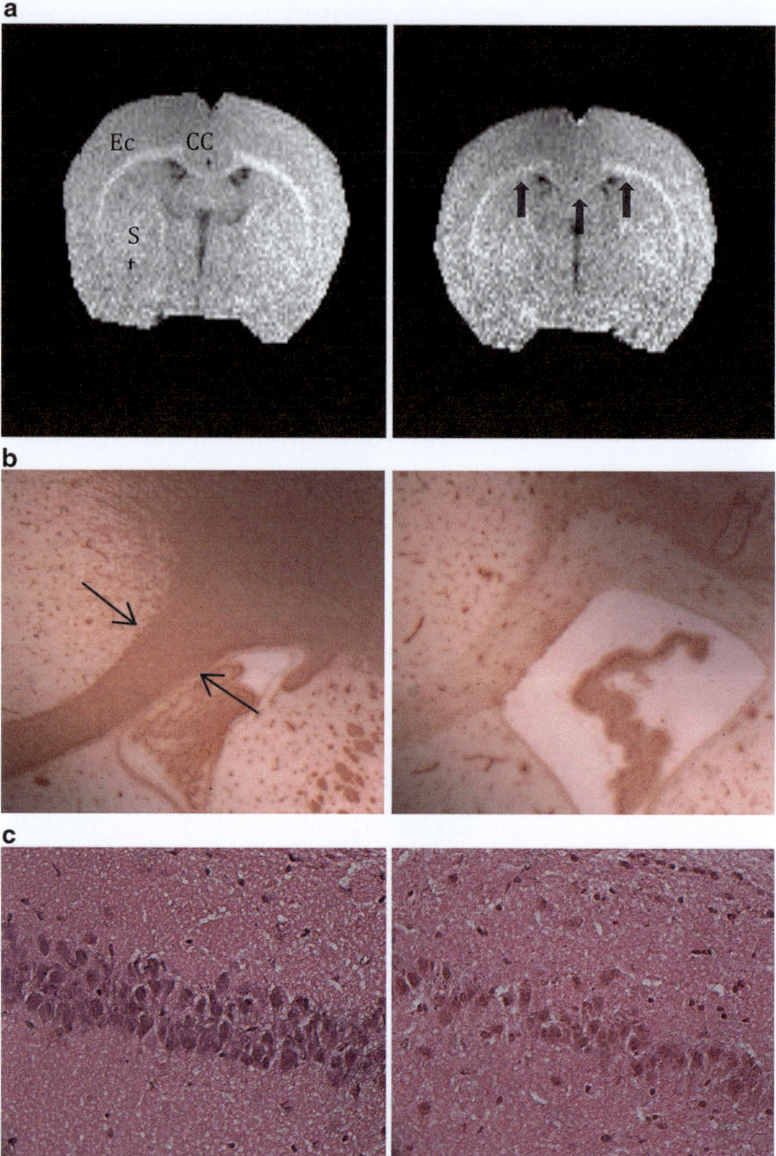

Fig. 5 (**a**) Coronal magnetization transfer ratio (MTR) image of Sham (*left*) and BUAL rat pup at postnatal day 21. The white matter tracts of the corpus callosum (CC) and external capsule (EC) are clearly visible in the Sham animal, but substantially thinned in the BUAL rat pup (*arrows*). (**b**) Photomicrograph of corpus callosum stained with myelin basic protein (*arrow*). On the *left*, the SHAM animal has normal myelin basic protein staining. The BUAL pup on the *right* has significantly diminished staining, an indication of poor myelination, and a pathologic hallmark of cerebral palsy. (**c**) Photomicrograph of cells in the CA_1 region of the hippocampus in SHAM (*left*) and BUAL (*right*) rat pups at 21 days of age. Note the marked decrease in hippocampal cells in those animals exposed to placental insufficiency based on the BUAL model

knots. The same applies to securing the skin. The reasoning behind this methodology is to significantly reduce the risk of herniation, as the dams will tend to chew their sutures.

7. While an absorbable vicryl suture should be used for the closure of the muscle, silk is the preferable choice for the skin. In our experience, dams find vicryl skin sutures more irritating than silk, and are therefore more likely to chew them off. The skin is thicker than the muscle, therefore using a reverse cutting needle is more effective.

8. The dam should be observed for a minimum of 6 h postoperatively. Before returning her to her usual housing facility, ensure that all sutures are intact. This can be performed by simple observation. However, it is imperative that the entire incision be visualized, and thus may require handling of the dam to do so. If this happens, place the dam under anesthetic and replace the sutures.

9. The damage of the brain observed in this model is in the mild to moderate category, and would be reflective of the majority of children with cerebral palsy.

10. The behavioral manifestations (abnormalities) appear to be more robust that the pathologic abnormalities, and therefore it is important to test both when determining your outcome parameters, based on the experimental paradigm.

11. The model does take some practice to perfect. Consistency in approach, utilizing two persons, instead of one during the operation, and maintaining appropriate sterilization are extremely important for success.

7 Conclusion

The bilateral uterine artery ligation model is extremely effective in depicting a model that is more uniform with the human condition. In this regard, the insult occurs antepartum, is associated with an animal that is non-precocial, and affiliates associated risk factors with the outcome. Phenotypically, the pups show evidence of delay in their early developmental reflexes, abnormalities of both fine and gross motor activities, dysfunctional cognitive capabilities reminiscent of early mental health issues, and persistence of these abnormalities into adulthood. The rodent, as a model, provides for the assessment of long term survival and outcomes and is therefore appropriate in the development of therapeutic interventions. Increasingly, animal models that more accurately reflect the human condition will be required for improving the translatability of pre-clinical findings to the bedside.

References

1. Robertson CM, Svenson LW, Joffres MR (1998) Prevalence of cerebral palsy in Alberta. Can J Neurol Sci 25(2):117–122

2. Robertson CM, Watt MJ, Dinu IA (2009) Outcomes for the extremely premature infant: what is new? And where are we going? Pediatr Neurol 40(3):189–196

3. Robertson CM, Watt MJ, Yasui Y (2007) Changes in the prevalence of cerebral palsy for children born very prematurely within a population-based program over 30 years. JAMA 297(24):2733–2740

4. Badawi N, Kurinczuk JJ, Keogh JM, Alessandri LM, O'Sullivan F, Burton PR et al (1998) Intrapartum risk factors for newborn encephalopathy: the Western Australian case-control study. BMJ 317(7172):1554–1558

5. Badawi N, Kurinczuk JJ, Keogh JM, Alessandri LM, O'Sullivan F, Burton PR et al (1998) Antepartum risk factors for newborn encephalopathy: the Western Australian case-control study. BMJ 317(7172):1549–1553

6. Jarvis S, Glinianaia SV, Torrioli MG, Platt MJ, Miceli M, Jouk PS et al (2003) Cerebral palsy and intrauterine growth in single births: European collaborative study. Lancet 362(9390):1106–1111

7. Jarvis S, Glinianaia SV, Blair E (2006) Cerebral palsy and intrauterine growth. Clin Perinatol 33(2):285–300

8. Blair E, Stanley F (1990) Intrauterine growth and spastic cerebral palsy. I. Association with birth weight for gestational age. Am J Obstet Gynecol 162(1):229–237

9. Blair E, Stanley F (1992) Intrauterine growth and spastic cerebral palsy II. The association with morphology at birth. Early Hum Dev 28(2):91–103

10. Shevell MI, Dagenais L, Hall N (2009) The relationship of cerebral palsy subtype and functional motor impairment: a population-based study. Dev Med Child Neurol 51(11):872–877

11. Shevell MI, Dagenais L, Hall N (2009) Comorbidities in cerebral palsy and their relationship to neurologic subtype and GMFCS level. Neurology 72(24):2090–2096

12. Morgane PJ, Mokler DJ, Galler JR (2002) Effects of prenatal protein malnutrition on the hippocampal formation. Neurosci Biobehav Rev 26(4):471–483

13. Smith GN, Flynn SW, McCarthy N, Meistrich B, Ehmann TS, MacEwan GW et al (2001) Low birthweight in schizophrenia: prematurity or poor fetal growth? Schizophr Res 47(2–3):177–184

14. Barker DJ (1995) Intrauterine programming of adult disease. Mol Med Today 1(9):418–423

15. Barker DJ, Clark PM (1997) Fetal undernutrition and disease in later life. Rev Reprod 2(2):105–112

16. Joss-Moore LA, Lane RH (2009) The developmental origins of adult disease. Curr Opin Pediatr 21(2):230–234, Pubmed Central PMCID: 2726974

17. Jacobs SE, Berg M, Hunt R, Tarnow-Mordi WO, Inder TE, Davis PG (2013) Cooling for newborns with hypoxic ischaemic encephalopathy. Cochrane Database Syst Rev 1: CD003311

18. Nelson KB (1988) What proportion of cerebral palsy is related to birth asphyxia? J Pediatr 112(4):572–574

19. Nelson KB (1989) Relationship of intrapartum and delivery room events to long-term neurologic outcome. Clin Perinatol 16(4):995–1007

20. Nelson KB (1991) Prenatal origin of hemiparetic cerebral palsy: how often and why? Pediatrics 88(5):1059–1062

21. Wigglesworth JS (1974) Fetal growth retardation. Animal model: uterine vessel ligation in the pregnant rat. Am J Pathol 77(2):347–350

22. Olivier P, Baud O, Evrard P, Gressens P, Verney C (2005) Prenatal ischemia and white matter damage in rats. J Neuropathol Exp Neurol 64(11):998–1006

23. Ke X, Schober ME, McKnight RA, O'Grady S, Caprau D, Yu X et al (2010) Intrauterine growth retardation affects expression and epigenetic characteristics of the rat hippocampal glucocorticoid receptor gene. Physiol Genomics 42(2):177–189

24. Schober ME, McKnight RA, Yu X, Callaway CW, Ke X, Lane RH (2009) Intrauterine growth restriction due to uteroplacental insufficiency decreased white matter and altered NMDAR subunit composition in juvenile rat hippocampi. Am J Physiol Regul Integr Comp Physiol 296(3):R681–R692

25. Black AM, Armstrong EA, Scott O, Juurlink BJH, Yager JY (2015) Broccoli Sprout Supplementation During Pregnancy Prevents Brain Injury in the Newborn Rat following Placental Insufficiency. Behavioural Brain Research 15;291:289–298.

Chapter 3

Perinatal Intracerebral Hemorrhage Model and Developmental Disability

Janani Kassiri and Marc Del Bigio

Abstract

Perinatal intracerebral hemorrhage, also known as germinal matrix hemorrhage (GMH), refers to the bleeding that arises from the sub-ependymal (or periventricular) germinal region of the immature brain. Intraventricular hemorrhage (IVH) refers to the bleeding that extends into the ventricles, usually as an extension of GMH. Clinical studies have shown that infants who experience GMH/IVH may develop hydrocephalus or suffer from long-term neurological dysfunctions, including cerebral palsy, seizures, and learning disabilities. Understanding the pathogenesis of subsequent brain damage is important for the prevention and management of GMH/IVH. Appropriate animal models are necessary to achieve this understanding. Rodent models of GMH/IVH are economical, homogenous within a strain, and suitable for studying some long-term outcomes. This chapter reviews rodent models of GMH/IVH and their neurobehavioral outcomes.

Key words Germinal matrix/intraventricular hemorrhage, Perinatal, Brain damage, Premature birth, Neurobehavioral outcome

1 Clinical Background of Perinatal Intracerebral Hemorrhage

Perinatal intracerebral hemorrhage or germinal matrix hemorrhage (GMH) refers to the bleeding that arises from the periventricular (or subependymal) germinal region of the immature brain. The ganglionic eminence, which lies over the head of the caudate nucleus, is the most prominent germinal zone in which bleeding arises. It develops in the human fetus during the second trimester of gestation and involutes at 34–36 weeks gestation. Intraventricular hemorrhage (IVH) refers to the bleeding that extends into the ventricles and is usually an extension of GMH [1]. Lateral extension of GMH is associated with destruction of maturing parenchyma, including the head of the caudate and internal capsule [2, 3].

GMH is a major problem in children born prematurely. Clinical studies have shown that infants who experience GMH/IVH may develop hydrocephalus or suffer from long-term neurological

Jerome Y. Yager (ed.), *Animal Models of Neurodevelopmental Disorders*, Neuromethods, vol. 104,
DOI 10.1007/978-1-4939-2709-8_3, © Springer Science+Business Media New York 2015

dysfunctions (~25 %), including cerebral palsy and seizures [1, 4, 5]. GMH is graded according to the extent of hemorrhage: Grade I indicates isolated subependymal hemorrhage (SEH); Grade II indicates GMH with IVH but without ventricle enlargement; Grade III indicates IVH with enlarged ventricles; and Grade IV indicates IVH and GMH extending into the brain beyond the ganglionic eminence [6].

2   Rodent Model of GMH/IVH

A number of animal species including rabbits, sheep, pigs, dogs, cats, and primates have been used to model GMH/IVH (see Table 1). For literature review that critically evaluates the animal models of GMH/IVH, please refer to Balasubramaniam and Del Bigio [7]. Newborn rodent brains are developmentally comparable to 24–26 week gestational age human brains, at birth. As in the preterm infant at the end of the second trimester, neuronal genesis is complete in most regions [8, 9]. A considerable literature concerning rodent brain development is available. Importantly, the germinal matrix region has been studied well in rodents, making them an attractive model for studying GMH [5, 10–12]. Moreover, rodents do not exhibit spontaneous brain hemorrhage at birth. Rats are highly suitable for studying long-term behavioral changes, as their neurological outcomes are well documented [13, 14].

3   Developmental Disability

GMH/IVH that occurs during a critical period of brain development (23–34 weeks) likely interrupts normal brain development and long-term neurological processes. Considering the fact that perinatal brain injury is among the most prevalent and costly forms of neurological disabilities [15, 16], it is important to understand the long-term consequences of GMH/IVH. A potential consequence of germinal matrix hemorrhage is a permanent neurological deficit, such as cerebral palsy or post-hemorrhagic hydrocephalus [17]. Cerebral palsy is a clinical term to describe a broad group of motor syndromes, and cognitive comorbidities resulting from damage to the developing brain [18]. Few good animal models mimic this process. In rodents, the postnatal period from day 1 to day 14 involves corresponding rapid changes in brain structure [19, 20]. Thus, it is likely that insults, such as blood injection, can interrupt the developmental processes necessary for timely acquisition of behavioral skills.

Table 1
Summary of non-rodent GMH/IVH models

Animal models	Experimental procedure	Advantage	Problems
Rabbits (Full term 21–32 days)	Induced hypertension [73] Hypotension by intraperitoneal injection of glycerol [74] Furosemide diuresis [38, 75] Sodium bicarbonate infusion [76]	Postnatal brain development and germinal matrix similar to human fetus. At 28 days gestation, preterm rabbit can survive outside uterus because respiratory system is matured.	Diffuse spontaneous hemorrhage in parenchyma
Sheep (Full term 147–150 days)	Arterial and venous hypertension with asphyxia [77] Hypovolemia, hypervolemia [78] Maternal hypotension [79]	The germinal layers of mid-gestation resemble human 26–30 weeks gestational age. Vascular pattern of germinal matrix resembles that of human infant.	Need to deliver animal prematurely but sheep fetus must remain connected via umbilical cord to the dam. Presence of carotid rete mirables.
Pigs (Full term 113–117 days)	Autologous blood injection [80]	Mature brain at birth.	No germinal matrix at term.
Dogs (Full term 60–63 days)	Hypertension [81] Hypercarbia [82] Hypovolemic hypotension followed by autotransfusion [83] Premature delivery and asphyxia [84]	Large size and extensive blood vessels around germinal matrix layer first days of life. Matrix matures and rapidly involutes by 4–10 days post birth.	Spontaneous hemorrhage. Presence of carotid rete mirables. Histologically 2–3-day-old pup germinal matrix is comparable to 32-week human GA.
Cats (Full term 60–63 days)	Intraperitoneal injection of sodium [85]	Postnatal neuronal development is well established.	Germinal matrix not well studied.
Primates (Full term 184–188 days)	Spontaneous hemorrhage at premature birth [86]	Germinal matrix of 100 day gestation age baboon and rhesus similar to 24 week human GA.	Expensive Diffuse bleeding And secondary white matter changes

4   Neurobehavior of Rodents

Ambulation, surface righting, and negative geotaxis are among the earliest motor developmental milestones in rats [14, 21–23] and mice [13]. By postnatal day 3–4, rodent pups are able to ambulate and pivot using their forelimbs, torsos, and heads. This behavior peaks at postnatal day 7 and disappears by postnatal day 15. Reflex behaviors, such as the righting response, emerge during the first 3 postnatal days and mature during the first week of postnatal development [22, 24]. Between postnatal days 9 and 11, rats develop negative geotaxis, gaining the ability to orient themselves on an inclined plane. The righting reflex (free fall) can be detected between postnatal days 9 and 18 [25, 26]. Rats and mice develop adult-like locomotion shortly after developing full vision, around postnatal day 16. Around the same period, rodents can stand on their hind limbs while leaning on their forelimbs for support [23]. Fine manual dexterity is difficult to assess in immature rodents. Various techniques, such as skilled reaching and wire climbing, are used to study fine motor tasks in more mature adult rats. Whishaw et al. developed a skilled reaching task, which involves single food pellet retrieval through a narrow slit. The trained rats have shown a range of limb movement and accurate ability to reach tiny pellets at different distances located in different directions [27, 28].

5   Autologous Blood Injection Rodent Model

Autologous blood injection model analyzes the effect that blood has on the surrounding brain tissue. We developed this model for use in newborn mice (1 day old) and rats (2–3 days old) by injecting autologous tail blood into the subependymal region [29]. In the present chapter, we will describe the development of rodent autologous blood infusion GMH/IVH model and discuss early and long-term neurobehavioral testing methods and findings in this model.

6   Methods

6.1   Autologous Blood Injection

The brains of both the 1-day-old mouse and the 2–3-day-old rat are at a level of maturity that is roughly equivalent to a 24–26 week gestation human fetus [5, 10]. Our studies use 1 day mice or 2–3 day rats. Initially we found that cooling newborn rodents on an ice bed provides a fast way of anesthetizing animals and allows for a quick recovery. Current animal protocols utilize isoflurane anesthetic delivered by nose cone.

Autologous blood must to be collected using a sterile syringe. The tail is placed in warm (30 °C) water for 1 min, then the skin is cleansed with 70 % alcohol. Using a sterile scalpel or razor blade, 1–2 mm should be cut off the tail tip [26]. Cutting more than this will result in stunted tail growth that may be associated with poor balance in adulthood. Unlike adult models, the tail vein [30, 31], central tail artery [32], or femoral artery [33] cannot be accessed in neonates. The blood (~30 µL) is drawn into a sterile syringe. Although some adult models use heparinized blood [31, 32, 34], we do not add anticoagulant to blood because it could modify the effect of blood. Gentle pressure on the tail with sterile gauze (optionally with antibiotic ointment) will stop the bleeding.

Neonatal rodents cannot be secured accurately in a stereotaxic frame because of their small size. Thus, blood injections are performed freehand with one hand securing the head and the other manipulating the syringe and needle. We developed a custom shield around the 28 gauge needle to stabilize against the scalp and ensure correct depth of penetration (2.5 mm deep to the skull surface). The surface coordinates for targeting the lateral periventricular region are 1.5 mm lateral to midline, and 0.5 mm posterior to the outer canthus of the right eye. The scalp should be cleansed with 70 % alcohol prior to cutting he tail; it is critical to perform the injections quickly to prevent blood clotting in the syringe. To avoid experimental bias, animal injections should be done consistently by one researcher while behavioral testing should be done by a different researcher.

In this model, blood (15 µL) is injected slowly over 2–3 min. The needle is left in the place for 10–20 s and then removed slowly. For sham controls, an equal volume of sterile saline is injected. Because needle insertion alone can cause brain damage, intact control animals subjected only to anesthesia should also be included in these experiments. The animals are warmed on a 25 °C blanket. Surface blood is washed away with damp gauze prior to return to the mother. Pups should be identified with toe tattoos while under anesthesia; we find that ear punches in the early neonatal period causes too much damage. Numberings can be converted to a permanent ear punch at day 10. Dams and pups are housed in standard cages, and they are on 12-h light–dark schedule with free access to chow and water.

It is important to assess the extent of blood spread immediately after the injection using an imaging modality. Magnetic resonance (MR) imaging of rodents is done 20–30 min after injecting blood or saline. Sham injections that result in substantial hemorrhage and misplaced blood injections (typically too lateral) are excluded [35]. Figure 1 shows a typical MR image. It is possible that high resolution computed tomography scanning would also be useful. In early pilot studies we sacrificed a series of animals immediately after imaging to ascertain the histopathologic features of

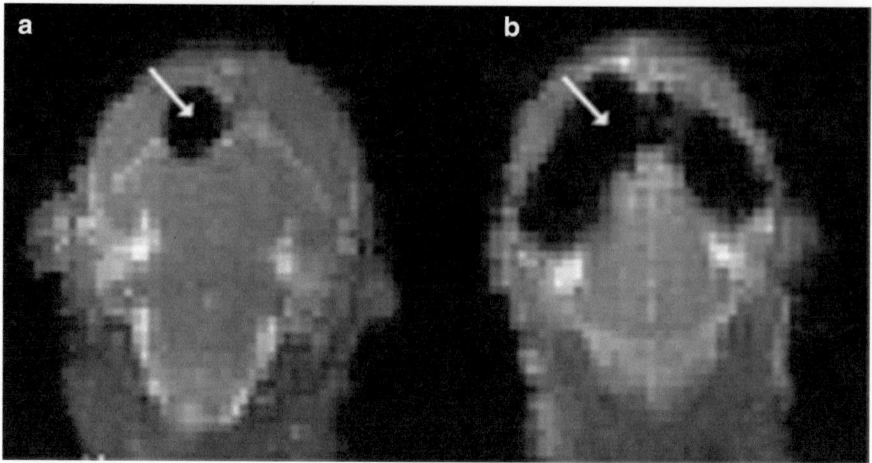

Fig. 1 T2-weighted magnetic resonance images showing horizontal slices of 1-day-old rat brain. Fresh blood collections have hypointense (*dark*) signal. Ten minutes after saline injection (15 μL) there is small blood collection due to needle damage (*arrow*) in the striatum (**a**). Ten minutes after autologous blood injection (15 μL) the hematoma spreads from the striatum (*arrow*) into the ventricles (**b**)

the acute brain damage [29]. Figure 2a, b illustrates typical rat brain sections stained with hematoxylin and eosin (H and E). It is recommended that all investigators do a set of similar pilot experiments so that the relationship between the image and the actual blood spread is clearly understood.

In terms of technique, the damage introduced by needle insertion is a confounding factor. Therefore, alternate noninvasive approaches, such as the induction of transient neonatal hypertension (with phenylephrine) following 24 h in an 8 % oxygen environment, can be utilized for some aspects of GMH/IVH research. The problem with this method is the anticipated variability, as reported in dog and rabbit models [36–38].

Rodents have also been used to study other disorders, such as post-hemorrhagic hydrocephalus, by injecting blood into the lateral ventricles of 7-day-old rats [39–41]. With respect to GMH, a potential limitation of these models is that 7-day-old rats are comparable to about 34–36-week gestation infants in which the GM has involuted. Moreover, our work [42], as well as others using different models of brain injury [43], clearly show that the neonatal rodent brain responds differently than do the young rodent brain or the mature brain.

6.2 Collagenase Model

Local injection of bacterial collagenase to induce intracerebral hemorrhage has been widely studied [44–46]. Bacterial collagenase disrupts basal lamina of cerebral blood vessels causing the blood to leak into surrounding tissue. All of the collagenase studies have been conducted on older rodents with involuted GM. When our lab attempted to inject collagenase into neonatal rats, we found

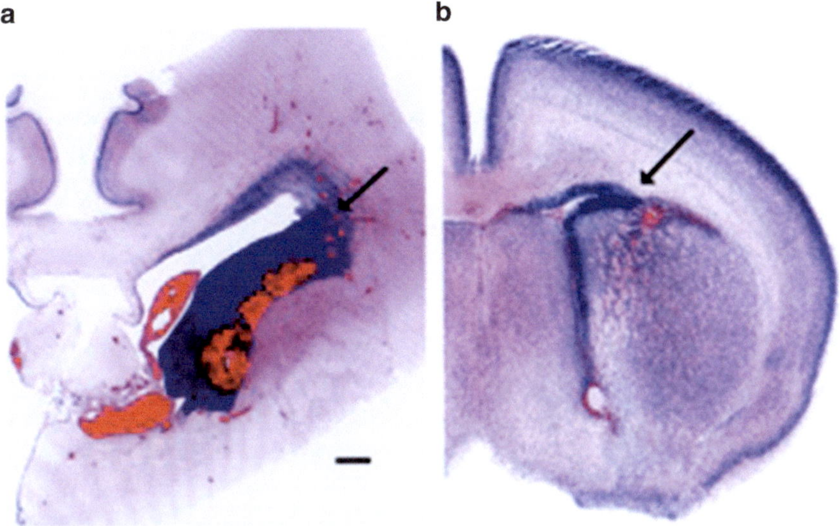

Fig. 2 Hematoxylin and eosin stained coronal brain sections of periventricular germinal tissue from a human fetus (~25 weeks gestational age) obtained after several hours survival (**a**), and a neonatal rat brain obtained 24 h after blood injection (**b**). The hematoma (*red*) is seen in the germinal matrix (*blue, arrow*) and in the lateral ventricles (*white space*) of both

widespread but not reproducible hemorrhages distal to the hematoma. In addition, other drawbacks include the significant inflammation that is seen in this model but not in human clinical pathology [47].

Mutant mice have also been used to address the hypotheses concerning specific GMH mechanisms. For example, alpha V integrin knockout mice develop spontaneous ICH in utero, likely because of impaired endothelial cell adhesion [48].

6.3 Behavioral Testing

For detailed GMH/IVH rodent behavioral testing and results please refer to Balasubramaniam et al. [49] and Xue et al. [42]. Of note, most of the neurobehavioral tests are well described in rats and not in mice likely because mice are small, fast, and difficult to train. Testing in the early period gives an indication of developmental delay. Testing in the adolescent to young adult period (i.e., after 6 weeks in mice or 8 weeks in rats) gives an indication of deficits that might be permanent.

6.3.1 Early Behavior Tests

The surface righting response is measured by gently placing pups onto their backs on a cotton sheet. The time to return to prone position and the directional preferences are recorded daily between days 4 and 15. The duration of the test is limited to 2 min and is done in triplicate [50].

Negative geotaxis test reflects the pups' preference to face upward on a sloped surface. The pup is placed facing head down on a slope inclined at 25° plane. The pup is held gently in the starting position for 5 s before being released. The time and direction are recorded when the pup turns to face up the incline. The maximum allotted time can be 2 min. The test is replicated three times.

Early ambulation can be assessed daily or on alternate days between days 4 and 15. Each pup placed on a clean surface (e.g., 50 cm×50 cm blotting paper) is observed for 5 min (triplicate trials). Movements can be categorized as follows: No movement = 0, asymmetric limb movements = 1, slow crawling = 2, and fast walking = 3.

To assess forelimb grip strength, pups are allowed to grasp a wire (1.5-mm diameter, 70 cm long between two poles at a height of 40 cm with a soft sponge placed underneath) with their forepaws beginning with day 11. The time to falling is measured with the maximum allotted time of 2 min.

Figure 3 shows a typical test assessing autologous blood infused in newborn rodent in this behavioral experiment. In the righting test, on days 4 and 8, blood infused rats were significantly slower compared to sham or intact control rats. However, by day 11, when all rats righted very quickly, no difference and no directional preferences were observed. In the same model, the average ambulation score of blood infused rats on days 4, 8, 11, and 15 was significantly ($p < 0.05$) lower compared to that of the age-matched intact and sham controls. In the negative geotaxis on day 4, most

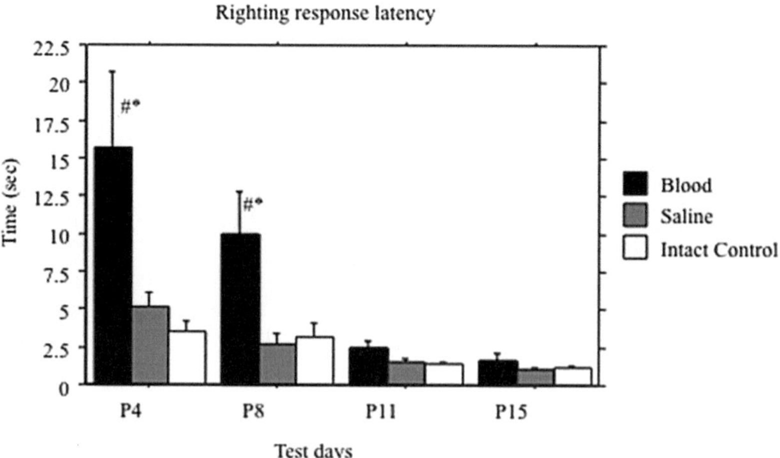

Fig. 3 Righting response in rats with PVH relative to saline and intact control rats assessed on days 4, 8, 11, and 15. Blood injected rats showed significantly longer latency than sham or intact control groups at day 4 ($F = 2.247$, $p = 0.0315$) and day 8 ($F = 2.254$, $p = 0.0161$). * indicates intact control vs. PVH (day 4 $p = 0.0315$ and day 8 $p = 0.0461$) and # indicates saline control vs. PVH (day 4 $p = 0.0762$ and day 8 $p = 0.0349$). Data are expressed as mean ± SEM and were analyzed with 2-way ANOVA followed by Scheffe's post hoc test

pups did not turn and received the maximum score. On day 8 and day 15, blood infused rats took significantly longer to turn to the incline position. These results are similar to the findings by Aquilina et al. in which IVH animals failed to improve on the negative geotaxis test at 2 weeks [41]. In our model, the wire-hanging test revealed no significant differences between the groups; however, the test may not be sensitive to very early injury models.

6.4 Adult Behavioral Testing

Pups should be housed with their mother until weaning, i.e., postnatal day 21. They should be then placed in standard housing plastic cages with 2–3 the same gender in each cage. Anytime after week 3 juveniles can be tested for spontaneous behaviors that do not require training.

To assess forepaw preference, rats are placed in a transparent cylinder (20 cm diameter, 30 cm height) for 5 min [51]. The natural response is to search for an escape route. Mirrors placed behind the cylinder at angles of 45° and 135° or video recording from below through a transparent floor enables the observer to record forelimb movements when the animal is turned away from camera. The number of times the rat rears and the paw that is initially used to support its body against the cylinder wall are documented on video recording and later analyzed frame by frame.

Ambulatory agility can be assessed using a rotating cylinder (7-cm diameter) ("rotorod"). First, endurance at a constant speed (e.g., 5 rpm) is assessed for a maximum of 2 min. Second, the ability to stay on the cylinder at an accelerating speed is tested (e.g., beginning at 2.5 rpm and increasing at a rate of 0.1 rpm every second) for up to 2 min. The time is measured from the moment the rat is placed on the cylinder until it falls off. Each animal can be given three trials.

After observing motor behaviors in forced novel situations, rats can receive complex training in skilled reaching behavior or ladder/beam walking [52]. For the skilled reaching test, food should be restricted to about 80 % of free feeding amount (~15 g/day) for up to 7 days to reduce the body weight by 5–15 %. Rats should be weighed daily and habituated to the special food (we use 45 mg 3.5 mm diameter; Rodent "Dustless" Precision Pellets; Bio-Serv, Frenchtown NJ) during this period. Daily 20–30-min skill reaching training of each rat lasts approximately 2 weeks. It is a tedious and long process; therefore, it should be done with manageable number of animals. Rats are placed in a Plexiglas box (25 × 25 × 30 cm high) that has a narrow opening (10 mm wide slot) extending from the floor to a height of 15 cm. On the outside wall, in front of the slot, is a 2 cm wide by 4 cm long shelf 3 cm above the floor. During the training period, rats are placed in the box until 20–25 pellets had been retrieved and eaten with the forepaw of their choice. During the second training week, the forelimb ipsilateral to the brain lesion of all rats (including sham and intact control) is wrapped with an adhesive bandage to force the rats to use the limb contralateral to the injection site [52]. As an alternative to wrapping the forelimb, the feeding box can be constructed with the slot 1–2 cm from the side wall (rather than in the center) of the limb to be tested; this position prevents comfortable use of the contralateral forelimb. The testing period (approximately 2 weeks after the training period) consists of exposure to similar feeding conditions for 7 consecutive days. A restricted number (~20) of pellets are offered individually. Reaches are only considered successful if the pellet was grasped and eaten on the first attempt. Failed attempts include all limb advances resulting in a dropped, missed, or displaced pellet. For the qualitative analysis, eating is documented on videotape and detail forelimb movements are scored according to Whishaw and coworkers [52, 53]. The single pellet retrieval task examines subtle impairments.

In our model, blood-infused rats retrieved significantly fewer pellets than did controls on the skilled reaching test. This is consistent with previous findings on other unilateral lesions [28, 54]. Qualitative evaluation of the reaching movement using frame-by-frame video analysis showed that rats exhibited marked impairment in advance, digit extension, grasp, supination, and release phases.

A potential alternative to the skilled reaching test is the staircase-feeding test, which assesses the independent use of forelimbs in skilled reaching and grasping tasks. This was described in detail by Montoya and coworkers [55]. They use a plexiglas box with a removable baited double staircase. Food pellets are placed on the staircase and presented bilaterally at 7 graded stages of reaching difficulty to provide objective measures of side bias, maximum forelimb extension, and grasping skill. The apparatus can be used to assess the reaching performance of rats following unilateral lesions of the sensorimotor cortex, unilateral lesions of the posterior cortex, or bilateral lesions of the olfactory bulbs. The task has the advantage of objective over rating measurement, and the simplicity of the apparatus permits many animals to be tested concurrently. The speed to climb and the number of foot slips are recorded [56].

The investigator must note that many of these tests were developed for use in adult animals and required training prior to injuring of the brain. In that circumstance premorbid side preference could be ascertained and post injury decrements could be easily detected. Use of these tests in animals that have sustained perinatal brain injury is associated with the caveat that there is no opportunity for premorbid training and the possibility that the training period might also have a rehabilitation effect that reduces the actual difference [57–59].

6.5 Long-Term Imaging and Histology

At the end of the behavioral testing, juvenile or adult animals can be imaged for structural and volumetric analysis. Subsequently, the animals may be euthanized for a broad range of investigations.

We have used enzyme linked immunosorbent assays (ELISA) to quantify proteins related to synapse (synaptophysin), astrocytes (glial fibrillary acidic protein), and myelin (myelin basic protein). We have also fixed brain tissue in aldehydes for histological or immunohistochemical assessments, Golgi impregnation to demonstrate neuronal dendrites, and for electron microscopy.

7 Notes

7.1 General Notes on Experimental Models of Germinal Matrix Hemorrhage

1. Developmental brain anatomy and physiology are not well documented in all species that are used to model GMH, making interspecies extrapolation of information difficult. One must consider carefully the state of maturity of the periventricular tissues if one desires to model GMH. Animal models should have a prominent GMH in locations comparable to that of humans, i.e., the ganglionic eminence. There are obvious advantages of studying species whose brains, compared to human brains, are less mature at the normal time of birth because their brains resemble those of premature humans who are at risk for GMH. However, there are discrepancies between the relative maturity of other organ systems, most obviously pulmonary, making one consider that other less obvious differences (e.g., endocrine/hormonal) might influence the presentation and pathogenesis of GMH in those animal models. Large animals with gyrencephalic brains probably have greater similarity to humans with respect to brain damage pathogenesis; however, they are expensive and more complex.

2. A general lack of understanding of the pathogenesis of brain damage follows GMH in the immature brain. Hemostasis and hemostatic proteins differ between the fetus (and premature infant) and the full-term infant [60–62]. Different blood components have different effects on surrounding tissues. Using cell culture studies we have shown that soluble blood components, particularly thrombin, have an adverse effect on maturing subventricular zone cells and oligodendrocyte progenitor cells derived from newborn rat brain [63]. Furthermore, in a blood infusion neonatal mouse model [35] and in a study of human infants and rats (unpublished data), reduced cellular proliferation has been observed in the germinal matrix near the site of the hematoma.

3. Experimental GMH/IVH is induced under extreme physiological conditions, which are imposed on healthy animals to mimic the clinical scenarios of premature infants. The physiological parameters are controlled in order to understand the variables; however, they may oversimplify the human situation. For instance, hematomas created by injection of blood into the

GM region do not mimic the human situation, but they do offer a simple way of understanding the pathogenesis of cellular brain damage that follows GMH.

4. Imaging and volumetric analysis become crucial in studies on structural and functional correlations. When possible, initial hemorrhage sites and sizes as well as final brain volumes and ventricular sizes should be documented and quantified by imaging modalities such as magnetic resonance or computed tomography. GMH produces motor disabilities and reduces cortical volume in human infants [64, 65]. Imaging studies of premature infants show cerebral volume reductions that correlate to some extent with late developmental outcome [65–69]. In the blood infused rats, we observed a significant decrease in right hemispheric brain volume and atrophy in the striatum independent of large cortical lesions. Additionally, we detected behavioral abnormalities in blood infused animals with large structural defects. This finding is consistent with other behavioral studies that have shown a relationship between severity of lesion and functional deficit [70]. In our model, the initial hematoma and the final brain volumes did not correlate strongly. This finding is consistent with human data showing that periventricular hemorrhage size is not a good predictor of functional outcome [71, 72].

7.2 Points to Consider Before Designing Behavioral Tests

1. It is important to choose behavioral tests that are age specific. Newborn animal behavioral tests are limited, as very few studies have looked at early neurological behaviors [14, 21]. In addition, older behavioral testing used animals that received injury in the adult age and not neonatal period. Thus, when an injury occurs very early on in immature animals, some of the available adult behavioral tests may not be sensitive enough to detect abnormalities. For example, in our study, tests used to detect limb use asymmetry such as cylinder exploration was not sensitive, although this test has been shown to be effective in other unilateral injuries of mature brain [51].

2. It is crucial to recognize the type and location of injury in a model before designing behavioral tests. For example, our model produces a predominantly unilateral lesion; therefore, we chose behavioral tests that are sensitive to lateralization. Furthermore, considering the location of injury, we chose tests that assess sensorimotor deficits with minimal adverse effect on learning and memory.

3. Training could affect brain recovery; therefore, behavioral tests should be organized so that non-trained spontaneous behaviors (e.g., cylinder exploration) are assessed first, followed by more complex behaviors that require minimal learning (walking on rotating cylinder and ladder) and finally behaviors that

require complex training (e.g., the skilled reaching test). It is important to note that it is not practical to test all rats when a test requires extensive training (e.g., skilled reaching test) because of the considerable training time required. Statistical analyses should be used to calculate the sufficient number of animals to detect statistically significant results. Moreover, tests that are done to explore subtle changes (e.g., cylinder exploration, ladder walking, and skilled reaching tests) should be video recorded and analyzed using a frame-by-frame analysis method because rodents move very fats and subtle deficits are easily missed [52].

8 Conclusion

It is clear that no single animal models are suitable for the study of all physiological, developmental, pathological, and molecular aspects of GMH/IVH. We acknowledge the physiological and anatomical differences between rodent and human brains are substantial, and it is very difficult to monitor the former. Nevertheless, the presence of a prominent GM in the postnatal period, its affordability and suitability for studying long-term behavioral outcomes makes it an attractive model. This model could have value for testing specific hypothesis concerning pathophysiology of perinatal intracerebral hemorrhage as well as therapeutic studies.

References

1. Volpe JJ (2001) Neurology of the newborn, 4th edn. W.B. Saunders, Philadelphia, PA

2. Del Bigio MR (2004) Hemorrhagic lesions. In: Golden JA, Harding BN (eds) Developmental neuropathology. International Society of Neuropathology Press, Basel, Switzerland, pp 150–155

3. Leviton A, Gilles FH, Dooling EC (1983) The epidemiology of ganglionic eminence hemorrhage. In: Gilles FH, Leviton A, Dooling EC (eds) The developing human brain. Growth and epidemiologic neuropathology. John Wright Inc., Boston, MA, pp 204–216

4. Volpe JJ (2001) Perinatal brain injury: from pathogenesis to neuroprotection. Ment Retard Dev Disabil Res Rev 7:56–64

5. Kakita A, Goldman JE (1999) Patterns and dynamics of SVZ cell migration in the postnatal forebrain: monitoring living progenitors in slice preparations. Neuron 23:461–472

6. Papile LA, Burstein J, Burstein R, Koffler H (1978) Incidence and evolution of subependymal and intraventricular hemorrhage: a study of infants with birth weights less than 1,500 gm. J Pediatr 92:529–534

7. Balasubramaniam J, Del Bigio MR (2006) Animal models of germinal matrix hemorrhage. J Child Neurol 21:365–371

8. Dobbing J, Sands J (1979) Comparative aspects of the brain growth spurt. Early Hum Dev 3:79–83

9. Romijn HJ, Hofman MA, Gramsbergen A (1991) At what age is the developing cerebral cortex of the rat comparable to that of the full-term newborn human baby? Early Hum Dev 26:61–67

10. Levers TE, Edgar JM, Price DJ (2001) The fates of cells generated at the end of neurogenesis in developing mouse cortex. J Neurobiol 48:265–277

11. Zhu Y, Li H, Zhou L, Wu JY, Rao Y (1999) Cellular and molecular guidance of GABAergic neuronal migration from an extracortical origin to the neocortex. Neuron 23:473–485

12. Sturrock RR, Smart IH (1980) A morphological study of the mouse subependymal layer from embryonic life to old age. J Anat 130:391–415

13. Fox WM (1965) Reflex ontogeny and behavioral development of the mouse. Anim Behav 13:234–241

14. Altman J, Sudarshan K (1975) Postnatal development of locomotion in the laboratory rat. Anim Behav 23:896–920

15. MacDonald BK, Cockerell OC, Sander JW, Shorvon SD (2000) The incidence and lifetime prevalence of neurological disorders in a prospective community-based study in the UK. Brain 123:665–676

16. Rubin RJ, Gold WA, Kelley DK, Sher JP (1992) The cost of disorders of the brain. National Foundation for Brain Research/Lewin-ICF, Washington, DC

17. Luu TM, Ment LR, Schneider KC, Katz KH, Allan WC, Vohr BR (2009) Lasting effects of preterm birth and neonatal brain hemorrhage at 12 years of age. Pediatrics 123:1037–1044

18. Shapiro BK (2004) Cerebral palsy: a reconceptualization of the spectrum. J Pediatr 145:S3–S7

19. Davison AN, Dobbing J (1966) Myelination as a vulnerable period in brain development. Br Med Bull 22:40–44

20. Eayrs JT, Goodhead B (1959) Postnatal development of the cerebral cortex in the rat. J Anat 93:385–402

21. Westerga J, Gramsbergen A (1990) The development of locomotion in the rat. Brain Res Dev Brain Res 57:163–174

22. Smart JL, Dobbing J (1971) Vulnerability of developing brain. VI. Relative effects of foetal and early postnatal undernutrition on reflex ontogeny and development of behaviour in the rat. Brain Res 33:303–314

23. Bolles R, Woods P (1964) The ontogeny of behavior in the albino rat. Anim Behav 12:427

24. Bignall KE (1974) Ontogeny of levels of neural organization: the righting reflex as a model. Exp Neurol 42:566–573

25. Unis AS, Petracca F, Diaz J (1991) Somatic and behavioral ontogeny in three rat strains: preliminary observations of dopamine-mediated behaviors and brain D-1 receptors. Prog Neuropsychopharmacol Biol Psychiatry 15:129–138

26. Vorhees CV, Acuff-Smith KD, Moran MS, Minck DR (1994) A new method for evaluating air-righting reflex ontogeny in rats using prenatal exposure to phenytoin to demonstrate delayed development. Neurotoxicol Teratol 16:563–573

27. Whishaw IQ (2000) Loss of the innate cortical engram for action patterns used in skilled reaching and the development of behavioral compensation following motor cortex lesions in the rat. Neuropharmacology 39:788–805

28. Ballermann M, Tompkins G, Whishaw IQ (2000) Skilled forelimb reaching for pasta guided by tactile input in the rat as measured by accuracy, spatial adjustments, and force. Behav Brain Res 109:49–57

29. Xue M, Balasubramaniam J, Buist RJ, Peeling J, Del Bigio MR (2003) Periventricular/intraventricular hemorrhage in neonatal mouse cerebrum. J Neuropathol Exp Neurol 62:1154–1165

30. Lee JC, Cho GS, Choi BO, Kim HC, Kim YS, Kim WK (2006) Intracerebral hemorrhage-induced brain injury is aggravated in senescence-accelerated prone mice. Stroke 37:216–222

31. Qu Y, Chen-Roetling J, Benvenisti-Zarom L, Regan RF (2007) Attenuation of oxidative injury after induction of experimental intracerebral hemorrhage in heme oxygenase-2 knockout mice. J Neurosurg 106:428–435

32. Rynkowski MA, Kim GH, Komotar RJ, Otten ML, Ducruet AF, Zacharia BE, Kellner CP, Hahn DK, Merkow MB, Garrett MC, Starke RM, Cho BM, Sosunov SA, Connolly ES (2008) A mouse model of intracerebral hemorrhage using autologous blood infusion. Nat Protoc 3:122–128

33. Zhao X, Sun G, Zhang J, Strong R, Dash PK, Kan YW, Grotta JC, Aronowski J (2007) Transcription factor Nrf2 protects the brain from damage produced by intracerebral hemorrhage. Stroke 38:3280–3286

34. Belayev A, Saul I, Liu Y, Zhao W, Ginsberg MD, Valdes MA, Busto R, Belayev L (2003) Enriched environment delays the onset of hippocampal damage after global cerebral ischemia in rats. Brain Res 964:121–127

35. Xue M, Balasubramaniam J, Del Bigio MR (2003) Brain inflammation following intracerebral hemorrhage. Clin Neuropharmacol 1:325–332

36. Johnson DL, Getson P, Shaer C, O'Donnell R (1987) Intraventricular hemorrhage in the newborn beagle puppy. A limited model of intraventricular hemorrhage in the premature infant. Pediatr Neurosci 13:78–83

37. Ment LR, Stewart WB, Duncan CC, Lambrecht R (1982) Beagle puppy model of intraventricular hemorrhage. J Neurosurg 57:219–223

38. Lorenzo AV, Welch K, Conner S (1982) Spontaneous germinal matrix and intraventricular hemorrhage in prematurely born rabbits. J Neurosurg 56:404–410

39. Cherian SS, Love S, Silver IA, Porter HJ, Whitelaw AG, Thoresen M (2003) Posthemorrhagic ventricular dilation in the neonate: development and characterization of a rat model. J Neuropathol Exp Neurol 62:292–303

40. Cherian S, Whitelaw A, Thoresen M, Love S (2004) The pathogenesis of neonatal post-hemorrhagic hydrocephalus. Brain Pathol 14:305–311

41. Aquilina K, Chakkarapani E, Love S, Thoresen M (2011) Neonatal rat model of intraventricular haemorrhage and post-haemorrhagic ventricular dilatation with long-term survival into adulthood. Neuropathol Appl Neurobiol 37:156–165

42. Xue M, Balasubramaniam J, Parsons KA, McIntyre IW, Peeling J, Del Bigio MR (2005) Does thrombin play a role in the pathogenesis of brain damage after periventricular hemorrhage? Brain Pathol 15:241–249

43. Kolb B, Tomie JA (1988) Recovery from early cortical damage in rats. IV. Effects of hemidecortication at 1, 5 or 10 days of age on cerebral anatomy and behavior. Behav Brain Res 28:259–274

44. Del Bigio MR, Yan HJ, Buist R, Peeling J (1996) Experimental intracerebral hemorrhage in rats. Magnetic resonance imaging and histopathological correlates. Stroke 27:2312–2320

45. Rosenberg GA, Estrada E, Kelley RO, Kornfeld M (1993) Bacterial collagenase disrupts extracellular matrix and opens blood-brain barrier in rat. Neurosci Lett 160:117–119

46. Clark W, Gunion Rinker L, Lessov N, Hazel K (1998) Citicoline treatment for experimental intracerebral hemorrhage in mice. Stroke 29:2136–2139

47. Xue M, Del Bigio MR (2003) Comparison of brain cell death and inflammatory reaction in three models of intracerebral hemorrhage in adult rats. J Stroke Cerebrovasc Dis 12:152–159

48. McCarty JH, Monahan-Earley RA, Brown LF, Keller M, Gerhardt H, Rubin K, Shani M, Dvorak HF, Wolburg H, Bader BL, Dvorak AM, Hynes RO (2002) Defective associations between blood vessels and brain parenchyma lead to cerebral hemorrhage in mice lacking alphav integrins. Mol Cell Biol 22:7667–7677

49. Balasubramaniam J, Xue M, Buist RJ, Ivanco TL, Natuik S, Del Bigio MR (2006) Persistent motor deficit following infusion of autologous blood into the periventricular region of neonatal rats. Exp Neurol 197:122–132

50. Schroeder H, Humbert AC, Koziel V, Desor D, Nehlig A (1995) Behavioral and metabolic consequences of neonatal exposure to diazepam in rat pups. Exp Neurol 131:53–63

51. Schallert T, Fleming SM, Leasure JL, Tillerson JL, Bland ST (2000) CNS plasticity and assessment of forelimb sensorimotor outcome in unilateral rat models of stroke, cortical ablation, parkinsonism and spinal cord injury. Neuropharmacology 39:777–787

52. Whishaw IQ, O'Connor WT, Dunnett SB (1986) The contributions of motor cortex, nigrostriatal dopamine and caudate-putamen to skilled forelimb use in the rat. Brain 109(Pt 5):805–843

53. Vergara-Aragon P, Gonzalez CLR, Whishaw IQ (2003) A novel skilled-reaching impairment in paw supination on the "good" side of the hemi-Parkinson rat improved with rehabilitation. J Neurosci 23:579–586

54. Whishaw IQ, Woodward NC, Miklyaeva E, Pellis SM (1997) Analysis of limb use by control rats and unilateral DA-depleted rats in the Montoya staircase test: movements, impairments and compensatory strategies. Behav Brain Res 89:167–177

55. Montoya CP, Campbell-Hope LJ, Pemberton KD, Dunnett SB (1991) The "staircase test": a measure of independent forelimb reaching and grasping abilities in rats. J Neurosci Methods 36:219–228

56. Metz GA, Whishaw IQ (2002) Cortical and subcortical lesions impair skilled walking in the ladder rung walking test: a new task to evaluate fore- and hindlimb stepping, placing, and coordination. J Neurosci Methods 115:169–179

57. Maclellan CL, Plummer N, Silasi G, Auriat AM, Colbourne F (2011) Rehabilitation promotes recovery after whole blood-induced intracerebral hemorrhage in rats. Neurorehabil Neural Repair 25:477

58. Auriat AM, Wowk S, Colbourne F (2010) Rehabilitation after intracerebral hemorrhage in rats improves recovery with enhanced dendritic complexity but no effect on cell proliferation. Behav Brain Res 214:42–47

59. Kolb B, Teskey GC (2012) Age, experience, injury, and the changing brain. Dev Psychobiol 54:311

60. Manco-Johnson MJ, Jacobson LJ, Hacker MR, Townsend SF, Murphy J, Hay W Jr (2002) Development of coagulation regulatory proteins in the fetal and neonatal lamb. Pediatr Res 52:580–588

61. Manco-Johnson MJ (2005) Development of hemostasis in the fetus. Thromb Res 115(Suppl 1):55–63

62. Kuhle S, Male C, Mitchell L (2003) Developmental hemostasis: pro- and anticoagulant systems during childhood. Semin Thromb Hemost 29:329–338

63. Juliet PA, Frost EE, Balasubramaniam J, Del Bigio MR (2009) Toxic effect of blood components on perinatal rat subventricular zone cells and oligodendrocyte precursor cell

proliferation, differentiation and migration in culture. J Neurochem 109:1285–1299

64. Pinto-Martin JA, Riolo S, Cnaan A, Holzman C, Susser MW, Paneth N (1995) Cranial ultrasound prediction of disabling and nondisabling cerebral palsy at age two in a low birth weight population. Pediatrics 95:249–254

65. Vasileiadis GT, Gelman N, Han VK, Williams LA, Mann R, Bureau Y, Thompson RT (2004) Uncomplicated intraventricular hemorrhage is followed by reduced cortical volume at near-term age. Pediatrics 114:e367–e372

66. Inder TE, Huppi PS, Warfield S, Kikinis R, Zientara GP, Barnes PD, Jolesz F, Volpe JJ (1999) Periventricular white matter injury in the premature infant is followed by reduced cerebral cortical gray matter volume at term. Ann Neurol 46:755–760

67. Peterson BS, Vohr B, Staib LH, Cannistraci CJ, Dolberg A, Schneider KC, Katz KH, Westerveld M, Sparrow S, Anderson AW, Duncan CC, Makuch RW, Gore JC, Ment LR (2000) Regional brain volume abnormalities and long-term cognitive outcome in preterm infants. JAMA 284:1939–1947

68. Weisglas-Kuperus N, Baerts W, Sauer PJ (1993) Early assessment and neurodevelopmental outcome in very low-birth-weight infants: implications for pediatric practice. Acta Paediatr 82:449–453

69. Whitaker AH, Van Rossem R, Feldman JF, Schonfeld IS, Pinto-Martin JA, Tore C, Shaffer D, Paneth N (1997) Psychiatric outcomes in low-birth-weight children at age 6 years: relation to neonatal cranial ultrasound abnormalities. Arch Gen Psychiatry 54:847–856

70. Tomimatsu T, Fukuda H, Endoh M, Mu J, Watanabe N, Kohzuki M, Fujii E, Kanzaki T, Oshima K, Doi K, Kubo T, Murata Y (2002) Effects of neonatal hypoxic-ischemic brain injury on skilled motor tasks and brainstem function in adult rats. Brain Res 926:108–117

71. Dubowitz LM, Levene MI, Morante A, Palmer P, Dubowitz V (1981) Neurologic signs in neonatal intraventricular hemorrhage: a correlation with real-time ultrasound. J Pediatr 99:127–133

72. Palmer P, Dubowitz LM, Levene MI, Dubowitz V (1982) Developmental and neurological progress of preterm infants with intraventricular haemorrhage and ventricular dilatation. Arch Dis Child 57:748–753

73. Coulter DM, LaPine T, Gooch WM 3rd (1984) Intraventricular hemorrhage in the premature rabbit pup. Limitations of this animal model. J Neurosurg 60:1243–1245

74. Conner ES, Lorenzo AV, Welch K, Dorval B (1983) The role of intracranial hypotension in neonatal intraventricular hemorrhage. J Neurosurg 58:204–209

75. Greene CS Jr, Lorenzo AV, Hornig G, Welch K (1985) The lowering of cerebral spinal fluid and brain interstitial pressure of preterm and term rabbits by furosemide. Z Kinderchir 40(Suppl 1):5–8

76. Sugimoto T, Yasuhara A, Matsumura T (1981) Intracranial hemorrhage following administration of sodium bicarbonate in rabbits. Brain Dev 3:297–303

77. Reynolds ML, Evans CA, Reynolds EO, Saunders NR, Durbin GM, Wigglesworth JS (1979) Intracranial haemorrhage in the preterm sheep fetus. Early Hum Dev 3:163–186

78. Ting P, Yamaguchi S, Bacher JD, Killens RH, Myers RE (1984) Failure to produce germinal matrix or intraventricular hemorrhage by hypoxia, hypo-, or hypervolemia. Exp Neurol 83:449–460

79. Wheeler AS, Sadri S, Gutsche BB, DeVore JS, David-Mian Z, Latyshevsky H (1979) Intracranial hemorrhage following intravenous administration of sodium bicarbonate or saline solution in the newborn lamb asphyxiated in utero. Anesthesiology 51:517–521

80. Stankovic MR, Fujii A, Maulik D, Boas D, Kirby D, Stubblefield PG (1998) Optical monitoring of cerebral hemodynamics and oxygenation in the neonatal piglet. J Matern Fetal Investig 8:71–78

81. Goddard J, Lewis RM, Armstrong DL, Zeller RS (1980) Moderate, rapidly induced hypertension as a cause of intraventricular hemorrhage in the newborn beagle model. J Pediatr 96:1057–1060

82. Goddard J, Lewis RM, Alcala H, Zeller RS (1980) Intraventricular hemorrhage – an animal model. Biol Neonate 37:39–52

83. Goddard-Finegold J, Armstrong D, Zeller RS (1982) Intraventricular hemorrhage, following volume expansion after hypovolemic hypotension in the newborn beagle. J Pediatr 100:796–799

84. Leuschen MP, Nelson RM Jr (1987) Effects of asphyxia on telencephalic microvessels of premature beagle pups. J Perinatol 7:93–99

85. Turbeville DF, Bowen FW Jr, Killam AP (1976) Intracranial hemorrhages in kittens: hypernatremia versus hypoxia. J Pediatr 89:294–297

86. Dieni S, Inder T, Yoder B, Briscoe T, Camm E, Egan G, Denton D, Rees S (2004) The pattern of cerebral injury in a primate model of preterm birth and neonatal intensive care. J Neuropathol Exp Neurol 63:1297–1309

Chapter 4

Preterm Rabbit Model of Glycerol-Induced Intraventricular Hemorrhage

Praveen Ballabh

Abstract

Intraventricular hemorrhage (IVH) remains a major complication of prematurity. We have developed a model of IVH in preterm rabbit pups. We deliver rabbit pups prematurely by C-section at 29-day gestational age (term = 32 days) and inject intraperitoneal glycerol at 3 h age to induce IVH. About 80 % of glycerol-treated pups develop IVH within 8 h age, which can be detected accurately with head ultrasound. These pups are reared in an infant incubator and gavage fed. At postnatal day 14, they exhibit hypomyelination, gliosis, and impaired neurological function. Hence, this is a novel animal model of IVH that can be used to test strategies to prevent post-hemorrhagic complication of IVH.

Key words Intraventricular hemorrhage, Glycerol, Germinal matrix, Hypomyelination, Gliosis, Head ultrasound

List of Abbreviations

FA   Fractional anisotropy
GFAP   Glial fibrillary acidic protein
IVH   Intraventricular hemorrhage
MRI   Magnetic resonance imaging
VEGF   Vascular endothelial growth factor

1   Background and Overview

About 12,000 premature infants develop intraventricular hemorrhage (IVH) every year in the USA alone [1, 2]. The incidence of IVH in very low birth weight infants (<1,500 g) has reduced from 40–50 % in the early 1980s to 20 % in the late 1980s [3]. However, over the last two decades the occurrence of IVH has remained almost stationary [4]. In extremely premature infants weighing 500–750 g, IVH occurs in about 45 % of infants [5]. IVH results in neurologic sequelae including cerebral palsy, mental retardation,

Jerome Y. Yager (ed.), *Animal Models of Neurodevelopmental Disorders*, Neuromethods, vol. 104, DOI 10.1007/978-1-4939-2709-8_4, © Springer Science+Business Media New York 2015

cognitive deficits, and post-hemorrhagic hydrocephalus. Since both the prematurity rate and survival of preterm infants have increased, IVH has remained a major public health concern.

IVH typically initiates in the periventricular germinal matrix (ganglionic eminence). The germinal matrix is located on the head of caudate nucleus and underneath ventricular ependyma. It is a highly vascular collection of glial and neuronal precursor cells. This brain region has selective propensity to hemorrhage in premature infants during the first 72 h of life. When the hemorrhage in the germinal matrix is considerable, the ependyma breaks and blood leaks into the ventricle, giving rise to IVH. Thus, IVH is a natural progression of germinal matrix hemorrhage. These premature infants are usually asymptomatic and the diagnosis is made on screening cranial ultrasound. Some infants exhibit subtle abnormalities in the level of consciousness, movement, tone, respiration, and eye movement; and uncommonly, there is a rapid deterioration, when the premature newborns present with stupor, coma, decerebrate posturing, generalized tonic seizures, and quadriparesis. The etiology of IVH is multifactorial and is primarily ascribed to the intrinsic fragility of the germinal matrix vasculature and disturbances in the cerebral blood flow [6]. The germinal matrix microvasculature is fragile because of an abundance of angiogenic blood vessels that exhibit a paucity of pericytes, immaturity of the basal lamina, and a deficiency of glial fibrillary acidic protein (GFAP) in the ensheathing astrocytic endfeet [7–11]. A rapid angiogenesis in the germinal matrix is attributed to the presence of high VEGF and angiopoietin-2 levels in this periventricular brain region [8]. Premature infants rarely develop IVH in utero. However, after birth they are exposed to a number of risk factors, including vaginal delivery, low Apgar score, severe respiratory distress syndrome, pneumothorax, hypoxia, hypercapnia, seizures, patent ductus arteriosus (PDA), infection, and others, which triggers IVH by inducing fluctuation in the cerebral blood flow [12–14].

A number of animal species, including rabbit, beagle, and rodents, have been used to model IVH [15]. We use prematurely delivered rabbit pups to model IVH because of certain close similarities of rabbits to the human conditions. These include; (1) perinatal brain development, (2) a relatively large size of the germinal matrix, (3) the blood supply to the brain by internal carotid and vertebral arteries, (4) completed maturation of the lungs a few days before term, making them capable of survival premature birth, and (5) a risk for spontaneous IVH. A number of interventions, including administration of hyperosmolar or hypertensive agents, induction of hypoxia, hypercapnia, or hypervolemia have been done to induce IVH in these animals [16–18]. In addition, blood has been injection of blood into the germinal matrix to mimic IVH [23]. We use intraperitoneal glycerol (hyperosmolar agent) for induction of IVH in premature rabbit pups because it neither causes direct injury to the brain by needle insertion nor confounds the model with unwanted metabolic changes in the neural cells such as hypoxic-ischemia or hypercapnia. Glycerol treatment results in intravascular dehydration and an elevation in serum osmolarity, which is attended by a decline in intracranial pressure and a consequential increase in transmural pressure across the vessel wall, causing rupture of the microvasculature [18]. Importantly, the development of a germinal matrix hemorrhage in this model is an effect of intracranial hypotension and not of the osmotic agent per se, as there is no hemorrhage when a fall in intracranial pressure is prevented by the intracisternal infusion of saline [19].

2   Equipment, Materials, and Methods

Timed pregnant New Zealand rabbits are used in all of our experiments. The pups are delivered by C-section at a gestational age of 29 days (full-term is 32 days). The pups are dried and kept warm in an infant incubator which is maintained at a temperature of 35 °C. We use a human infant incubator (Ohmeda, Ohio Care Plus incubator) to house the premature pups. The pups are fed 1–2 mL Esbilac (PETAG Inc) at 4 h of age and then 2 mL every 12 h (100–120 mL/kg/day) using a 3.5 French feeding tube. After day 2, we advance feeds to 125 mL/kg, 150 mL/kg, 200 mL/kg, 250 mL/kg, and 280 mL/kg on postnatal days 3, 5, 7, 10, and 14, respectively. At the age of 3–5 h, the pups are treated with an intraperitoneal injection of 50 % glycerol (6.5 g/kg). Injection of glycerol is done in the right flank about 2 cm lateral to the umbilicus. Head ultrasound (Acuson Sequoia C256 system, Siemens) is performed at 6 and 24 h of postnatal age to assess for the presence and severity of IVH. Brains without gross hemorrhage are evaluated for microscopic hemorrhages after H and E staining of the sections. IVH is classified as (1) mild, no gross signs but microscopic hemorrhage detected in hematoxylin and eosin stained brain sections; (2) moderate, gross hemorrhage into lateral ventricles (two separate lateral ventricles discerned); or (3) severe, gross IVH leading to fusion of lateral ventricles into a common chamber (Fig. 1) [20]. We also quantify IVH by measuring dimensions of the bleed in coronal and sagittal views (horizontal width × vertical width × depth) of the head ultrasound.

C-section in pregnant rabbit: We perform C-section on pregnant rabbits (E29) under inhaled isoflurane anesthesia. When the animal is unresponsive to toe pinch, a 5 cm midline incision is made on the lower abdomen and uterus is exposed. Uterine incision is made to deliver the pups. The dam is humanely sacrificed with pentobarbital 100 mg/kg IV.

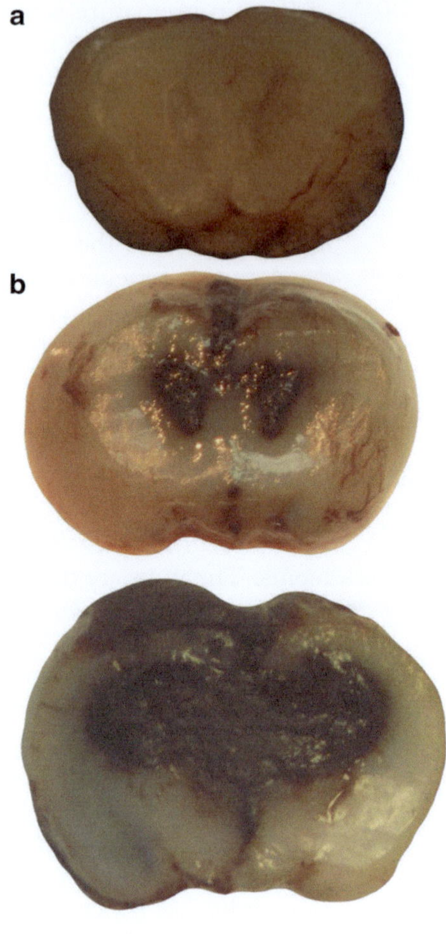

Fig. 1 GM hemorrhage in rabbit pups. (**a**) Coronal brain section of a premature rabbit pup brain at the level of midseptal nucleus shows a normal slit-like ventricle (*arrowhead*) indicating no IVH (*upper panel*). (**b**) Coronal brain section of two pups at the level of midseptal nucleus treated with i.p. glycerol shows moderate (*middle panel*) and severe IVH (*lower panel*). Scale bar, 1 cm

Neurobehavioral assessment: We perform neurobehavioral examinations [21] and grade each item as follows:

1. *Consciousness*: Comatose (0), sleepy (1), alert (2), active alert (3)

2. *Cranial Nerves*:

    (a) *Smell* (Peppermint and 100 % ethanol): no response (0), doubtful response (1), some response (2), definite response (3).

    (b) *Startle to clapping*: no response (1), doubtful response (2), some response (3), definite response (3).

    (c) *Suck and swallow*: No movement of jaw, milk dribbles out completely (0), some sucking and swallowing, some milk appears in nose (1), good suck and swallow, no milk in nose or dribbling outside, finishes 3 mL milk in >3 min (2), good suck and swallow, no milk in nose or dribbling outside, finishes 3 mL milk in <3 min (3).

(d) *Head turn during feeding (Rooting)*: No movement away (0), Occasional movement of head (1), distinct movement of head to one side (2), distinct movement of head to both sides (3).

(e) *Eye blink test*: (approaching eye with the cotton-tipped applicator to eliciting blink of the eye). Percentage of normal will be recorded.

(f) *Visual cliff test*: latency to edge in seconds

3. *Motor examination*

(a) *Posture*: Lies supine or on side (0), Maintains prone position, but wobbly (1), Prone position with trunk close to surface and inability to raise head (2), Prone position with legs coiled (3), Sits on buttocks and brings front leg in air (4).

(b) *Tone in foreleg, hind leg* (Based on Ashworth scale). *If hypertonia, grade the hypertonia*: No increase in tone (0), slight increase in tone giving a catch when limb is moved in flexion or extension (1), more marked increase in tone, but limb easily flexed (2), considerable increase in tone, passive movement difficult (3), limb is rigid in flexion or extension (4). *If hypotonia, grade hypotonia*: No hypotonia (0), slight decrease in tone, some resistance when limb is moved in flexion or extension (1), considerable decrease in tone (2), complete paralysis (3).

(c) *Motor activity in head, foreleg, and hind leg*: No movement (0), some movement (1), considerable movement (2), normal movement (3).

(d) *Locomotion: 30′ inclined surface* (Observe for 2 min) Does not walk on inclined surface (0), Walks on inclined surface for less than 8″ (1), Walks on inclined surface for 9–18″ (2), Walks very well on inclined surface beyond 18″(3).

(e) *Speed of walking*: Walking straight for 20 s; walks <50 in. in 20 s (0), walks 50–100 in. in 20 s (1), walks 100–200 in. in 20 s (2), walks >200 in. in 20 s (3).

(f) *Righting reflex*: Number of times turns prone when placed in supine position within 2 s out of 5 tries.

0 out of 5 tries (0), 2 out of 5 tries (1), 4 out of 5 tries (2), 5 out of 5 tries (3).

(g) *Jumping*: Scale of 0–3. 3 is the best and 0 is the worst.

(h) Ability to stay at a ramp pitched at a slope of 60° inclined (Geotaxis): (test for muscle strength and coordination) Latency to slip down the slope.

4. *Gait*:

Level V: Normal walking

Level IV: Walks without restriction, can perform running, jumping, propels the body using synchronously the back legs, walks on inclined surface and crosses the fence, but limitation in speed, balance, and coordination manifesting as either *asymmetry or clumsiness in gait.*

Level III: Walks taking alternate steps, trunk low; cannot propel its body using synchronously the hind legs, but walks on inclined surface.

Level II: Walks taking alternate steps, trunk touching the ground but can neither propel its body using synchronously the back legs nor walk on inclined surface.

Level I: Crawls with trunk touching the ground for 2–4 steps and then rolls over.

Level 0: No independent mobility.

5. *Sensory examination*: examine face, trunk and extremities for

(a) Crude touch and pressure (Touch test: North coast medical Inc.): Scale of 0–3, 3 is the best and 0 is the worst.

(b) Hot plate test: Pain sensitivity was evaluated using a hot Plate at 50 °C: Latency in seconds showing discomfort as lifting paws or body movement.

We have also performed open field test and novel object recognition test (NOR) using video tracking system (ANY-maze, Stoelting). We perform open field test in pups at postnatal day 14, 21 and 30 using an arena of $100 \times 100 \times 30$ in. dimension. We record total distance travelled by the animal along the walls, sides, and center of the arena is recorded over 10 min. For NOR, rabbits at postnatal day 30 are placed in an arena (100×100 in.) with two identical objects for 3 min, followed by 1 h rest in home cage. This is repeated three times so that pup gets accustomed to two objects. After 1 h rest, the pups are placed in the arena for 3 min with one of the known object replaced with novel object. We analyze the amount of time the rabbit spends with novel vs. known object.

3 Notes

Occurrence of IVH: The premature rabbit pups (E29) are injected with 50 % intraperitoneal glycerol (6.5 g/kg) at 2–3 h of age. To determine the incidence and severity of IVH, the rabbit pups are killed at 48-h age and brains are evaluated for IVH.

Among glycerol treated pups, about 10–15 % pups develop mild (microscopic) IVH, 40 % moderate, and 40 % severe IVH; and only 10 % pups do not develop IVH. The premature rabbit pups can develop mild to moderate IVH (5–10 %) spontaneously. The source of ventricular blood is difficult to discern in pups with moderate and severe IVH since part of the choroid plexus and germinal matrix is typically destroyed. However, in pups with mild IVH (microscopic hemorrhage in GM), we noted congestion and some hemorrhage in the choroid plexus in about one third of such cases on histology. To identify the presence and severity of hemorrhage, we perform head ultrasounds on glycerol treated pups at 6- and 24-h postnatal age using Acuson Sequoia C256 system (Siemens). However, microscopic (mild) hemorrhage can be missed by head ultrasound and should be diagnosed histologically.

Survival and growth of pups: The pups with glycerol-induced IVH display some lethargy and poor activity. About 10–20 % of all pups with moderate to severe IVH develop seizures, which manifest as retraction of the neck, stiffening and jerking of all four extremities, and associated with a lack of responsiveness to external stimuli. The duration of these episodes is from 1 min to several minutes.

We have further determined the motor and sensory capabilities of 24 and 72-h-old premature pups delivered at E29. At 24-h of age, pups suck the milk delivered by a plastic pipette and swallow it without nasal regurgitation of the milk. On touching the angle of mouth with the pipette, both groups of pups move their head away from the stimulus. They are not responsive to light (flashlight) and sound (ringing of bell, clapping) stimuli at either 24- and 72-h age. All pups typically show an aversive response to alcohol swabs held closed to their nostrils. Motor activity and the ability to turn prone from supine position (righting reflex) are similar in the two groups of pups—with and without IVH—at both 24- and 72-h postnatal age. To determine the tone of the extremities, we perform flexion and extension of the joints and do not observe any difference in tone between pups with and without IVH at both time points. Of note, control pups without IVH at 72-h age are more active and display better ability to walk at 30° inclination compared to pups with IVH. However, this comparison is not significant at 24-h age. At 24-h postnatal age, pups with and without IVH are more wobbly and unstable during walking and are noted to tip over to their side with almost every step. However, at 72-h age, pups without IVH are less wobbly and rolled over to sides less frequently compared to pups with IVH. Those pups that are treated with glycerol and do not develop IVH are similar to pups without IVH on neurobehavioral evaluation at 24 and 72 h of age.

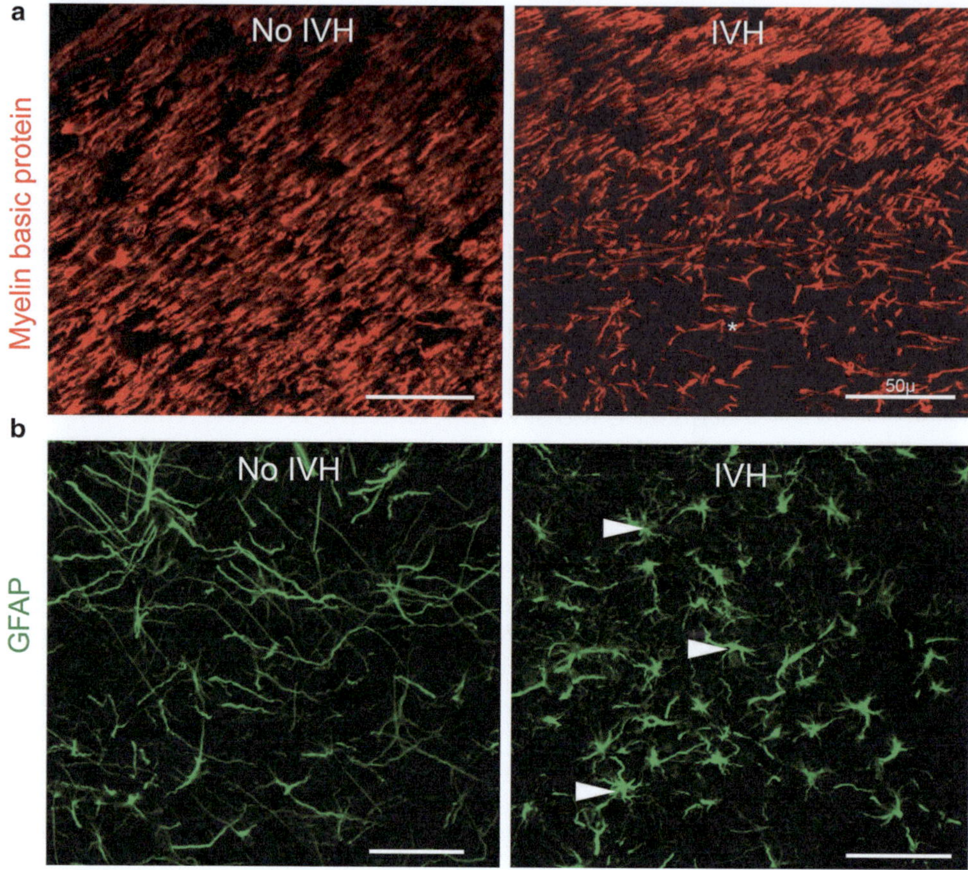

Fig. 2 (a) Representative immunofluorescence of cryosections from 14 day old pups labeled with myelin basic protein (MBP) specific antibody. Note reduced immuno-reactivity to MBP (*asterisk*) in periventricular white matter (corona radiata) of pups with IVH compared to control without IVH. (**b**) Representative labeling of cryosections from 14 day old rabbit pups labeled with GFAP antibody. Note normal looking astrocytes in the white matter of rabbit pups without IVH and abundant hypertrophic astrocytes—with large cell body and numerous processes making a dense network—in the periventricular white matter of pups with IVH (*arrowhead*)

Development of hypomyelination and gliosis: The pups with IVH at day 14, compared to controls without, show reduced myelination of the white matter on myelin basic protein immuno-labeling and Western blot analyses of brain homogenates of the forebrain (Fig. 2) [22]. The hypomyelination is noted in the periventricular corona radiata and corpus callosum, but not in the internal capsule. A relative sparing of internal capsule can be attributed to the distant anatomical location of this brain region from the blood filled ventricle. Diffusion tensor imaging (MRI of the fixed brain) of pups with IVH at day 14 show that the FA is significantly decreased in the corpus callosum, corona radiata, and fimbria–fornix compared with controls, but not in the internal capsule [22]. In addition, the specific white matter areas—the corpus callosum, fimbria–fornix, and corona radiata—are significantly reduced in size in pups with IVH compared with controls, but not in the internal capsule. The rabbit pups with IVH at 2 week of age also exhibit gliosis in the brain region around the ventricle that can be visualized by GFAP immunolabeling of brain sections (Fig. 2) In conclusion, rabbit pup model of IVH exhibits hypomyelination and gliosis at 2 weeks of age after the development of hemorrhage.

Other models of IVH: Several species including rats, mice, pigs and dogs have been used to model IVH [15]. Injection of autologous blood or collagenase in the periventricular germinal matrix of neonatal rats, mice, and pigs have been performed by several investigators [23, 24]. The demerit of the needle injection in the brain is that this can potentially traumatize the brain parenchyma despite ultrasound or MRI monitoring. In addition, it is difficult to inject precisely in the germinal periventricular germinal matrix of mouse and rat which is a microscopic structure in the newborn brain. IVH has been induced in beagle pups by altering hemodynamic parameters, including blood pressure, circulating blood volume, serum osmolarity, pCO_2, or O_2 levels [16, 17]. These interventions result in metabolic changes in the brain that can potentially confound the molecular and histological changes—cell death, inflammatory cell infiltration, and others—that follows IVH. The best model of IVH is one that mimics premature infants with IVH with respect to etiology, pathology, and clinical manifestations. The author believes that the glycerol model of IVH closely resembles premature infants with ventricular hemorrhage.

Acknowledgements

Supported by NIH/NINDS grant RO1 NS071263 (PB) and American Heart Association grant-in-aid # 09GRNT2310147(PB).

References

1. Guyer B, Hoyert DL, Martin JA, Ventura SJ, MacDorman MF, Strobino DM (1999) Annual summary of vital statistics—1998. Pediatrics 104:1229–1246

2. Heuchan AM, Evans N, Henderson Smart DJ, Simpson JM (2002) Perinatal risk factors for major intraventricular haemorrhage in the Australian and New Zealand Neonatal Network, 1995–97. Arch Dis Child Fetal Neonatal Ed 86:F86–F90

3. Philip AG, Allan WC, Tito AM, Wheeler LR (1989) Intraventricular hemorrhage in preterm infants: declining incidence in the 1980s. Pediatrics 84:797–801

4. Jain NJ, Kruse LK, Demissie K, Khandelwal M (2009) Impact of mode of delivery on neonatal complications: trends between 1997 and 2005. J Matern Fetal Neonatal Med 22:491–500

5. Wilson-Costello D, Friedman H, Minich N, Fanaroff AA, Hack M (2005) Improved survival rates with increased neurodevelopmental disability for extremely low birth weight infants in the 1990s. Pediatrics 115:997–1003

6. Ballabh P (2010) Intraventricular hemorrhage in premature infants: mechanism of disease. Pediatr Res 67(1):1–8

7. Ballabh P, Hu F, Kumarasiri M, Braun A, Nedergaard M (2005) Development of tight junction molecules in blood vessels of germinal matrix, cerebral cortex, and white matter. Pediatr Res 58:791–798

8. Ballabh P, Xu H, Hu F, Braun A, Smith K, Rivera A, Lou N, Ungvari Z, Goldman SA,

Csiszar A, Nedergaard M (2007) Angiogenic inhibition reduces germinal matrix hemorrhage. Nat Med 13:477–485

9. Xu H, Hu F, Sado Y, Ninomiya Y, Borza DB, Ungvari Z, Lagamma EF, Csiszar A, Nedergaard M, Ballabh P (2008) Maturational changes in laminin, fibronectin, collagen IV, and perlecan in germinal matrix, cortex, and white matter and effect of betamethasone. J Neurosci Res 86:1482–1500

10. Braun A, Xu H, Hu F, Kocherlakota P, Siegel D, Chander P, Ungvari Z, Csiszar A, Nedergaard M, Ballabh P (2007) Paucity of pericytes in germinal matrix vasculature of premature infants. J Neurosci 27:12012–12024

11. El-Khoury N, Braun A, Hu F, Pandey M, Nedergaard M, Lagamma EF, Ballabh P (2006) Astrocyte end-feet in germinal matrix, cerebral cortex, and white matter in developing infants. Pediatr Res 59:673–679

12. Antoniuk S, da Silva RV (2000) Periventricular and intraventricular hemorrhage in the premature infants. Rev Neurol 31:238–243

13. Kenny JD, Garcia-Prats JA, Hilliard JL, Corbet AJ, Rudolph AJ (1978) Hypercarbia at birth: a possible role in the pathogenesis of intraventricular hemorrhage. Pediatrics 62:465–467

14. Volpe JJ (1989) Intraventricular hemorrhage in the premature infant—current concepts. Part I. Ann Neurol 25:3–11

15. Balasubramaniam J, Del Bigio MR (2006) Animal models of germinal matrix hemorrhage. J Child Neurol 21(5):365–371

16. Ment LR, Stewart WB, Duncan CC, Lambrecht R (1982) Beagle puppy model of intraventricular hemorrhage. J Neurosurg 57:219–223

17. Goddard J, Lewis RM, Armstrong DL, Zeller RS (1980) Moderate, rapidly induced hypertension as a cause of intraventricular hemorrhage in the newborn beagle model. J Pediatr 96:1057–1060

18. Conner ES, Lorenzo AV, Welch K, Dorval B (1983) The role of intracranial hypotension in neonatal intraventricular hemorrhage. J Neurosurg 58:204–209

19. Luttrell CN, Finberg L (1959) Hemorrhagic encephalopathy induced by hypernatremia. I. Clinical, laboratory, and pathological observations. AMA Arch Neurol Psychiatry 81:424–432

20. Georgiadis P, Xu H, Chua C, Hu F, Collins L, Huynh C, Lagamma EF, Ballabh P (2008) Characterization of acute brain injuries and neurobehavioral profiles in a rabbit model of germinal matrix hemorrhage. Stroke 39:3378–3388

21. Vinukonda G, Csiszar A, Hu F, Dummula K, Pandey NK, Zia MT, Ferreri NR, Ungvari Z, LaGamma EF, Ballabh P (2010) Neuroprotection in a rabbit model of intraventricular haemorrhage by cyclooxygenase-2, prostanoid receptor-1 or tumour necrosis factor-alpha inhibition. Brain 133:2264–2280

22. Chua CO, Chahboune H, Braun A, Dummula K, Chua CE, Yu J, Ungvari Z, Sherbany AA, Hyder F, Ballabh P (2009) Consequences of intraventricular hemorrhage in a rabbit pup model. Stroke 40:3369–3377

23. Xue M, Del Bigio MR (2005) Injections of blood, thrombin, and plasminogen more severely damage neonatal mouse brain than mature mouse brain. Brain Pathol 15:273–280

24. Stankovic MR, Maulik D, Rosenfeld W (1999) Real-time optical imaging of experimental brain ischemia and hemorrhage in neonatal piglets. J Perinat Med 27:279–286

Models of Perinatal Brain Injury in Premature and Term Newborns Resulting from Gestational Inflammation Due to Inactivated Group B Streptococcus (GBS), or Lipopolysaccharide (LPS) from *E. coli* and/or Immediately Postnatal Hypoxia-Ischemia (HI)

Julie Bergeron, Marie-Julie Allard, Clémence Guiraut, Mathilde Chevin, Alexandre Savard, Djordje Grbic, Marie-Elsa Brochu, and Guillaume Sébire

Abstract

It is known that gestational and/or perinatal inflammation combined or not with hypoxia-ischemia (HI) is a risk factor for brain injuries, but the mechanisms underlying are still unclear. This chapter discusses about animal models mimicking those conditions, allowing scientists to uncover mechanisms involved and to study the adverse effects on the offspring. Here is presented a model of maternal inflammation induced by inactivated Group B Streptococcus (Sect. 2) and two experimental designs using LPS. One explores the effects of prenatal LPS administration and/or immediately postnatal HI (Sect. 3) and the second one, the immediately postnatal exposure to inflammation induced by LPS and/or HI (Sect. 4). For each animal model, the rationale supporting the model is exposed, followed by the procedures and the results obtained, allowing experimenter to reproduce and use these presented animal models.

Key words Gestational inflammation, Perinatal inflammation, Animal model, Group B streptococcus, LPS, Hypoxia-ischemia

1 Introduction

Preterm human newborns are at high risk of brain injuries [1–4]. Perinatal brain injury affects about 50 % of very preterm human neonates and also a fraction of late preterm and term newborns [1]. Perinatal brain injury leads to lifelong neurobehavioral disabilities such as cerebral palsy, learning impairments, and autism spectrum disorders (ASD) [2, 5]. Decreased oxygen and other blood nutrient supply to the brain, remote pathogen exposure, or both combined and their associated neuroinflammatory responses

Jerome Y. Yager (ed.), *Animal Models of Neurodevelopmental Disorders*, Neuromethods, vol. 104, DOI 10.1007/978-1-4939-2709-8_5, © Springer Science+Business Media New York 2015

are the most important perinatal risk factors associated with such brain injuries [6, 7]. The incidence of neonatal brain damage is inversely proportional to gestational age and thus higher in preterm than in term newborns. In addition, the type and distribution of brain lesions differ markedly between preterm and term newborns [8]. This is attributed to both different levels of brain maturity and vulnerability to injuries due to regional and age-specific metabolic needs.

Postnatal bacterial infections—that are more frequent in preterm than term newborns—often combined with hypoxia-ischemia (HI), are also associated with an increased risk to develop brain injuries [7, 9–13]. In this regard, both prenatal infection [14, 15] and postnatal infection [16–18] have been associated with an increase in the incidence of aberrant newborn development. The perinatal inflammatory response has also been hypothesized to be linked with later onset mental health difficulties [19], and is maturationally and time dependent [20].

To uncover potential differences in the mechanisms of brain injuries according to the stage of brain development and to intrauterine versus extrauterine stage of life, we have designed rat models of brain damage induced at different levels of development. In rats, postnatal day 1 (P1) corresponds, in terms of brain development, to the level of development of very preterm human brain, i.e., 26–32 weeks of gestation. This allows for rat newborns to be a pertinent model for the study of brain injuries affecting the premature human newborn [8].

The causal link between pathogen-induced gestational infection/inflammation and fetal or neonatal brain injury was established, thanks to preclinical animal models. It was also supported by human epidemiological and observational evidence [2, 12, 21]. However, the distinct effects on fetal brains of the different pathogens involved in maternal inflammation, as well as the mechanistic pathways whereby they exert their deleterious actions, remain largely unsettled. In addition, the respective part of fetal inflammatory response syndrome (FIRS) and of hypoxia-ischemia (HI), either from placental origin or from perinatal asphyxia, continues to require further study [6, 13].

Beyond prenatal infection and inflammation, the influence of bacterial components in modulating HI-induced perinatal brain injuries has been mostly reported in models of postnatal infectious components exposures. However, there have been few animal models of the likely sequence of events taking place in human newborns within the context of perinatal brain lesions, namely, prenatal intrauterine infection/inflammation, chief among them chorioamnionitis, immediately followed by postnatal HI.

2 Preclinical Rat Model of Maternal Inflammation Triggered by Inactivated Group B Streptococcus (GBS)

The effect of GBS-induced gestational inflammation on the placenta and fetal brain and the neurological outcome of these offspring had not been studied, even though GBS (*Streptococcus agalactiae*, a gram-positive bacterium) is the most frequently encountered pathogen in the fetal environment. GBS colonize the genital tract in 10–30 % of pregnant women. GBS infections, such as urinary infections or chorioamnionitis, are major causes of maternal infection, especially at the end of gestation [22].

2.1 Experimental Design

Lewis dams at gestational day 14 (G14) were acclimatized to our animal facility prior to experimental manipulation [22]. Rats are reared at 20–23 °C with a 12 h light/dark cycle and with access to food and water ad libitum. This experimental design is approved by the Animal review board of the *Université de Sherbrooke* and was in full agreement with the ad hoc guidelines from the Canadian Council on Animal Care.

Dams received intraperitoneal (ip) injections of 10^9 CFU of inactivated GBS (suspended in 100 µL of sterile saline every 12 h) from G19 to the end of gestation, and were compared to dams receiving ip injections/12 h of 100 µL of saline also from G19 to the end of gestation (G22) (Fig. 1). GBS was inactivated by suspension in formaldehyde 10 % for 24 h. Inactivated GBS was thoroughly washed with sterile saline prior to use. Because our goal was to establish a model for gestational inflammation mimicking GBS induced inflammation occurring at the end of human pregnancy, we chosed to expose dams to GBS from G19 to G22. This choice was based on human epidemiological studies showing that GBS infections mainly happen during the third tier of gestation. Following GBS exposure, we did not observe mortality in the dams, but they often showed diminished weight gain (6 %) during gestation as compared with controls (10 %). Caesarean sections were performed on G22 to collect the pups and placentas. Dams were placed under general isoflurane anesthesia (5 % for induction, 2 % for maintenance) and we performed an incision through the abdominal tissues to uncover the uterine horn (Fig. 2). Then, we carefully collected placentas. Dams were euthanized at the end of the surgery. Other dams gave birth naturally on G23 and we performed behavioral testing with offspring to mainly characterize abnormal functional traits, which in our circumstances, are felt to reflect Autism Spectrum Disorder. Offspring were euthanized on postnatal day 40 and brains were collected to assess neuropathological outcome and damage [22].

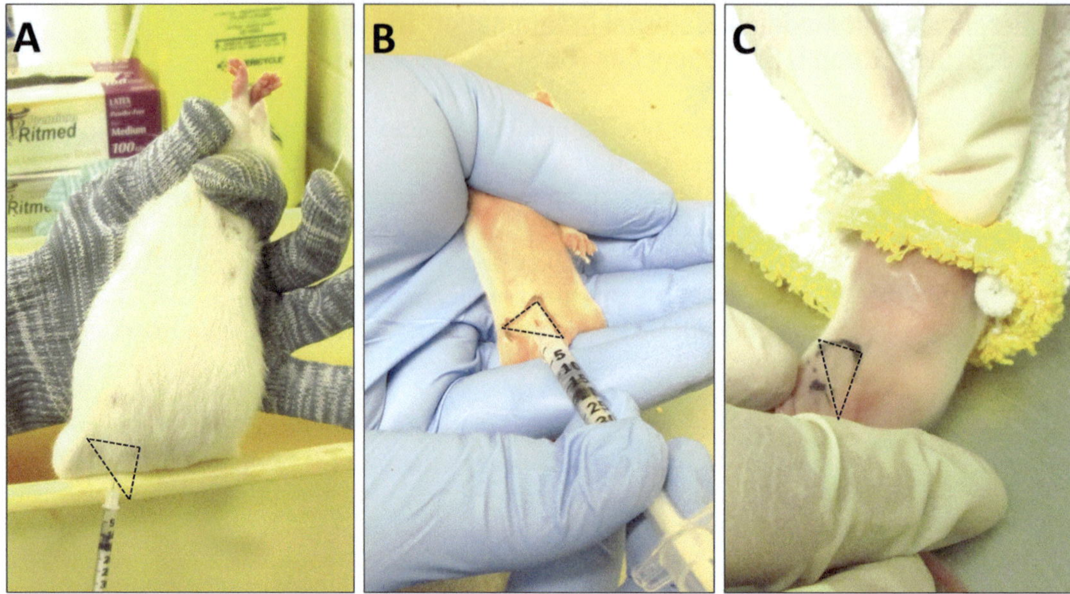

Fig. 1 Intraperitoneal (ip) injection of dams (**a**) and rat pups (**b**, **c**): the targeted ip region to be injected is delimited by *dotted triangle*; the injection is done with a needle angle of 45°. Front paws of dams are maintained crossed on its body (under its bottom jaw) with one hand, while the back paws are gently pressed against the edge of the cage to stretch the dams' body (**a**). Pups are held with one hand: pup's head under the thumb and back paws between other fingers (**b**) or wrapped in a towel (**c**)

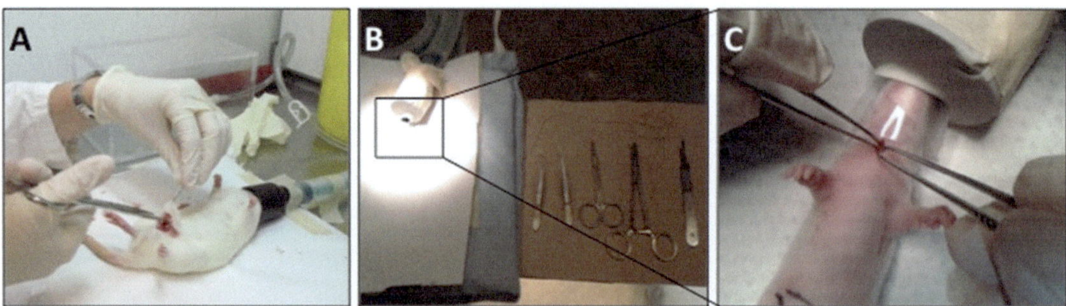

Fig. 2 Surgical set up and element of procedures used for C-section of dams at G20 (**a**) and carotid ligation of rat pups at P1 or P12 (**b**, **c**). Before starting the incision, evaluate the consciousness state of the animal by pinching the tail (dams) or back paws (dams or rat pups) with forceps. Dams are under isoflurane (2 %) inhalation while opening the abdominal cavity (T-shape incision) until reaching the uterine horn. Perform smallest incision as possible to keep the rat fetuses warm and moist inside the abdomen (**a**). Surgical set up for the carotid ligation and minimum surgical instruments, including a heating pad to optimize rat pup's recovery after surgery (**b**). Rat pup (P12) are under isoflurane (2 %) anesthesia while reaching the carotid. Perform smallest incision as possible to limit the number of stitches and the risk of infection (**c**)

2.2 GBS-Induced Chorioamnionitis and Neurobehavioral Impairments

GBS-exposed placentas displayed a peculiar pattern of chorioamnionitis featured by cystic lesions and polymorphonuclear infiltration mainly located within the decidual side of the placenta (referring as the maternal side of the placenta), contrasting with macrophagic infiltration affecting the labyrinth/fetal side of the placenta following end gestational LPS-induced maternal inflammation [22]. Offspring brain injuries, studied at P40, were characterized by lateral ventricular enlargement, more prominent in the males, reduced thickness of external capsules, oligodendrocyte dysmaturation (blockage at the olig-2 stage with a lack of CC-1 positive mature oligodendrocytes), and disorganization of fronto-parietal white matter architecture without any glial proliferation [22] (Fig. 1). Autistic-like traits were found in the GBS-exposed offspring: exemplified by deficits in motor behavior, social interaction, and communication tools. Interestingly, only male offspring presented these combined autistic-like hallmarks [22]. Thus, these results show that GBS-induced maternal inflammation plays a role in the induction of placental and brain injuries reproducing some cardinal features of human preterm neonatal brain injuries, as well as autistic-like features such as gender dichotomy and typical neurobehavioral traits. Unlike other models of prenatal inflammatory brain damage—induced for instance by interactions between viral components and toll-like receptor 3, or gram-negative components such as LPS and toll-like receptor 4—maternal inflammation resulting from the interaction between GBS and, mainly, toll-like receptor 2 induce a distinctive pattern of chorioamnionitis and remote cerebral injuries [22]. These results also show that, beyond genetic influences, modifiable environmental factors might be involved in both the occurrence of autism and its gender imbalance. To our knowledge, these preclinical may establish a link between GBS-induced maternal inflammation and the male autistic-like behavior in an animal model [22].

We also showed using this model [22], and others [23–26], that magnetic resonance imaging (MRI) of the placenta and of the offspring brain was feasible. MRI was performed on rats using a small-animal 7T MRI system. Dams are scanned as early as G17. They were anesthetized during the entire procedure (isoflurane 2 %) and imaged on a ventral position. Vital signs and temperature were monitored during the scan. Two types of images were acquired: T2-weighted respiration-gated images using a fast-spin echo pulse sequence and T1-weighted images to assess placental perfusion using a contrast agent (gadolinium, 0.143 mmol, injected into the tail vein) (Fig. 3). We also performed MRI on rats, as early as P20. The animal's vital signs and temperature were monitored throughout the MRI procedure. T2-weighted respiration-gated images were acquired using a fast spin echo pulse sequence (TR/TEeff: 2,500/48 ms, 8 echoes, FOV: 3.2 × 3.2 cm 2, matrix: (256) 2, NA: 8, 25 slices of 1.0 mm). These MRI images allowed us to

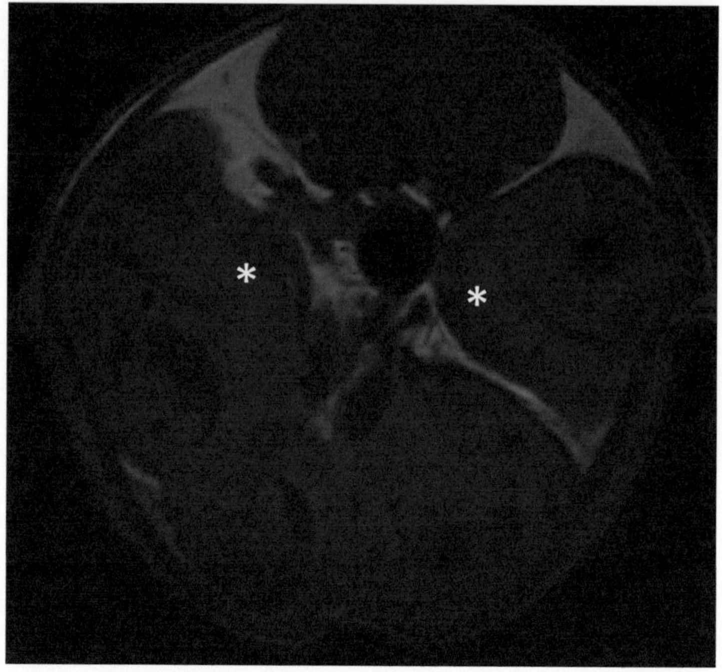

Fig. 3 In vivo MRI of placentas (*) in a control dam: T1-weighted images showing placentas within the uterine cavity

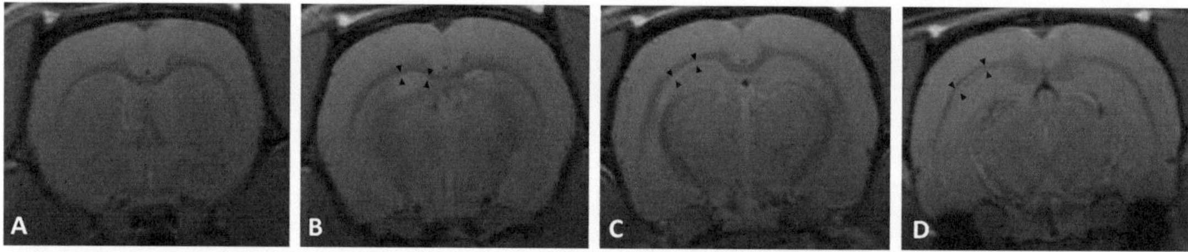

Fig. 4 (**a**, **b**) T2-weighted brain MRI of 5-month-old male rat in utero-exposed to GBS showing a unilateral ventricular hypersignal (*arrowheads*) (**c**, **d**) reflecting an asymmetrical enlargement of a lateral cerebral ventricle which is secondary to the atrophy of the adjacent external capsule/white matter

visualize in vivo chorioamnionitis at G20 [24, 26], brain white matter injuries and ventricle enlargement in the GBS animal model, and cerebral lesions into the LPS animal model at P20 (see hereunder experimental design) [22] (Fig. 4).

3   Experimental Design to Explore Effects of Prenatal LPS Administration and/or Immediately Postnatal HI

Pregnant rats were injected ip with LPS (200 μg/kg; *Escherichia coli*, 0127:B8) diluted in saline (100 μL), or with saline (100 μL) [1, 8, 21, 24, 25, 27]. These ip injections were given every 12 h starting 2–3 days before the end of gestation (4–6 injections).

We used a standard protocol of ip injection as showed in Fig. 1. Dams' mortality was unusual. Dams presented ruffled fur, transient weight loss and hypo-activity, without any drop of blood pressure. Twelve to thirty-six hours after birth (P1), pups of both sexes and from different litters were exposed to HI using slightly modified experimental methods originally described by Rice et al. [28] (*See* Chap. 1 by Nguyen et al.). Thus, to induce ischemia, the right common carotid artery of these pups is exposed following a small incision of the skin (1.5 cm) and permanently ligated under isoflurane anesthesia (5 % for induction (2 min), 2 % for maintenance) (Fig. 2). The skin is sutured using a 4.0 Surgilon yarn. Pups were kept at normal temperature (36–38 °C) on a heat pad. The duration of the surgical procedure was less than 5 min. The full duration of anesthesia was about 8 min. Mortality rate associated with the surgical procedure was about 1 %. The pups were then returned to the dams. Thirty minutes later, these pups were exposed to 8 % O_2/N_2 balance, in a hypoxic chamber at ambient temperature of 36 °C for 3.5 h. All pups were immediately returned to the dams. All pups survive hypoxia.

Using this model, we showed (1) that prenatal exposure to LPS rapidly induced a chorioamnionitis and subsequent placental hypo-perfusion likely associated with fetal brain hypo-perfusion [23, 25]; (2) that the extent of injury in the brains of rats exposed to postnatal HI was potentiated by prenatal exposure to LPS [29]; and (3) that blocking the maternal or neonatal interleukin-1 (IL-1) pathway—using a natural antagonist of IL-1, (namely, IL-1 receptor antagonist) alleviated placental and/or neonatal brain damage and subsequent behavioral anomalies arising from prenatal LPS exposure or postnatal HI [24–26]. This animal model thus provides a precious experimental means particularly designed for investigating the role of the immune-inflammatory system in the pathogenesis of perinatal brain damage.

4   Models Representing Postnatal Exposure to LPS and/or HI

Postnatal bacterial infections—that are more frequent in preterm than term newborns—are associated with an increased risk to develop neonatal brain injuries. *E. coli* and GBS are the most common pathogen involved in neonatal infections. To uncover potential differences between term and preterm neuroinflammatory responses to neonatal insults, we have also used rat models of brain damage induced at different stages of brain development. P1 corresponds, in term of brain development, to the very preterm human brain, whereas P12, corresponds to the level of development of the term human neonate brain. We compared the intra-cerebral profiles of pro- and anti-inflammatory cytokines responses, resulting chemokines responses, and related immune cell recruitments at both developmental stages [8, 30].

4.1 Experimental Design

Lewis dams were purchased between G14 and 18 and give birth naturally in our animal facility. At P1 or P12, pups were injected ip with either LPS (200 µg/kg, *Escherichia coli*, 0127:B8) diluted in 20 (P1) or 50 µL (P12) of saline, or saline alone (Fig. 1). HI is induced 4 h after the LPS injection by permanent ligation of the right common carotid artery (Fig. 2) followed by exposure to 8 % O_2/N_2 for 180 min at P1 or 90 min at P12 in a chamber at 36 °C [8, 30]. The duration of hypoxia was decreased at P12 because of the higher mortality rates of pups in this more mature age range during hypoxia than at P1. P12 rat pups showed generalized tremor and circling behavior, whereas P1 pups did not show any of these patterns. The P1 and P12 pups were then randomized into five groups: (1) control, (2) sham (surgical procedure without ligation), (3) HI, (4) LPS, (5) LPS+HI. This protocol was approved by the appropriate institutional Animal Research Ethics Board and conducted in accordance with all applicable laws and regulations [8, 30].

4.2 Results

LPS, HI and LPS+HI trigger a pro-inflammatory oriented immune response within both the preterm- and term-like forebrains, with a maximum inflammatory reaction induced by the combination of LPS+HI. The profiles of these neuroinflammatory responses present striking variations according to the level of development: no or downregulated anti-inflammatory responses associated with mainly IL-1β release are seen in the preterm-like brains (P1), whereas the term-like brains (P12) present a strong anti- as well as pro-inflammatory response, including both IL-1β and TNF-α releases, and blood–brain-barrier leakage [8, 30]. These developmental-dependent variations of neuroinflammatory response could play a key role in the differential pattern of brain lesions observed across gestational ages in humans. These results highlight the necessity to take into account the maturation stage, of both brain and immune systems, in order to develop new anti-inflammatory neuroprotective strategies.

5 Notes

- To model gestational or neonatal inflammation, intraperitoneal (ip) injection of pathogen components to the dam or rat pups must be performed as described in Fig. 1 to prevent dams' mortality and morbidity.

- Inactivated pathogen (lipopolysaccharide (LPS), or group B Streptococcus (GBS)) administered to the dam induces about 50 % fetal mortality, significantly reducing the litters' number; this has to be taken into account in the experimental design: sample calculation, end-points ....

- Following pathogen exposure and common carotid ligation, the tolerance of rat pups to the duration of hypoxia (8 %) decreases with postnatal age: mortality is less than 5 % at postnatal day (P)1 (3 h of hypoxia), but reaches 40 % at P12 (1 h of hypoxia).

- Systemic inactivated pathogen administration to the dam rapidly induces chorioamnionitis. This inflammatory response has to be taken into account in these rat models when assessing the pathophysiological mechanisms linking gestational inflammation and fetal brain injuries.

- Dams' chorioamnionitis is detectable in vivo/in utero, and can be monitored, thanks to magnetic resonance imaging (MRI) (Fig. 3).

- Cytokines induced by inactivated pathogen and/or HI play a causal role prenatally and postnatally, in placental and offspring brain injuries, and therefore have to be considered among other factors in such perinatal pathophysiological processes and their modelling.

- The timing and onset of fetal inflammation clearly play an important role in the outcome following this form of insult.

Acknowledgments

This work was supported by scholarships from Fonds de la Recherche du Québec-Santé (FRQ-S), a grant from the Canadian Institutes of Health Research (CIHR), and a grant from the Foundation of Stars. G.S. is a member of the Centre de Recherche Clinique Etienne Le Bel du CHUS, of the Centre de Neurosciences de l'Université de Sherbrooke, and of the Centre de Recherche Mère & Enfant de l'Université de Sherbrooke. We thank Sylvie Girard for her contribution to the design of the LPS, and LPS + HI models.

References

1. Girard S, Kadhim H, Roy M et al (2009) Role of perinatal inflammation in cerebral palsy. Pediatr Neurol 40:168–174

2. Salmaso N, Jablonska B, Scafidi J, Vaccarino FM, Gallo V (2014) Neurobiology of premature brain injury. Nat Neurosci 17: 341–346

3. Kadhim H, Tabarki B, De Prez C, Sebire G (2003) Cytokine immunoreactivity in cortical and subcortical neurons in periventricular leukomalacia: are cytokines implicated in neuronal dysfunction in cerebral palsy? Acta Neuropathol 105:209–216

4. Kadhim H, Tabarki B, Verellen G, De Prez C, Rona AM, Sebire G (2001) Inflammatory cytokines in the pathogenesis of periventricular leukomalacia. Neurology 56:1278–1284

5. Kuzniewicz MW, Wi S, Qian Y, Walsh EM, Armstrong MA, Croen LA (2014) Prevalence and neonatal factors associated with autism spectrum disorders in preterm infants. J Pediatr 164:20–25

6. Schendel D, Nelson KB, Blair E (2012) Neonatal encephalopathy or hypoxic-ischemic encephalopathy? Ann Neurol 72:984–985

7. Girard S, Kadhim H, Beaudet N, Sarret P, Sebire G (2009) Developmental motor deficits induced by combined fetal exposure to lipopolysaccharide and early neonatal hypoxia/ischemia: a novel animal model for cerebral palsy in very premature infants. Neuroscience 158:673–682

8. Brochu ME, Girard S, Lavoie K, Sebire G (2011) Developmental regulation of the neuroinflammatory responses to LPS and/or hypoxia-ischemia

between preterm and term neonates: an experimental study. J Neuroinflammation 8:55

9. Jonsson M, Agren J, Norden-Lindeberg S, Ohlin A, Hanson U (2014) Neonatal encephalopathy and the association to asphyxia in labor. Am J Obstet Gynecol 211:667

10. Ellenberg JH, Nelson KB (2013) The association of cerebral palsy with birth asphyxia: a definitional quagmire. Dev Med Child Neurol 55:210–216

11. Kadhim H, Khalifa M, Deltenre P, Casimir G, Sebire G (2006) Molecular mechanisms of cell death in periventricular leukomalacia. Neurology 67:293–299

12. Nelson KB, Bingham P, Edwards EM et al (2012) Antecedents of neonatal encephalopathy in the Vermont Oxford Network Encephalopathy Registry. Pediatrics 130:878–886

13. Shevell A, Wintermark P, Benini R, Shevell M, Oskoui M (2014) Chorioamnionitis and cerebral palsy: lessons from a patient registry. Eur J Paediatr Neurol 18:301–307

14. Wu YW, Colford JM Jr (2000) Chorioamnionitis as a risk factor for cerebral palsy: a meta-analysis. JAMA 284:1417–1424

15. Wu YW, Escobar GJ, Grether JK, Croen LA, Greene JD, Newman TB (2003) Chorioamnionitis and cerebral palsy in term and near-term infants. JAMA 290:2677–2684

16. Chau V, Poskitt KJ, McFadden DE et al (2009) Effect of chorioamnionitis on brain development and injury in premature newborns. Ann Neurol 66:155–164

17. Chau V, McFadden DE, Poskitt KJ, Miller SP (2014) Chorioamnionitis in the pathogenesis of brain injury in preterm infants. Clin Perinatol 41:83–103

18. Glass HC, Bonifacio SL, Chau V et al (2008) Recurrent postnatal infections are associated with progressive white matter injury in premature infants. Pediatrics 122:299–305

19. Hagberg H, Gressens P, Mallard C (2012) Inflammation during fetal and neonatal life: implications for neurologic and neuropsychiatric disease in children and adults. Ann Neurol 71: 444–457

20. Rees S, Harding R, Walker D (2011) The biological basis of injury and neuroprotection in the fetal and neonatal brain. Int J Dev Neurosci 29:551–563

21. Roy M, Girard S, Larouche A, Kadhim H, Sebire G (2009) TNF-alpha system response in a rat model of very preterm brain injuries induced by lipopolysaccharide and/or hypoxia-ischemia. Am J Obstet Gynecol 201:493.e1–10

22. Bergeron JD, Deslauriers J, Grignon S et al (2013) White matter injury and autistic-like behavior predominantly affecting male rat offspring exposed to group B streptococcal maternal inflammation. Dev Neurosci 35:504–515

23. Girard S, Sebire G, Kadhim H (2010) Proinflammatory orientation of the interleukin 1 system and downstream induction of matrix metalloproteinase 9 in the pathophysiology of human perinatal white matter damage. J Neuropathol Exp Neurol 69:1116–1129

24. Girard S, Tremblay L, Lepage M, Sebire G (2010) IL-1 receptor antagonist protects against placental and neurodevelopmental defects induced by maternal inflammation. J Immunol 184:3997–4005

25. Girard S, Sebire H, Brochu ME, Briota S, Sarret P, Sebire G (2012) Postnatal administration of IL-1Ra exerts neuroprotective effects following perinatal inflammation and/or hypoxic-ischemic injuries. Brain Behav Immun 26:1331–1339

26. Girard S, Tremblay L, Lepage M, Sebire G (2012) Early detection of placental inflammation by MRI enabling protection by clinically relevant IL-1Ra administration. Am J Obstet Gynecol 206:358.e1–9

27. Girard S, Larouche A, Kadhim H, Rola-Pleszczynski M, Gobeil F, Sebire G (2008) Lipopolysaccharide and hypoxia/ischemia induced IL-2 expression by microglia in neonatal brain. Neuroreport 19:997–1002

28. Rice JE 3rd, Vannucci RC, Brierley JB (1981) The influence of immaturity on hypoxic-ischemic brain damage in the rat. Ann Neurol 9:131–141

29. Larouche A, Roy M, Kadhim H, Tsanaclis AM, Fortin D, Sebire G (2005) Neuronal injuries induced by perinatal hypoxic-ischemic insults are potentiated by prenatal exposure to lipopolysaccharide: animal model for perinatally acquired encephalopathy. Dev Neurosci 27:134–142

30. Savard A, Lavoie K, Brochu ME et al (2013) Involvement of neuronal IL-1beta in acquired brain lesions in a rat model of neonatal encephalopathy. J Neuroinflammation 10:110

Chapter 6

Fetal Brain Activity in the Sheep Model with Intrauterine Hypoxia

Bryan S. Richardson and Brad Matushewski

Abstract

The fetal and early neonatal period is a time of rapid brain growth and development, dependent upon adequate oxygenation. Fetal hypoxia, whether chronic due to placental insufficiency or intermittent cord compression, or acute during labor, may give rise to aberrant development and neurologic sequelae. This chapter reviews the metabolic activities and behavioral states of the brain throughout the late fetal period, and the methods for studying them using the chronically instrumented ovine fetus. Also outlined are methods for studying hypoxia in the ovine fetus through embolization of the placenta, and occlusion of the umbilical cord.

Key words Fetal hypoxia, Cerebral metabolism, Cerebral blood flow, Behavioral states, Umbilical cord occlusions, Growth restriction

1 Introduction

1.1 Perinatal Hypoxia and Brain Development/ Injury

Human clinical studies sampling umbilical cord blood at birth, indicate an increased risk for newborn encephalopathy with pH values <7.00, which in turn may result in long term neurologic sequelae including cerebral palsy, although the majority of these infants will still be without noted complications [1–4]. However, among newborn encephalopathy infants subsequently categorized as being "non-impaired," a proportion will test one or more grade levels below expected suggesting lesser degrees of neurologic derangement [5]. Furthermore, in utero hypoxia has been ongoing antenatally in some of these individuals, predisposing them to further hypoxia/acidemia during labor.

Fetal intrauterine growth restriction (IUGR), which is well associated with hypoxemia throughout the latter part of pregnancy [6, 7], is a primary risk factor for newborn encephalopathy [8, 9] and in turn for cerebral palsy [10, 11]. Moreover, there is increasing evidence that IUGR, whether in isolation or in combination with acidemia at birth, may lead to other neurologic complications

Jerome Y. Yager (ed.), *Animal Models of Neurodevelopmental Disorders*, Neuromethods, vol. 104, DOI 10.1007/978-1-4939-2709-8_6, © Springer Science+Business Media New York 2015

including cognitive impairment, attention deficit disorders and possibly schizophrenia and autism [12–16].

Another strong predictor of hypoxemia is variable fetal heart rate decelerations, suggestive of umbilical cord compression, which has been associated with an increased incidence of neonatal acidosis, impaired oxygen delivery to the fetus, and nuchal cord at birth [17–20]. Infants with a symptomatic nuchal cord at birth with non-reassuring fetal heart rate decelerations have been shown to have subclinical neurodevelopmental deficits at 1 year of age [21]. In infants born at a normal weight, the presence of a tight nuchal cord at delivery is the most common hypoxic condition associated with infant cerebral palsy [22].

1.2 Fetal Cerebral Metabolic Rate

Measurements of the metabolic activity of the brain in the unanesthetized fetus are largely limited to sheep, an animal that is relatively neuroanatomically and electrophysiologically mature at birth [23, 24]. Studies near term (term = 145 days gestation) using the microsphere technique for blood flow determination and Fick methodology, have revealed cerebral metabolic rates for oxygen ($CMRO_2$) of ~160 µmol/100 g/min [25]. Glucose is the major metabolic substrate which accounts for all oxidative metabolism, with uptake rates of ~28 µmol/100 g/min [25]. This substrate uptake provides the energy required for tissue maintenance, functional activity and growth processes, and for the provision of substrate materials for tissue growth.

Energy requirements for tissue maintenance are likely lower in the developing brain owing to a lesser extent of cell process branching, and lower surface-to-volume ratios; however, energy demands for tissue growth are increased, requiring high efficiency of substrate delivery [26]. The latter is facilitated by a relatively high cerebral blood flow, compared to that of the neonatal lamb or adult sheep, with reported values of ~160 ml/100 g/min [25]. Regional blood flow differences are evident, with brainstem and subcortical flows being substantially higher than flows to other brain regions [25]. Given the tight coupling between blood flow and metabolic rate reported for brain tissue, this differential blood flow likely reflects regional differences in metabolic activity within the brain [27].

1.3 Fetal Behavioral States

The study of electrocortical performance status (electrocorticography, or ECOG) provides for electrophysiologic measures of brain development and offers insight into associated control circuitries. For the ovine fetus, such studies using stainless steel wire electrodes implanted biparietally on the dura, reveal well-differentiated ECOG patterns from ~120 days gestation, with a temporal relationship to episodic muscle and breathing activity indicative of behavioral states [24, 28, 29]. Studies of the maturation of ECOG patterns of behavioral activity indicate a similarity in the development of sleep–wakefulness patterns in humans and other mammals, with the degree of development at birth well correlated with the neuroanatomical development of the brain [25].

Initially, low-voltage/high frequency (LV/HF) activity is prominent, with rapid eye movements (REM) for 40–50 % of the time, whereas 30–40 % of the time is marked by high-voltage/ low-frequency (HV/LF) activity without rapid eye movements (NREM). Periods of apparent wakefulness are brief. Thereafter, there is a progressive decrease in the incidence of LV/HF-REM to ~40 % time by term, with a continued fall off in the incidence of LV/HF-REM sleep postnatally.

A behavioral state effect on cerebral metabolic rate is evident for the ovine fetus, with a ~20 % increase in cerebral blood flow and O_2 and glucose uptake during the LV/HF-REM state, when compared to that of the HV/LF-NREM state [30, 31]. The early prominence in LV/HF-REM and timing across species in relation to brain growth, as well as the increase in cerebral metabolic rate at this time, support a role for behavioral/sleep state activity in stimulating growth processes thereby contributing to the brain's development [24].

1.4 Hypoxia and Fetal Cerebral Metabolic Rate/ Behavioural State

In the ovine fetus, sustained hypoxemia sufficient to result in on-setting metabolic acidemia leads to a marginal decrease in $CMRO_2$, which may reflect a decrease in nonessential metabolic activity, thus lowering oxidative needs [32]. Fetal ECOG activity is likewise little changed unless hypoxemia is severe enough to result in associated acidemia with a decrease in the incidence of the metabolically active LV/HF-REM state noted [33]. Studies with prolonged and graded hypoxemia demonstrate similar findings, with $CMRO_2$ and LV/HF-REM activity initially marginally decreased, and only markedly lowered with worsening acidemia [34, 35]. While the decrease in $CMRO_2$ and LV/HF-REM with chronic hypoxemia is only marginal, it may prove detrimental to growth processes in the long term.

A study in the ovine fetus through 60 s of cord occlusion has shown that $CMRO_2$ is maintained, but requires an increase in fractional O_2 extraction thereby necessitating a further lowering of the brain's oxygenation [36]. Behavioral state is disrupted with a flattening of ECOG activity by 90 s of cord occlusion, with an overall decrease in LV/HF-REM following repetitive insults [37]. Cord occlusion of 2 min duration results in decreased O_2 availability to the brain, with the low PO_2 gradient now rate limiting for $CMRO_2$. This, in turn, signals a shift to anaerobic metabolism, suppression in ECOG activity, and the probable shutdown of other energy-using processes [38]. These findings in conjunction with a study examining the neuropathological changes following cord occlusion, has indicated that the near-term ovine fetal brain is able to tolerate severe cord occlusion insults of several minutes duration with little, if any, overt cortical necrotic cell injury, despite a marked decrease in estimated energy production within the brain [38, 39]. However, this disruption in energy-dependent processes, although it is insufficient to induce cell death, may lead to altered development within the brain with repetitive cord occlusion insults over time.

Studies of the developing brain and responses to hypoxic/asphyxia insults in utero, primarily in sheep, have reported on most of the mechanisms outlined for overt neuropathology, with cellular necrosis either primary or delayed in the adult [40–50]. However, these studies have usually been short term with severe acidemia and in relation to cell death as the measure of brain injury; hence, while relevant for modeling intrapartum processes leading to brain injury, they are less applicable for understanding the antenatal conditioning of brain development. There has been limited research of less severe insults with chronic or intermittent hypoxia over longer periods. Such insults might impact growth processes operating through different mechanisms including altered growth factors, stimulus cues and protein synthesis.

2   Methods

2.1   Fetal Cerebral Metabolic Rate

Cerebral metabolic rate in the chronically catheterized ovine fetus has mostly been studied using Fick methodology with closely timed measurements of cerebral arteriovenous differences for the substrate of interest, and cerebral blood flow using the labeled microsphere technique.

2.1.1   Animal Preparation

For studies in near term fetal sheep, we use polyvinyl catheters constructed from V4 (0.72 mm I.D. × 1.22 mm O.D.) inner tubing and reinforced with V11 (1.68 mm I.D. × 2.39 mm O.D) outer tubing (Scientific Commodities Inc., Lake Havasu City, AR). The catheters are placed in the axillary artery at the elbow and advanced 5 cm toward the brachiocephalic trunk for sampling pre-ductal arterial blood, and in the hind limb pedal vein and advanced 18 cm toward the inferior vena cava for injection of labeled microspheres (Fig. 1). A 2.3 mm spherical drill bit is used to remove the bone and expose the dura over 1 cm of the superior sagittal sinus just caudal to the coronal suture. A similar polyvinyl catheter is then placed in the superior sagittal sinus through an 18 gauge needle hole approximately midway between the points where the coronal and lambdoid sutures intersect the sagittal suture (Figs. 1 and 2). The catheter is advanced posteriorly 1.5–2.0 cm, such that the tip lays near the confluens sinus, and after confirming patency for free flowing blood withdrawal, is cemented at its entry through the dura with tissue adhesive (Krazy Glue, Columbus, OH). This catheter then provides for sampling venous blood draining from the brain.

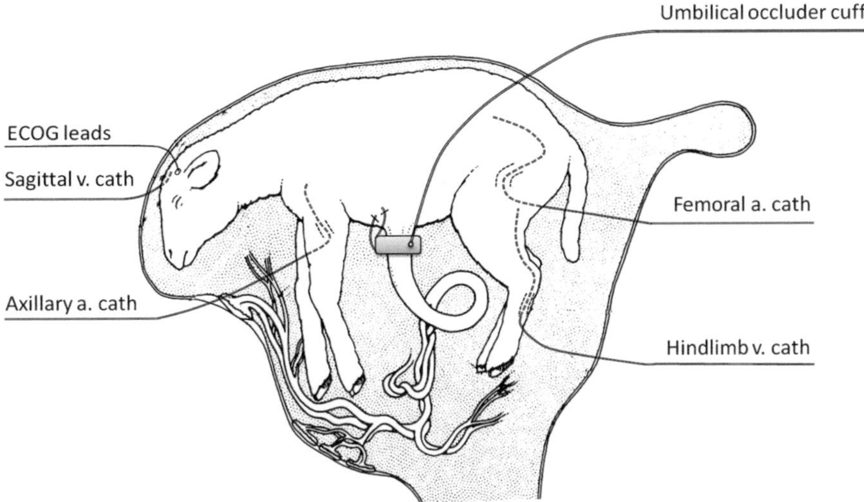

Fig. 1 Chronically instrumented ovine fetus showing the placement of polyvinyl catheters in the axillary artery, in the hind limb vein, in the superior sagittal sinus and in the femoral artery, electrocortical leads implanted biparietally, and an occluder cuff around the proximal portion of the umbilical cord with retaining ties sutured to the abdominal skin

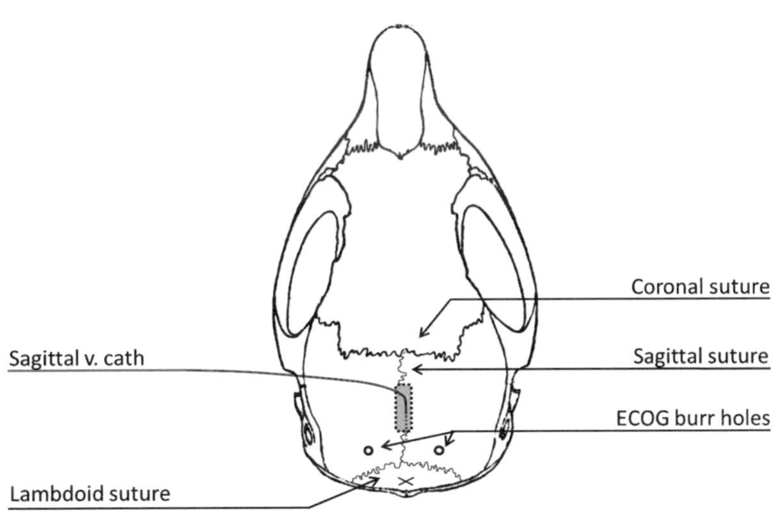

Fig. 2 Outline of the ovine fetal skull from above showing placement of the superior sagittal sinus catheter midway between the points where the coronal and lambdoid sutures intersect the sagittal suture with the catheter advanced posteriorly such that the tip lays near the confluens sinus; and of the cortical electrodes implanted biparietally on the dura through small burr holes in the skull bone 1–1.5 cm lateral and slightly rostral to the junction of the sagittal and lambdoid sutures. The *shaded rectangular area* indicates where skull bone is removed to facilitate placement of the sagittal vein catheter, while *X* indicates the location of the reference electrode in the loose connective tissue overlying the occipital bone at the back of the skull

<table>
<tr><td>2.1.2 Cerebral
Arteriovenous Substrate
Measurements</td><td>

Paired blood samples are taken simultaneously from the axillary arterial catheter and from the superior sagittal sinus catheter. Blood is drawn slowly over 20–30 s into 1 or 2 cm^3 plastic syringes which have been rinsed with heparinized saline and placed on ice. These whole blood samples are placed back on ice and then immediately analyzed for the metabolites of interest including oxygen saturation, hemoglobin, glucose, and lactate. We use an ABL-725 blood gas analyzer (Radiometer, Copenhagen, Denmark) with temperature corrected to that of the ovine fetus, namely, 39.5 °C, and oxygen content is calculated using an oxygen capacity of 1.36 ml/g hemoglobin. Any remaining whole blood can be centrifuged using a cold desk top centrifuge set at 4 °C for 4 min at 9,000g, after which time the plasma may be pipetted off, divided into required aliquots, and then frozen at −80 °C until subsequent measurement for other metabolites.

</td></tr>
<tr><td>2.1.3 Cerebral Blood
Flow Measurements</td><td>

Cerebral blood flow (CBF) measurements have most commonly been made using the labeled microsphere technique, initially using different radioactively labeled microspheres and more recently different-colored or fluorescent microspheres. These are commercially available and made of inert plastic with the label incorporated and related to sphere volume.

Flow measurements are made by the distribution of the microspheres within tissues after injection into the fetal circulation. Immediately following injection into the circulation, the microspheres must be well mixed and evenly distributed and approximate red blood cell distribution. They are then largely trapped in the arteriolar or capillary vascular system on the first circulation after injection depending on their size. It has been calculated and confirmed experimentally, that to have blood flow distribution variability within 10 % of the mean distribution at the 95 % confidence level, about 400 microspheres need to be present in the organ, or portion of organ of interest [51, 52]. Accordingly, it is important that the number of microspheres injected be carefully calculated to satisfy these criteria for the organ or portion of organ with the lowest flow that is to be measured.

We use 15 μm diameter microspheres a size which satisfies both conditions of distribution similar to red blood cells and absence of significant non-entrapment [51]. These are prepared according to the manufacturer's specifications with attention to minimizing microsphere aggregation by placing the container vial in a vortexer for 20 s and then passing the microsphere suspension to be injected through a #26 gauge needle. Using the "reference sample technique" and its application for the measurement of upper body organ blood flow, pre-ductal arterial blood is collected at a constant rate immediately before, during, and after injection of the microspheres. For this we use an axillary artery catheter attached via a blunt needle onto a previously weighed heparinized

</td></tr>
</table>

10-ml glass syringe then placed in a Harvard withdrawal pump set at a withdrawal rate of 2–2.5 ml/min. The withdrawal is commenced and after about 5 s, when it is apparent that blood is being withdrawn without difficulty, 1–1.5 million microspheres are injected over 15 s into the hind limb pedal vein catheter. This is followed by a saline flush to insure all microspheres have been cleared from the catheter line. The withdrawal is continued for 2 min after the microsphere injection is completed, for a total of 2.5 min. The syringe is then detached and reweighted to calculate the volume of blood withdrawn (blood weight/specific gravity of blood), as a check against the accuracy of the withdrawal rate. The blood is then transferred into counting vials and the syringe flushed with distilled water washing out any residual blood and microspheres and hemolyzing the blood.

Upon completion of experimental study, the ewe and fetus are euthanized, and necropsy performed to validate catheter and electrode placements. The fetal brain is weighed and then dissected into the regions and sub-regions of interest. These tissues are weighted separately and then prepared for microsphere counting according to the manufacturer's specifications and published literature [51, 53–55], a detailed accounting of which is beyond the scope of this review.

2.1.4 Cerebral Metabolic Rate Calculations

Cerebral arteriovenous substrate differences for each time point of interest are measured and reported in mmol/L. CBF for each time point is calculated from the labeled counts attributed to that respective microsphere injection found within the prepared brain tissues according to the following formula, where Blood flow ref = the reference sample withdrawal rate:

$$\text{Blood flow}\left(\text{ml}/\text{min}/100\,\text{g tissue}\right) = \frac{\text{labeled counts}/\text{min}/100\,\text{g tissue} \times \text{Blood flow ref ml}/\text{min}}{\text{labeled counts ref}/\text{min}}$$

Cerebral metabolic rate for given substrates of interest is then calculated as the product of cerebral substrate arteriovenous differences and respective CBF measurements. For these measurements, we only include blood flow to the cerebral hemispheres, since this is the brain region primarily drained by the superior sagittal sinus. Cerebral substrate delivery is calculated as the product of arterial substrate concentration in mmol/L and respective CBF measurements. Cerebral substrate fractional extraction, namely, the rate of substrate consumption as a fraction of substrate delivery, is calculated by dividing the substrate arteriovenous differences by the respective arterial substrate concentrations. Cerebral substrate metabolic quotients representing the ratio of substrate uptake to oxygen uptake, expressed stoichiometrically, for complete oxidation of that substrate, are calculated according to the following formula:

$$\text{Substrate metabolic quotient} = \frac{\text{arterio} - \text{venous substrate difference} \left(\text{mmol} / \text{L}\right) \times k}{\text{arterio} - \text{venous oxygen difference} \left(\text{mmol} / \text{L}\right)}$$

where k = number of moles of O_2 required for complete oxidation of 1 mol of substrate. k values are lactate, 3; and glucose, 6.

2.2 Fetal Behavioral States

2.2.1 Animal Preparation

For these studies in near term fetal sheep, we use cortical electrodes made by pulling three lengths of 36 gauge teflon-coated multi-stranded stainless-steel wire (Cooner Wire, Chatsworth, CA) through V11 polyvinyl tubing (Scientific Commodities Inc., Lake Havusu City, AR) with 10 cm of free wire at each end. Both ends of the catheter tubing are hermetically sealed by forcing silastic rubber into the lumen for a distance of 4 cm. Pliable plastic disks (1 mm thick, 5 mm diameter) are punctured with a 25 gauge needle and slid on two of the wires 2 cm from their free ends. A knot is tied in the wire, just above and below the disk to secure it in place, with the remaining free end then stripped of its insulation. The third wire is used as the reference electrode with a single knot tied 6 cm from the end, and a 0.5 mm length of insulation removed immediately beyond it. The opposite free ends of these three wires are marked with a corresponding number of knots to facilitate their connections at the time of experimental study.

The cortical electrodes are implanted biparietally on the dura through small burr holes in the skull bone ~2 mm in diameter made with a battery-powered hand held drill. These are placed 1–1.5 cm lateral to the junction of the sagittal and lambdoid sutures with care taken to avoid puncturing the dura (Figs. 1 and 2). The bared portion of wire to each active electrode is rolled into a small ball and then inserted into each burr hole to rest on the dura with the plastic disk then covering each burr hole. This is then thinly coated with tissue adhesive (Krazy Glue, Columbus, OH) and held in position against the skull bone surrounding each burr hole until firmly adherent. The reference electrode is then placed in the loose connective tissue in the midline overlying the occipital bone at the back of the skull such that the exposed portion of the wire beyond the knot is imbedded (Fig. 2).

2.2.2 ECOG Acquisition and Analyses

We presently use a PowerLab system for data acquisition and analysis (Chart 6 For Windows, AD Instruments Pty Ltd, Castle Hill, Australia). ECOG is recorded with a band pass filter of 0.3–30 Hz at 1,000 Hz sampling rate and then down-sampled to 100 Hz for offline analysis. Subsequently, ECOG amplitude and frequency components as well as the 95 % spectral edge frequency (SEF), ie., the ECOG frequency below which 95 % of spectral power is found, are calculated in 4 s non-overlapping sliding window intervals (Spektralparameter, GJB Datentechnik CmbH, Langewiesen, Germany). The frequency components are calculated as the relative spectral powers in the following frequency bands: 1–4 Hz (delta), 4–8 Hz (theta), 8–12 Hz (alpha) and 12–25 Hz (beta). For each ECOG recording, the ECOG amplitude, SEF and relative spectral power in each frequency band are synchronized and saved in a single file, where they are smoothed with a 1.5 min sliding Bartlett window to reduce noise and movement artifacts.

ECOG activity is scored using automated analysis of amplitude and frequency components (Matlab, MathWorks, Natick, MA) as we have recently reported [56] to distinguish LV/HF and HV/LF state epochs, along with indeterminate voltage/frequency (IV/F) and transition period activities (Fig. 3). Briefly, LV/HF and HV/LF activities are classified using individual animal cutoff values which are established for each experimental day for both primary (ECOG amplitude and 95 % SEF) and secondary (relative spectral power in the delta and beta bands) parameters. This is done by selecting uniform periods of clearly differentiated LV/HF and HV/LF activities that are a minimum of 3 min in duration as determined by visual examination of ECOG amplitude and 95 % SEF. These periods are then used to calculate LV/HF and HV/LF activity related means ± 2 SD for each ECOG parameter. Duration criteria for LV/HF and HV/LF state epochs have been established using normoxic control animals at 134 days of gestation, since ECOG state activity is clearly delineated and the duration of LV/HF and HV/LF state epochs are longer and stabilized by this gestational age [29]. From these animals all LV/HF and HV/LF activity periods were ranked by duration and the 25 percentiles were determined to be 32 s and 28 s, respectively, which were then used as the duration cutoff values for scoring ECOG state epochs. Accordingly, all LV/HF and HV/LF activity <32 s and <28 s, respectively, is reclassified as IV/F activity, while periods of IV/F activity occurring between epochs of LV/HF and HV/LF states are reclassified as transition periods.

2.3 Umbilical Cord Occlusion *2.3.1 Animal Preparation*	We have used an inflatable occluder cuff, 14 HD for preterm fetal sheep at 110 days gestation and 16 HD for near-term fetal sheep at 130 days gestation (In Vivo Metric, Healdsburg, CA), positioned around the proximal portion of the umbilical cord and secured to the abdominal skin (Fig. 1). The volume required for complete occluder cuff inflation is checked at surgery.
2.3.2 Experimental Protocol	We have generally used complete cord occlusion of varying duration and frequency, since this degree of cord occlusion is consistent and reproducible and does not require monitoring of umbilical blood flow. For modeling antenatal cord compression, we use intermittent complete cord occlusions sufficiently spaced so as to give rise to acute, yet limited fetal hypoxemia without cumulative acidosis, thus ensuring animal survival and study relevance for longer term development. We have studied both preterm and near

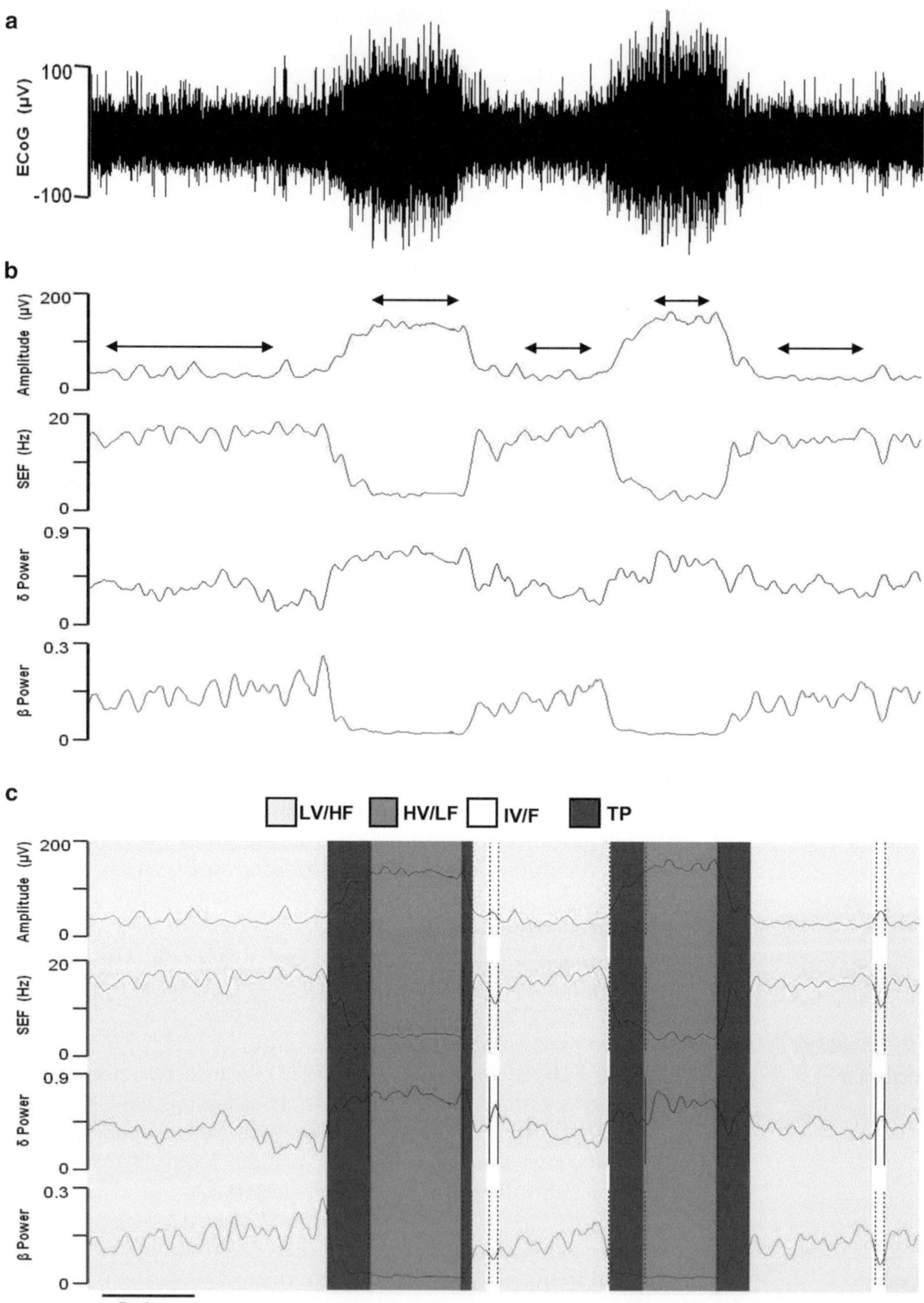

Fig. 3 (**a**) Forty-five minutes of representative ECOG recording and (**b**) respective periods (*arrows*) of uniform ECOG Low-Voltage/High Frequency (LV/HF) and High-Voltage/Low Frequency (HV/LF) activities used to establish individual cutoff values for each parameter. (**c**) Automated assignment of ECOG activity based on the primary (amplitude and 95 % spectral edge frequency (SEF)) and secondary (relative spectral power in the delta and beta frequency bands) parameters. *Light grey* represents: LV/HF; *Medium grey*: HV/LF; *White*: Indeterminate Voltage/ Frequency (IV/*F*); *Dark grey*: transition periods (TP). Reproduced from 'Keen AE, Frasch MG, Sheehan MA, Matushewski BJ, Richardson BS. Electrocortical activity in the near-term ovine fetus: automated analysis using amplitude and frequency components. Brain Res, 2011 Jul 21;1402:30–7, with permission from Elsevier

term fetal sheep with intermittent complete cord occlusion of 90 s duration every 30 min for 4 h each day over four successive days [37, 57]. This cord occlusion insult gives rise to a large decline in fetal arterial PO_2 of 15 Torr, and pH from 7.36 to 7.30, as well as a rise in PCO_2 of 8 Torr as measured at the end of each cord occlusion, but with no cumulative pH change following successive cord occlusions.

For modeling intrapartum cord compression, we have used repetitive complete cord occlusions of increasing frequency leading to fetal heart rate decelerations with worsening acidemia as might be seen in human labor [58]. During the first hour a mild umbilical cord occlusion series has been performed consisting of occlusion lasting for 1 min and repeating every 5 min. The overall effect achieved is that of a mild fetal acidemia with arterial pH falling from 7.36 to 7.32. During the second hour, a moderate occlusion series is performed consisting of cord occlusion for 1 min duration and repeating every 3 min which overall gives rise to a modest degree of fetal acidemia with pH falling to 7.20. During the third hour and continued as needed, a severe occlusion series is performed consisting of cord occlusion for 1 min duration, repeated every 2 min until a targeted degree of severe fetal acidemia has been attained with arterial pH <7.00.

We have also studied intermittent cord occlusions of 4 min duration occurring every hour for 4 h as might be seen during human labor with repeated episodes of fetal bradycardia [38]. These result in marked fetal hypoxemia, hypercapnia and acidemia as measured at the end of each of these UCO insults, and a modest interim pH change from 7.34 to 7.30 as measured over the 4 h of study.

2.4 Placental Embolization	We have chronically instrumented animals at 110 days gestation and used polyvinyl catheters constructed from V4 inner tubing and reinforced with V11 outer tubing (Scientific Commodities Inc., Lake Havasu City, AR). These are placed in a fetal femoral artery for microsphere embolizations and advanced 8 cm into the descending abdominal aorta such that the catheter tip is situated distal to the renal arteries and 1–2 cm above the common umbilical artery (Fig. 1). The correct position of this catheter is confirmed at postmortem examination.
2.4.1 Animal Preparation	
2.4.2 Experimental Protocol	On the initial day of experimental study, animals assigned to placental embolization receive non-radiolabeled carbonized latex 15 μm microspheres (Fisher Scientific, Ottawa, OH) diluted with sterile saline solution injected in boluses of one to two million every 15 to 20 min over a 2-h period through the fetal femoral artery catheter. During this time 0.5 ml of fetal blood is sampled, usually from the contralateral femoral artery which is also catheterized, every 15 min to measure arterial blood gases and oxygen content. Bolus embolizations are continued until oxygen content has been lowered by a targeted amount and maintained for two

consecutive blood samplings, approximately 30–40 % below control values. Care is taken to measure fetal arterial oxygen content in the absence of uterine activity. Microspheres are then injected again daily in a similar manner with the number of microspheres adjusted to lower oxygen content by the targeted amount. This induction of chronic fetal hypoxemia is continued for the study duration of interest, usually ranging between 10 and 21 days [59–63].

Although embolized fetuses always become hypoxemic with chronic repetitive embolization, there is partial recovery in fetal oxygenation the first several days requiring repeat embolizations, after which time fetal oxygen content usually remains near the targeted level with limited need for further embolizations.

3   Notes

3.1   Fetal Cerebral Metabolic Rate

3.1.1   Animal Preparation

Only the V4 inner tubing is advanced into the vessel and then secured to it with a 2.0 silk ligature, while the V11 outer reinforcing tubing is secured distally along the vessel if ligated (axillary artery and hind limb pedal vein) as well as externally to the skin with silk sutures.

We have found it beneficial to place the superior sagittal sinus catheter on a slow infusion of heparinized saline immediately postoperative to maintain vessel/catheter patency at the time of animal experimentation. All vascular catheters are otherwise flushed daily with heparinized saline immediately postoperative until animal experimentation to preserve vessel patency.

We often catheterize both axillary arteries to further insure that at least one of these will be patent at the time of animal experimentation, and to provide a second arterial catheter for the simultaneous monitoring of fetal arterial blood pressure if needed.

3.1.2   Cerebral Arteriovenous Substrate Measurements

We attempt to obtain at least two paired cerebral arteriovenous substrate measurements for each CBF measurement. This is done to reduce methodologic variance due to the inherent error in measuring substrate levels, which can become large relative to the arteriovenous difference during induced hypoxia when CBF is increased, and to reduce biologic variance due to transient fluctuations in CBF which will then impact on cerebral arteriovenous substrate differences.

We have measured the cerebral arteriovenous difference of other substrates from plasma including leucine and [14C] leucine [64]. However, since the uptake of these substrates by the fetal brain is relatively low a greater number of measurements/experiments may be needed to show significant uptakes and/or differences.

*3.1.3 Cerebral Blood
Flow Measurements*

The volume of blood withdrawn for the reference sample is not critical, so long as it is withdrawn at a constant known rate and gives rise to more than 400 microspheres in the sample. The duration of withdrawal is important and must be long enough to insure that flow is occurring to the reference sample, i.e., the withdrawal syringe, for as long as microspheres are circulating after their injection. We have found in our ovine fetal studies that by injecting 1.0–1.5 million microspheres with the withdrawal rate at 2–2.5 ml/ min and continuing for 2 min after completion of the injection, that no microspheres are continuing to circulate thereafter and there are always more than 400 microspheres in the reference blood sample. This number of microspheres injected is also sufficient to result in more than 400 microspheres for most brain regions and sub-regions when all brain tissue is in fact processed and counted [30–32, 36, 38, 65].

Fifteen micrometer microspheres are satisfactory for brain blood flow measurements since we and others [51] have shown that these are not detected in brain venous blood post-injection indicating complete entrapment within the brain. Injection of microspheres into the fetal hind limb pedal vein insures adequate mixing within the heart after shunting across the foramen ovale, and prior to their distribution to upper body tissues. Comparisons of the number of microspheres per gram of tissue in the left and right cerebral hemispheres, also provides an internal check of the adequacy of mixing.

On completion of cerebral arteriovenous blood sampling and related reference sample withdrawal for CBF determination, we routinely transfuse the fetus with an equivalent amount of maternal blood, to compensate for estimated blood loss from sampling.

A single continuous reference withdrawal can be used for sequential microsphere injections thereby allowing for the study of sequential changes in CBF over several minutes in relation to an event of interest, for example occlusion of the umbilical cord [38, 65]. A slower reference sample withdrawal rate is then used to minimize blood loss, but with a longer duration for sampling to insure the entrapment of all circulating microspheres.

Although the microsphere technique has proven to be reliable and allows for regional as well as global measurements of CBF, only a limited number of measurements can be made, each representing an integrated flow over the period of time the microspheres are circulating. As such, a continuous measurement of blood flow to the brain cannot be determined, which might provide for better temporal resolution in relation to both biological and pathological events and for use in long term study. A transit time flow probe has been placed on the external carotid artery [66] and although this vessel does supply the majority of flow to the ovine brain, a significant proportion also supplies cranial structures which cannot be isolated from the cerebral circulation without extensive surgical ligation of extracerebral vessels. The coupled thermojunction technique has also been used [67], and provides continuous, but only relative changes in CBF obtained indirectly through measurement of fluctuations in temperature from a discrete brain region. More recently, laser-Doppler flowmetry has been studied as a continuous measure of CBF [68], but again, is only able to measure relative changes in cerebral perfusion within a small region of brain tissue, which may not be linearly related to actual arterial inflow to that region. The use of a transit time flow probe on the superior sagittal sinus has been validated as an accurate and quantitative measure of CBF in the newborn lamb under resting conditions [69]. We have similarly shown this technique to provide a relatively stable and continuous measure of CBF in the near term ovine fetus under resting conditions [70]. However, with induced hypoxia superior sagittal sinus flow increases much less than arterial inflow to the cerebral cortex measured with microspheres [38] suggesting that alternative venous drainage is occurring.

3.1.4 Cerebral Metabolic Rate Calculations

An advantage of cerebral substrate fractional extractions and substrate/oxygen quotients is that these measures of cerebral metabolism can be quantified without requiring measurement of CBF.

Cerebral substrate metabolic quotients calculate a hypothetical stoichiometric amount of oxygen that would be required to completely oxidize a given substrate, but do not imply that the given substrate is indeed completely oxidized.

The Fick method as used in the study of cerebral metabolism in the ovine fetus is rather insensitive to small changes in cerebral metabolic rate. This is due to the measurement error in oxygen content and other substrates which becomes large relative to the respective arteriovenous differences during induced hypoxia and to a lesser extent to the estimated error of 10 % with the use of the microsphere technique [52]. Together these contribute to a standard deviation of measurement in oxygen consumption on the order of 30 % during hypoxic study with the need therefore for a greater number of experiments to show statistical differences.

Cerebral metabolic rate in the ovine fetus has also been studied using the [^{14}C] deoxyglucose method [71]. This provides for an integrated measure of cerebral glucose utilization over a prolonged period, i.e., 20–30 min, while [^{14}C] deoxyglucose is circulating after administration, and for regional measurements down to a spatial resolution of 100–200 μm.

3.2 Fetal Behavioral States

3.2.1 Animal Preparation

In fetuses where the ECOG has differentiated into LV/HF and HV/LF activities, the difference between the two states is greatest when recorded bilaterally from parietal electrodes [72].

3.2.2 ECOG Acquisition and Analyses

Individual animal cutoff values for classifying LV/HF and HV/LF activity need to be established since each parameter shows some variance across animals. This is possibly due to individual differences in brain maturation as well as slight differences in the placement of the recording electrodes and thereby in the summated neuronal activity represented.

Automated analysis of ECOG activity in the near term ovine fetus as described allows for the processing and quantitative evaluation of large amounts of data rapidly and objectively. We have utilized both ECOG amplitude and frequency components to more fully characterize ECOG activity and have established individual parameter cutoff values as well as population based duration cutoff values to distinguish LV/HF and HV/LF state epochs, along with IV/F and transition period activities. With this analysis, we have shown that the incidence of the predominant LV/HF and HV/LF activity states is comparable to that previously reported; however, the duration of these state epochs are considerably shorter due to the detection of brief periods of IV/F activity using automated analysis techniques which would be difficult to capture using visual analysis [56].

3.3 Umbilical Cord Occlusion

3.3.1 Animal Preparation

We have found the In Vivo Metric HD occluder cuffs to perform well under study conditions with repetitive cuff inflations over many days [73] and to allow for reuse in subsequent animal experiments after gas sterilization.

3.3.2 Experimental Protocol

In studies of intermittent umbilical cord occlusion in preterm and near term fetal sheep with acute, but limited hypoxemia and no cumulative acidosis, we have shown minimal effects on measures of cellular necrosis and apoptosis within the brain [74, 75]. However, this results in a selective reduction in structural proteins which may involve BDNF and its high-affinity receptor TrkB [76, 77].

We have studied repetitive umbilical cord occlusion in near term fetal sheep leading to severe acidemia with pH <7.00 and shown an inflammatory response both systemically and locally within the brain as a possible contributor for brain injury with severe acidemia at birth [78].

3.4 Placental Embolization

3.4.1 Animal Preparation

It is important that the tip of the femoral artery catheter is distal to where the renal arteries branch off the descending aorta to avoid embolization of the kidneys, but above the common umbilical artery to insure placental embolization.

3.4.2 Experimental Protocol

These microsphere embolizations have been shown to chronically occlude the placental arteries and arterioles, which decreases umbilical blood flow and the number of perfused villi and causes a reduction in the surface area available for gas exchange, thus leading to chronic fetal hypoxemia [61, 79].

Fetal placental embolizations reproduce many of the features of placental insufficiency including chronic fetal hypoxemia, abnormal umbilical artery Doppler flow velocity waveforms, elevated placental vascular resistance and asymmetrical growth restriction [59, 61, 62, 79].

Other experimental techniques which have been used in fetal sheep to induce chronic hypoxemia and growth restriction include pre-pregnancy removal of endometrial caruncles, maternal embolization of the uteroplacental circulation, and single umbilical artery ligation [80–82].

4   Summary

The sheep model is the most appropriate animal model for the study measurements and experimental techniques outlined. The ovine fetal size allows for chronic instrumentation and the study of responses to induced hypoxia, behavioral state development is similar to that of humans, and much is already known in this species from the standpoint of cerebral metabolic and anatomical development. Study during the last 4 weeks of ovine pregnancy would be equivalent to the last 8 weeks of human pregnancy, which is most clinically relevant for the timing of concerning hypoxia during human pregnancy and clinical management decisions. It should be noted that the anatomical development of the sheep brain as a "prenatal developer," is in advance of that of the human brain as a "perinatal developer" and the rodent brain as a "postnatal developer," making the ovine fetus a better model for studying in utero conditioning [23]. Nevertheless, the sheep brain is proportionally smaller than the human brain and requires a lower portion of cardiac output [83], and therefore may not be as readily "damaged" with a given hypoxic insult. Thus, there is a need to characterize these insults metabolically, both systemically and for the brain.

References

1. Goldaber KG, Gilstrap LC III, Leveno KJ, Dax JS, McIntire DD (1991) Pathologic fetal acidemia. Obstet Gynecol 78:1103–1106

2. American College of Obstetricians and Gynecologists (1996) Use and abuse of the Apgar Score. American College of Obstetricians and Gynecologists, Washington, DC, The College; Opinion 174

3. Low JA, Galbraith RS, Muir DW, Killen HL, Pater EA, Karchmar EJ (1984) Factors associated with motor and cognitive deficits in children after intrapartum fetal hypoxia. Am J Obstet Gynecol 148:533–539

4. Winkler CL, Hauth JC, Tucker JM, Owen J, Brumfield CG (1991) Neonatal complications at term as related to the degree of umbilical artery acidemia. Am J Obstet Gynecol 164:637–641

5. Robertson CMT, Finer NN, Grace MGA (1989) School performance of survivors of neonatal encephalopathy associated with birth asphyxia at term. J Pediatr 114:753–760

6. Soothill PW, Nicolaides KH, Campbell S (1987) Prenatal asphyxia, hyperlacticaemia, hypo-glycaemia, and erythroblastosis in growth retarded fetuses. Br Med J 294:1051–1053

7. Cox WL, Daffos F, Forestier F et al (1988) Physiology and management of intrauterine growth retardation: a biologic approach with fetal blood sampling. Am J Obstet Gynecol 159:36–41

8. Low JA, Panagiotopoulos C, Derrick EJ (1995) Newborn complications after intrapartum asphyxia with metabolic acidosis in the preterm fetus. Am J Obstet Gynecol 172:805–810

9. Low JA, Boston RW, Pancham SR (1972) Fetal asphyxia during the intrapartum period in intrauterine growth-retarded infants. Am J Obstet Gynecol 113:351–357

10. Badawi N, Watson L, Petterson B et al (1998) What constitutes cerebral palsy? Dev Med Child Neurol 40:520–527

11. Nelson KB, Ellenberg JH (1986) Antecedents of cerebral palsy: multivariate analysis of risk. N Engl J Med 315:81–86

12. Low J, Handley-Derry MH, Burke SO et al (1992) Association of intrauterine fetal growth retardation and learning deficits at age 9 to 11 years. Am J Obstet Gynecol 167:1499

13. Berg AT (1989) Indices of fetal growth-retardation, perinatal hypoxia-related factors and childhood neurological morbidity. Early Hum Dev 19:271–283

14. Dalman C, Thomas HV, David AS, Gentz J, Lewis G, Allebeck P (2001) Signs of asphyxia at birth and risk of schizophrenia. Br J Psychiatry 179:403–408

15. Cannon M, Jones PB, Murray RM (2002) Obstetric complications and schizophrenia: historical and meta-analytic review. Am J Psychiatry 159:1080–1092

16. Hultman CM, Sparén P, Cnattingius S (2002) Perinatal risk factors for infantile autism. Epidemiology 13:417–423

17. Anyaegbunam A, Brustnam L, Divon M, Langer O (1986) The significance of antepartum variable decelerations. Am J Obstet Gynecol 155:707–710

18. Hoskins IA, Frieden FJ, Young BK (1991) Variable decelerations in reactive nonstress tests with decreased amniotic fluid index predict fetal compromise. Am J Obstet Gynecol 165:1094–1098

19. Dawes GS, Lobb MO, Mandruzzato G, Moulden M, Redman CWG, Wheeler T (1993) Large fetal heart rate decelerations at term associated with changes in fetal heart rate variation. Am J Obstet Gynecol 168:105–111

20. Osak R, Webster K, Bocking A, Campbell K, Richardson B (1997) Nuchal cord evident at birth impacts on fetal size relative to that of the placenta. Early Hum Dev 49:193–202

21. Clapp JF, Lopez B, Simonean S (1999) Nuchal cord and neurodevelopment performance at 1 year. J Soc Gynecol Investig 6:268–272

22. Nelson KB, Grether JK (1998) Potentially asphyxiating conditions and spastic cerebral palsy in infants of normal birth weight. Am J Obstet Gynecol 179:507–513

23. McIntosh GH, Baghurst KI, Potter BJ, Hetzel BS (1979) Foetal brain development in the sheep. Neuropathol Appl Neurobiol 5:103–114

24. Richardson B, Gagnon R (2008) Behavioural state activity and fetal health and development. In: Creasy RK, Resnik R (eds) Maternal-fetal medicine, 6th edn. WB Saunders Co., Philadelphia, PA

25. Richardson BS (1991) Metabolism of the fetal brain: biological and pathological development. In: Hanson M (ed) The fetal neonatal brainstem. Cambridge University Press, Cambridge

26. Abrams RM, Hutchinson AA (1985) Energy metabolism of developing brain. J Reprod Med 30:318–323

27. Sokoloff L (1981) Relationships among local functional activity, energy metabolism and blood flow in the cerebral nervous system. Fed Proc 40:2311–2316

28. Ruckebusch Y, Gaujoux M, Eghbali R (1977) Sleep cycles and kinesis in the foetal lamb. Electroencephalogr Clin Neurophysiol 42:226–237

29. Szeto HH, Hinman DJ (1985) Prenatal development of sleep-wake patterns in sheep. Sleep 8:347–355

30. Richardson BS, Patrick JE, Abduljabbar H (1985) Cerebral oxidative metabolism in the fetal lamb: relationship to electrocortical state. Am J Obstet Gynecol 153:426–431

31. Richardson BS, Carmichael L, Homan J, Gagnon R (1989) Cerebral oxidative metabolism in lambs during perinatal period: relationship to electrocortical state. Am J Physiol 257: R1251–R1257

32. Richardson BS, Rurak D, Patrick JE, Homan J, Carmichael L (1989) Cerebral oxidative metabolism during sustained hypoxaemia in fetal sheep. J Dev Physiol 11:37–43

33. Bocking AD (1992) Fetal behavioural states: pathological alteration with hypoxia. Semin Perinatol 16:252–257

34. Richardson BS, Carmichael L, Homan J, Patrick JE (1993) Cerebral oxidative metabolism in fetal sheep with prolonged and graded hypoxemia. J Dev Physiol 19:77–83

35. Richardson BS, Carmichael L, Homan J, Patrick JE (1992) Electrocortical activity, electroocular activity, and breathing movements in fetal sheep with prolonged and graded hypoxemia. Am J Obstet Gynecol 167:553–558

36. Richardson BS, Carmichael L, Homan J, Johnston L, Gagnon R (1996) Fetal cerebral, circulatory and metabolic responses during heart rate decelerations with umbilical cord compression. Am J Obstet Gynecol 175: 929–936

37. Kawagoe Y, Green L, White S, Richardson B (1999) Intermittent umbilical cord occlusion in the ovine fetus near term: effects on behavioural

state and cardiovascular measurements. Am J Obstet Gynecol 181:1520–1529

38. Kaneko M, White S, Homan J, Richardson B (2003) Cerebral blood flow and metabolism in relation to electrocortical activity with severe umbilical cord occlusion in the near term ovine fetus. Am J Obstet Gynecol 188:961–972

39. Mallard EC, Wiliams CE, Johnston BM, Gunning MI, David S, Gluckman PD (1995) Repeated episodes of umbilical cord occlusion in fetal sheep lead to preferential damage to the striatum and sensitize the heart to further insults. Pediatr Res 37:707–713

40. Roohey T, Raju TN, Moustogiannis AN (1997) Animal models for the study of perinatal hypoxic-ischemic encephalopathy: a critical analysis. Early Hum Dev 47:115–146

41. Williams CE, Gluckman PD, Gunn AJ, Tan WK, Dragunow M (1990) Perinatal asphyxic brain damage - the potential for therapeutic intervention. In: Dawes GS, Zacutti A, Borruto f, Zacutti A Jr (eds) Fetal autonomy and adaptation. John Wiley & Sons Ltd, New York, NY

42. Williams CE, Gunn AJ, Gluckman PD (1991) The time course of intracellular edema and epileptiform activity following prenatal cerebral ischemia in sheep. Science 22:516–521

43. De Haan HH, Gunn AJ, Williams CE, Gluckman PD (1997) Brief repeated umbilical cord occlusions cause sustained cytotoxic cerebral edema and focal infarcts in near-term fetal lambs. Pediatr Res 41:96–104

44. Hagberg H, Andersson P, Kjellmer I, Thiringer K, Thordstein M (1987) Extracellular overflow of glutamate, aspartate, GABA and taurine in the cortex and basal ganglia of fetal lambs during hypoxia-ischemia. Neurosci Lett 78:311–317

45. Penning DH, Chestnut DH, Dexter F, Hardy J, Poduska D, Atkins B (1995) Glutamate release from the ovine fetal brain during maternal hemorrhage. Anesthesiology 82:521–530

46. Kjellmer I, Andine P, Hagberg H, Thiringer K (1989) Extracellular increase of hypoxanthine and xanthine in the cortex and basal ganglia of fetal lambs during hypoxia-ischemia. Brain Res 478:241–247

47. De Haan HH, Van Reempts JLH, Vles JSH, De Haan J, Hasaart THM (1993) Effects of asphyxia on the fetal lamb brain. Am J Obstet Gynecol 169:1493–1501

48. Ikeda T, Murata Y, Quilligan EJ, Parer JT, Doi S, Park SD (1998) Brain lipid peroxidation and antioxidant levels in fetal lambs 72 hours after asphyxia by partial umbilical cord occlusion. Am J Obstet Gynecol 178:474–478

49. Wagner KR, Ting P, Westfall MV, Yamaguchi S, Bacher JD, Myers RE (1986) Brain metabolic correlates of hypoxic-ischemic cerebral necrosis in mid-gestational sheep fetuses: significance of hypotension. J Cereb Blood Flow Metab 6:425–434

50. Edwards AD, Yue X, Cox P, Hope PL, Azzopardi DV, Squier MV, Mehmet H (1997) Apoptosis in the brains of infants suffering intrauterine cerebral injury. Pediatr Res 42:684–689

51. Heymann MA, Payne BD, Hoffman JIE, Rudolph AM (1977) Blood flow measurements with radionuclide-labeled particles. Prog Cardiovasc Dis 20(1):55–78

52. Buckberg GD, Luck JC, Payne DB, Hoffman JIE, Archie JP, Fixler DE (1971) Sources of error in measuring regional blood flow with radioactive microspheres. J Appl Physiol 31:598–604

53. Prinzen FW, Glenny RW (1994) Developments in non-radioactive microsphere techniques for blood flow measurement. Cardiovasc Res 28:1467–1475

54. Tan W, Riggs KW, Thies RL, Rurak DW (1977) Use of an automated fluorescent microsphere method to measure regional blood flow in the fetal lamb. Can J Physiol Pharmacol 75:959–968

55. Walter B, Bauer R, Gaser E, Zwiener U (1997) Validation of the multiple coloured microsphere technique for regional blood flow measurements in newborn piglets. Basic Res Cardiol 92:191–200

56. Keen AE, Frasch MG, Sheehan MA, Matushewski BJ, Richardson BS (2011) Electrocortical activity in the near-term ovine fetus: automated analysis using amplitude and frequency components. Brain Res 1402:30–37

57. Green LR, Kawagoe Y, Fraser M, Challis JRG, Richardson BS (2000) Activation of the hypothalamic-pituitary-adrenal axis with repetitive umbilical cord occlusion in the preterm ovine fetus. J Soc Gynecol Investig 7:224–232

58. Frasch MG, Mansano RZ, McPhaul L, Gagnon R, Richardson BS, Ross MG (2009) Measures of acidosis with repetitive umbilical cord occlusions leading to fetal asphyxia in the near-term ovine fetus. Am J Obstet Gynecol 200:200–207

59. Gagnon R, Challis J, Johnston L, Fraher L (1994) Fetal endocrine responses to chronic placental embolization in the late-gestation ovine fetus. Am J Obstet Gynecol 170: 929–938

60. Murotsuki J, Challis JRG, Johnston L, Gagnon R (1995) Increased fetal plasma prostaglandin E_2 concentrations during fetal placental embolization in pregnant sheep. Am J Obstet Gynecol 173:30–35

61. Murotsuki J, Gagnon R, Matthews SG, Challis JRG (1996) Effects of long-term hypoxemia on pituitary-adrenal function in fetal sheep. Am J Physiol 271:E678–E685 (Endocrinol Metab 34)

62. Murotsuki J, Challis JRG, Han VKM, Fraher LJ, Gagnon R (1997) Chronic fetal placental embolization and hypoxemia cause hypertension and myocardial hypertrophy in fetal sheep. Am J Physiol 272:R201–R207 (regulatory Integrative Comp Physiol 41)

63. Keen AE, Frasch MG, Sheehan MA, Matushewski B, Richardson BS (2011) Maturational changes and effects of chronic hypoxemia on electrocortical activity in the ovine fetus. Brain Res 1402:38–45

64. Czikk MJ, Sweeley JC, Homan JH, Milley JR, Richardson BS (2003) Cerebral leucine uptake and protein synthesis in the near-term ovine fetus: relationship to fetal behavioural state. Am J Physiol Regul Integr Comp Physiol 284:R200–R207

65. Richardson BS, Caetano H, Homan J, Carmichael L (1994) Regional brain blood flow in the ovine fetus during transition to the low-voltage electrocortical state. Dev Brain Res 81:10–16

66. Gratton R, Carmichael L, Homan J, Richardson B (1996) Carotid arterial blood flow in the ovine fetus as a continuous measure of cerebral blood flow. J Soc Gynecol Investig 3:60–65

67. Abrams RM, Gerhardt DJ, Burchfield DJ (1991) Behavioural state transition and local cerebral blood flow in fetal sheep. J Dev Physiol 15:283–288

68. Lan J, Hunter CJ, Murata T, Power GG (2000) Adaption of Laser-Doppler flowmetry to measure cerebral blood flow in fetal sheep. J Appl Physiol 89:1065–1071

69. Grant DA, Franzini C, Wild J, Walker AM (1995) Continuous measurement of blood flow in the superior sagittal sinus of the lamb. Am J Physiol 269(2 Pt 2):R274–R279

70. Czikk MJ, Totten S, Homan J, White SE, Richardson B (2011) Sagittal sinus blood flow in the ovine fetus as a continuous measure of cerebral blood flow: relationship to behavioural to state activity. Dev Brain Res 131:103–111

71. Abrams R, Hutchison AA, Jay TM, Sokoloff L, Kennedy C (1988) Local cerebral glucose utilization non-selectively elevated in rapid eye movement sleep of the fetus. Dev Brain Res 40:65–70

72. Clewlow F, Dawes GS, Johnston BM, Walker DW (1983) Changes in breathing, electrocortical and muscle activity in unanaesthetized fetal lambs with age. J Physiol 341:463–476

73. Czikk MJ, Green LR, Kawagoe Y, McDonald TJ, Hill DJ, Richardson BS (2001) Intermittent umbilical cord occlusion in the ovine fetus: effects on blood glucose, insulin, and glucagon and on pancreatic development. J Soc Gynecol Investig 8(4):191–197

74. Rocha E, Hammond R, Richardson B (2004) Necrotic cell injury in the preterm and near-term ovine fetal brain and the effect of intermittent umbilical cord occlusion. Am J Obstet Gynecol 191:488–496

75. Falkowski A, Hammond R, Han V, Richardson B (2002) Apoptosis in the preterm and near-term ovine fetal brain and the effect of intermittent umbilical cord occlusion. Dev Brain Res 136:165–173

76. Rocha E, Totten S, Hammond R, Han V, Richardson B (2004) Structural proteins during brain development in the preterm and near-term ovine fetus and the effect of intermittent umbilical cord occlusion. Am J Obstet Gynecol 191:497–506

77. Nishigori H, Mazzuca DM, Nygard KL, Han VK, Richardson BS (2008) BDNF and TrkB in the preterm and near-term ovine fetal brain and the effect of intermittent umbilical cord occlusion. Reprod Sci 15(9):895–905

78. Prout AP, Frasch MG, Veldhizen RAW, Hammond R, Ross MG, Richardson BS (2010) Systemic and cerebral inflammatory response to umbilical cord occlusions with worsening acidosis in the ovine fetus. Am J Obstet Gynecol 202:82.e1–9

79. Gagnon R, Johnston L, Murotsuki J (1996) Fetal placental embolization in the late-gestation ovine fetus: alternations in umbilical blood flow and fetal heart rate patterns. Am J Obstet Gynecol 175:63–72

80. Robinson JS, Kingston EJ, Jones CT, Thorburn GD (1979) Studies on experimental growth retardation in sheep. The effect of removal of endometrial caruncles on fetal size and metabolism. J Dev Physiol 1:379–398

81. Creasy RK, Barrett CT, De Swiet M, Kahanpaa KV, Rudolph AM (1972) Experimental intra-uterine growth retardation in the sheep. Am J Obstet Gynecol 112(4):566–573

82. Miller SL, Supramaniam VG, Yawno T et al (2009) Neuropathology in intrauterine growth restricted (IUGR) lambs is associated with delayed attainment of behavioural milestones in the newborn period. Reprod Sci 16(3 Suppl):109A

83. Sheldon RE, Peeters LLH, Jones MD Jr, Makowski EL, Meschia G (1979) Redistribution of cardiac output and oxygen delivery in the hypoxemic fetal lamb. Am J Obstet Gynecol 135:1071–1078

Chapter 7

Studies of Perinatal Asphyxial Brain Injury in the Fetal Sheep

Paul P. Drury, Laura Bennet, Lindsea C. Booth, Joanne O. Davidson, Guido Wassink, and Alistair Jan Gunn

Abstract

In order to develop more effective ways of identifying, managing, and treating perinatal asphyxial brain injury, stable experimental models are essential. Although the outcome of clinical asphyxia is highly variable, modern imaging studies have distinguished two major patterns of injury in term infants, involving primary damage in either the parasagittal cortex or in the basal ganglia respectively. The present review describes the experimental preparation in detail, and the key experimental factors that determine the pattern and severity of brain injury in chronically instrumented fetal sheep, including the depth ("severity"), duration, and repetition of the insult, the maturity, and condition of the fetus. These models are valuable to dissect the pathogenesis of key clinical patterns of brain injury in a stable thermal and biochemical environment, and to test therapeutic interventions.

Key words Fetal sheep, Perinatal asphyxia, Hypoxic-ischemic encephalopathy, Hypotension, Cerebral blood flow

1 Introduction

Acute neonatal encephalopathy associated with asphyxia remains a significant cause of death and long-term disability [1]. Early onset neonatal encephalopathy is highly associated, on the one hand with evidence of asphyxia as shown by non-reassuring fetal heart rate (FHR) tracings and severe metabolic acidosis on umbilical cord blood [2, 3], and on the other with subsequent neurodevelopmental impairment [4]. However, as previously reviewed, the relationship between the biochemical severity of asphyxia or need for resuscitation and subsequent injury is relatively weak, inferring that many other factors are involved [5]. Experimental models are essential to help us understand the complex factors that determine whether the fetus will go on to develop neural injury, and once properly characterized, to test the effects of different methods of managing the asphyxiated fetus and newborn, and to investigate

Jerome Y. Yager (ed.), *Animal Models of Neurodevelopmental Disorders*, Neuromethods, vol. 104, DOI 10.1007/978-1-4939-2709-8_7, © Springer Science+Business Media New York 2015

proposed novel therapeutic interventions. A range of models is required, because no one model can satisfactorily reproduce the wide range of distinct patterns of cerebral damage seen at postmortem or using modern imaging techniques [6, 7].

1.1 What Initiates Neuronal Injury?

Before discussing the experimental approach, it is helpful to reflect on the triggers of asphyxial injury to the brain and other organs. Fundamentally, injury requires a period of insufficient oxygen and substrate delivery (glucose and other substances such as lactate) such that neurons (and glia) cannot maintain internal homeostasis and become depolarized [8–10]. At least broadly, the extent of injury is related to the duration of tissue depolarization [11]. Since organ perfusion is essential for maintaining oxygen and substrate delivery to tissues, not surprisingly, there is now considerable evidence to suggest that brain perfusion is the key factor that determines whether or not neural injury occurs after severe hypoxia/asphyxia. For example, in term fetal sheep, the extent of brain injury is strongly associated with the degree or duration of systemic hypotension during asphyxia (e.g., Fig. 1) [12–15]. Conversely, although adaptation to moderate hypoxia has been extensively studied, it is critical to appreciate that the fetus can fully adapt to

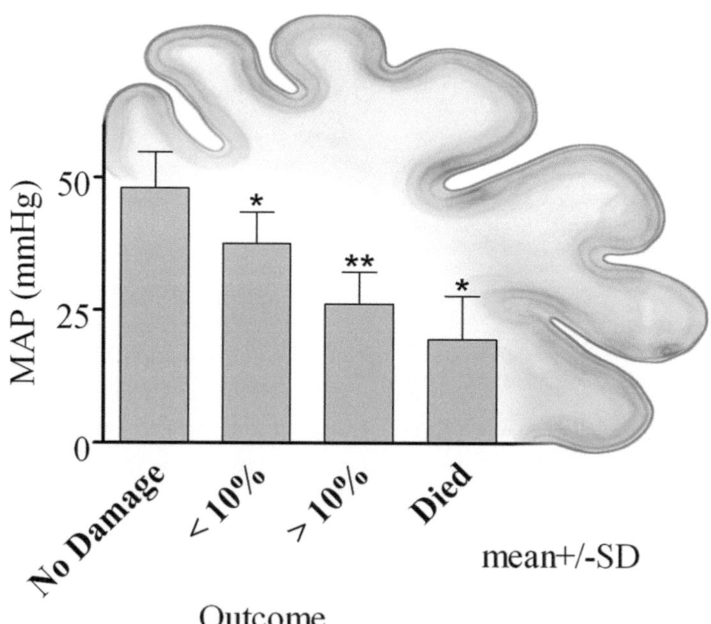

Fig. 1 The relationship between hypotension and neuronal damage. The severity of fetal systemic hypotension during asphyxia, induced by partial common uterine artery occlusion, is closely related to the degree of neuronal loss in the near-term fetal sheep (mean arterial pressure (MAP), image modified from ref [12])

mild to moderate reductions in oxygen tension without injury, from normal values of greater than 20 mmHg down to 10–12 mmHg [16, 17]. Thus, in general, hypotension during a severe hypoxic challenge is a central requirement to reliably induce asphyxial neural injury.

Compression of the umbilical cord is a known potential cause of fetal asphyxia. Normally, the presence of amniotic fluid protects the umbilical cord during uterine contractions. However, knots or entanglements of the cord can lead to compression of the cord before or during birth. Further, if amniotic fluid is reduced, e.g., in the growth retarded fetus, or lost after rupture of the amniotic membranes in an early stage of labor, or even worse, if the cord becomes trapped in the birth canal, physiologic uterine contractions can compress the umbilical cord between the fetus and uterine wall, giving rise to fetal hypoxemia, anaerobic glycolysis and acidemia, with repeated or fixed deceleration of the fetal heart rate [18].

In view of these considerations, the present review focuses on induction of severe asphyxia using umbilical cord occlusion in term-equivalent sheep, with a brief note on the effect of prematurity. Fetal responses to moderate hypoxemia without acidosis are addressed in Chap. 6 by Bryan Richardson. Note that acute fetal asphyxia can also be induced by interrupting perfusion of the uterus by occlusion of the uterine artery, but the degree of occlusion that can be obtained, with this method, is relatively variable due to residual anastomotic blood flow [12].

1.2 Advantages of the Sheep Preparation

The relatively large size of the sheep fetus enables extensive implantation of instrumentation (e.g., as outlined in Fig. 2) which allows for continuous measurement of fetal heart rate, blood pressure (arterial and venous) [19], behavior (body movements) [20], blood flow to the brain and periphery, and brain metabolism [21], cerebral oxygenation [22], sympathetic activity, temperature [23, 24], and more. Catheters in the ewe and fetus also allow access for blood sampling, treatments, and euthanasia. This comprehensive approach permits a significant physiological assessment of multiple organ systems in utero in a stable environment, without the confounding effects of anesthesia or changes in temperature.

It is important to appreciate that sheep are a highly precocial species, whose neural development around 0.8–0.85 of gestation is broadly similar to the term human fetus [25, 26]. Earlier gestations have also been studied; the 0.7 gestation fetus is broadly equivalent to the late preterm infant at 30–34 weeks, before the onset of cortical myelination, while at 0.6 gestation the sheep fetus is similar to the 26–28 week gestation human.

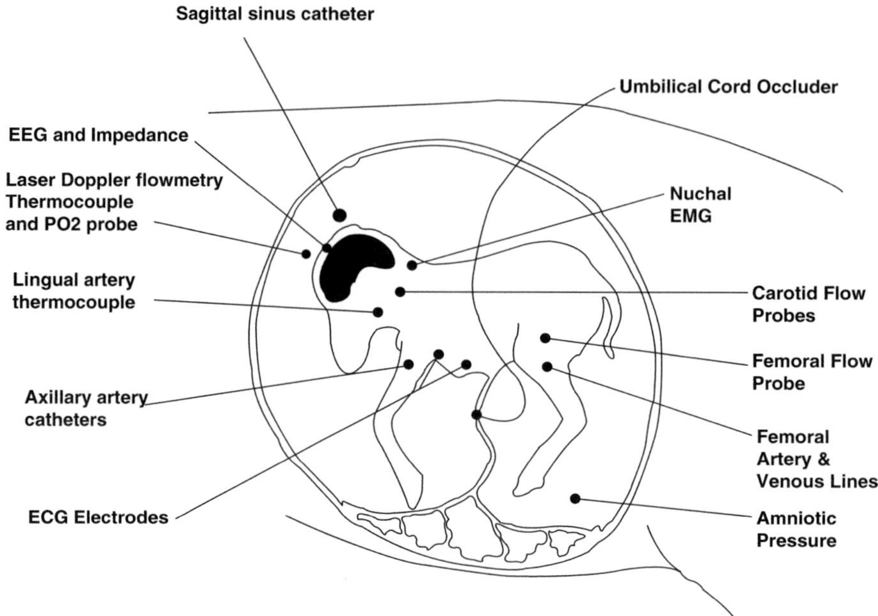

Fig. 2 Schematic representation of the chronically instrumented fetal sheep

2 Materials

2.1 Catheters

Fetal vascular catheters are nearly always custom-made. In our laboratory, vascular and amniotic catheters are made from 1.4 m lengths single lumen dural polyvinylchloride (PVC) tubing (2.0 mm outer diameter (OD) and 1.0 mm inner diameter (ID), Biocorp Australia Pty Ltd., AU). The distal end is briefly immersed in hot water (60–80 degree Celsius (°C)) to soften the PVC material and allow insertion of a 1.5 cm long 18 gauge (G) blunt needle (Health Support Ltd., Auckland, NZ), which allows connection to a three-way lockable luer stopcock (Onelink, Auckland, NZ). Two lengths (1.0 mm lengths, 2–3 mm distance) of slightly wider tubing (2.5 mm OD, 1.5 mm ID) are slipped over the vascular catheters, approximately 2 mm from the proximal end and fixed with cyclohexanone (Ajax Chemicals Ltd., Auckland, NZ). These cuffs serve as an attachment hub through which the catheter can be secured into a blood vessel or onto the skin. For amniotic catheters four small circular holes are cut at the proximal end to prevent catheter blockage. To accommodate different vascular sizes 20 cm lengths of smaller tubing diameters are fitted and secured with cyclohexanone at the proximal end: size 015 catheter (0.90 mm OD, 0.50 mm ID), size 025 catheter (1.20 mm OD, 0.80 mm ID) or size 030 catheter (1.27 mm OD, 0.86 mm ID). Infusion lines are used to provide the connection between the catheter-stopcock assembly and infusion syringes. These are made of 1.2 m lengths of tubing (2.0 mm OD, 1.0 mm ID), fitted with 18 G blunt needles at either end and capped with a white cap at the distal end and a male–male connector (Vygon, Swindon, UK) and red cap (Healthcare Logistics, Auckland, NZ) at the proximal end.

2.2 Stainless Steel Electrodes

All electrodes are custom-made in our laboratory. Electrocardiogram (ECG) electrodes are made from 1.65 m lengths of three-stranded stainless steel wire protected by an outer metal sheath and Teflon layer (AS633-3SSF, Cooner Wire Company, Chatsworth, USA). Forty-six centimeters of Teflon insulation is stripped from the proximal end and all three strands are isolated from the outer metal sheath. Thermal fabric discs (0.5 mm thick, 6 mm diameter, Centrepoint, Auckland, NZ) are threaded onto two wires and knotted at 40 cm from the proximal end. The Teflon coating below these knots is then removed and the bare wires rolled into balls, creating electrodes with a high electrical conductance. From the distal end of the electrode the same two wires are isolated at 3 cm and identified with a multi-meter. The tips of these wires are stripped of the Teflon coating (1 mm lengths), treated with soldering flux and soldered onto a bucket 9-pin adapter (R.S. Components, Auckland NZ). The outer metal sheath is covered with heat shrink tubing (3.2 mm ID, R.S. Components) to prevent contact between bare wires and soldered onto the bucket.

Electroencephalograph (EEG) electrodes consist of 1.4 m lengths of five-stranded (or seven-stranded if required) stainless steel wire protected by an outer metal sheath and Teflon layer (AS633-5SSF, Cooner Wire Company). Fifteen centimeters of Teflon is stripped from the proximal end and all strands are isolated from the outer metal sheath, which is kept as a reference electrode. Marked thermal fabric discs are threaded onto four wires and knotted at 10 cm from the proximal end. The Teflon coating is removed from below the knots and the bare wires rolled into balls. The final wire, acting as a second reference, is cut back to 10 cm and knotted at 5 cm. No more than 5 mm of Teflon is removed from below the knot. From the distal end of the electrode all wires are isolated at 7 cm, identified with a multi-meter and soldered onto a bucket 36-pin adapter (R.S. Components) as described above. The electrode discs are individually marked to allow correct placement at surgery.

2.3 Umbilical Cord Occluders

Inflatable silicone rubber occluders (In Vivo Metric, Healdsburg, USA) are extended to a length of 1.4 m with silicone tubing of an identical diameter (2.16 mm OD, 1.02 mm ID, Bamford Ltd., Lower Hutt, NZ) and distally fitted with an 18 G metal blunt cannula (Covidien, Dublin, Ireland). Tubing of a larger diameter (6 mm OD, 4.5 mm ID, 1 cm length) is then slipped over the exposed joints and fixed in place with flowable silicone sealant

(Electropar, Auckland, NZ). Two lengths (8 cm) of heavy duty umbilical tape (Johnson & Johnson Medical Pty Ltd., Sydney, AU) are then threaded through the eyelets of the umbilical cuff so that the cuff can be securely closed in surgery.

3   Experimental Population and Procedures

Many potentially suitable breeds of sheep are available, although there is little information on whether this affects responses to asphyxia. For reference, our studies are currently conducted using pregnant Romney ewes, time-mated with Suffolk rams at approximately 3–4 years of age. Ultrasound scans are performed at 30–35 days and again at 60 days of gestation to confirm the ewe's pregnancy. All ewes are kept on the farm's paddock and allowed to roam free before being transferred to a dedicated feedlot area for a period of 2 weeks to acclimatize to human handling and grow accustomed to the concentrated pellet feed (University C Mix, Camtech Nutrition Ltd., Hamilton, NZ) that is used at the University's Animal Resource Unit.

3.1   Transportation and Acclimatization

Following acclimatization on the farm to the laboratory feed, healthy ewes are transported to the Animal Resource Unit of the Faculty of Medical and Health Sciences, and transferred to individual metabolic cages that allows free access to concentrated pellet feed and water and are housed together in designated animal holding units. The large animal facilities are climate-controlled (ambient temperature 16 ± 1 °C, 50 ± 10 percent (%) humidity) with a 12 h (h) light–dark cycle (light hours 07:00–19:00). Ewes are acclimatized to these laboratory conditions for a period of 5 days before surgery.

3.2   Preoperative Preparation and Anesthesia

Food, but not water, is withdrawn 12–18 h before surgery to reduce the risk of vomiting which may lead to inhalation of gastric material during anesthesia. On the day of surgery the ewe is transferred to the animal preparation room in a specially designed portable sheep crate and weighed. The ewe is then restrained in a recumbent position and administered oxytetracycline (20 mg/kg, Phoenix Pharm, Auckland, New Zealand) by intramuscular injection 30 minutes before the start of surgery. The left front brachial vein is cannulated with a 20 G intravenous (i.v.) cannula and anesthesia induced by injection of propofol (5 mg/kg; AstraZeneca Limited, Auckland, New Zealand).

The ewe is placed in a V-shaped trough on the surgery table in a supine position with its legs tied caudally and cranially to prevent rotation during surgery. Intubation is performed with a 9 mm inflatable endotracheal tube (Health Support Ltd.) that is then connected to the anesthetic machine (CIG Midget 3 anesthetic apparatus, Commonwealth and Industrial Gases Pty Ltd., Preston,

AU). A 3–5 % isoflurane in oxygen mixture (Merial Ltd., Auckland, NZ) is administered until the ewe is completely anesthetized and then anesthesia is maintained using 2–3 % isoflurane in oxygen (2 L/min). Ewes typically breathe spontaneously under anesthesia, but are mechanically ventilated if necessary (Ivent Research Ltd., Auckland, NZ). Isoflurane readily crosses the placenta, and accordingly, the fetus is under anesthesia at all times during the surgical period. The depth of anesthesia, maternal heart rate, and respiration are monitored continuously by trained staff.

The ewe's wool is shorn from the primary surgical fields (abdomen, flank and leg) with industrial clippers and shaved with a size-40 razor blade (Southern Veterinary Supplies). The skin is then thoroughly cleansed with an antiseptic soap-water solution followed by the application of a 10 % povidone–iodine solution (Betadine solution, F.H. Faulding and Co., Mulgrave, AU) which is rinsed off with a 2 % hibitane (Microshield 5-chlorhexidine concentrate, Johnson & Johnson Medical Pty Ltd.) in 70 % isopropyl alcohol solution (Orica Chemnet, Auckland, NZ). Finally, a concentrated povidone–iodine solution is sprayed onto the cleaned areas. The ewe is then moved into the surgical theater suite and a constant isotonic saline drip (250 mL/h) connected via the brachial cannula is started to maintain maternal fluid balance. At this time the surgical team prepares for surgery, scrubbing hands thoroughly with a povidone–iodine surgical scrub brush (Health Support Ltd.) and donning masks, hats, sterile gowns and surgical gloves. The preparation is completed with five sterile linen drapes placed to allow an abdominal midline incision, and a sigmoidoscopy drape placed on the maternal flank.

3.3 Fetal and Maternal Surgery

The fetal sheep are instrumented using strict aseptic techniques, at appropriate gestational ages as noted in the introduction. Instruments, equipment, catheters, and electrodes are all sterilized using appropriate methods, including gamma irradiation, ethylene oxide gas, and autoclave [27]. A 10–12 cm abdominal midline skin incision is made with a scalpel, approximately 5 cm anterior to the mammary glands. Small blood vessels are cauterized to prevent bleeding, with a high-frequency diathermy (Electromedical Systems Inc., Danvers, USA) and the maternal fat is dissected. A laparotomy is then performed along the *linea alba* and the pregnant uterus palpated, to confirm the number of fetuses. At this stage, a maternal flank incision is made with a size-12 French trochar and cannula (Surgical Access Pty Ltd., Bibra Lake, USA) and the proximal ends of the catheters, electrodes, and flow-probes are introduced into the maternal abdominal cavity via the cannula. The leads are then pulled through, above the midline incision, and the cannula removed. All catheters are color coded for later identification and flushed with isotonic saline through stopcock attachments at the distal end.

The position of the fetus is palpated and a hysterectomy performed on the uterus in an area free of cotyledons and parallel to any major uterine vessels. Non-traumatic Babcock clamps are placed along the edges of the incision to prevent further tearing of the uterine or amniotic membranes and loss of amniotic fluid. The posterior half of the fetus must be carefully exteriorized to prevent any umbilical cord twisting, placed in an upright position (to allow easy surgical access) and wrapped in moist abdominal swabs (Health Support Ltd.) to prevent dehydration of fetal tissue.

An incision is made on the interior surface of the left fetal hindlimb. The femoral artery and vein are identified, exposed by blunt dissection and ligated distally. The respective vessels are then micro-incised with Vannas scissors (Cooper Medical Ltd.) and the proximal ends of the catheters inserted through the small incision in the top third of the blood vessels. A proximal ligature secures the catheter within the vessel. These catheters are used to measure fetal mean arterial pressure (MAP), and venous pressure. After insertion, the patency of each vascular catheter is briefly checked by drawing back blood, and then flushing with sterile saline. The incisions are closed and the catheters tightly secured to the skin with non-traumatic, non-cutting 2.0 silk sutures (Resorba Wundversorgung, Nürnberg, Germany) to minimize catheter migration. The right femoral artery is then dissected and an ultrasound blood flow probe (size 2.5PSS at 0.6 and 0.7 ga and 3PSS at 0.85 ga, Transonic Systems Inc., Ithaca, New York, USA) fitted around the vessel for the measurement of femoral blood flow (FBF). The probe is secured by suturing the exposed fascia on opposing sides of the wound over the probe head. The flow probe is irrigated with sterile saline to help expel air and so minimize the risk of acoustic error, the wound closed and flow probe tail secured to the skin. The fetus is then returned to the uterus in its original position, and the uterus repaired in two layers with a continuous non-locking suture, making sure that all chorionic and amniotic membranes are included. Non-traumatic, non-cutting 2.0 silk sutures are used for all fetal suturing, wound and uterine repairs.

The uterus is again palpated and the fetal head, chest, and forelimbs exteriorized through a second uterine incision. This two incision approach helps reduce the risk of twisting the fetus and thus the umbilical cord. As described previously for vascular catheters, polyvinyl catheters are placed in the right brachial or axillary artery (to allow pre-ductal blood sampling) and brachial vein respectively (for drug administration if required). ECG electrodes are placed subcutaneously over the right shoulder and chest at the apex level (fifth intercostal space) and sewn across the chest to record FHR. An insulated section of the ECG lead is then sutured over the fetal chest and again over the fetal shoulder to reduce tension on the electrodes. The amniotic catheter is attached to the second suture. The fetus is briefly retracted to abdominal level and the inflatable silicone occluder loosely fitted around the umbilical cord, for induction of asphyxia, and secured with umbilical cotton tape. The occluder lead is sutured to the skin in order to prevent tension on the umbilical cord. The fetus is then carefully replaced into the uterus and the uterus and membranes loosely tightened around the neck by readjusting the Babcock clamps to prevent further amniotic fluid loss.

The fetal head is flexed backwards, a 2 cm incision parallel to the trachea (0.5 mm distance) is made and the left carotid artery located by blunt dissection. An ultrasound blood flow probe (size 3PSS for all ages) is fitted around the carotid vessel to measure carotid blood flow (CaBF) and secured. The wound is closed and the catheters and flow probe secured to the skin. Next, EEG electrodes are implanted on the dura. The fetal scalp is incised along the sagittal plane and reflected back with blunt forceps. Bleeding capillary vessels are cauterized and four cranial burr holes (1 mm diameter) drilled with a standard foot-operated dental drill (Gunz Dental, Auckland, NZ). Two pairs of EEG electrodes are placed on the dura over the parasagittal parietal cortex (at term: 10 and 20 mm anterior to bregma and 10 mm lateral) and secured with cyanoacrylate glue (Selleys, Auckland, NZ). Two reference electrodes are sutured to the skin over the occiput. The fetal scalp is then replaced and secured to the skull with cyanoacrylate glue.

Once all fetal surgery is complete the fetus can then be gently returned to the uterus along with a generous length of the leads to allow free movement without fetal entanglement. The antibiotic Gentamycin (80 mg Gentamicin, Pfizer, NY, USA) is administered into the amniotic sac as well as sterile saline (warmed to 37 °C) to replace amniotic fluid lost during surgery. The uterus is then repaired in two layers. All leads are exteriorized through the maternal flank and the fetal vascular catheters flushed with heparinized sterile saline (10 U/mL, Health Support Ltd.) after which all the stopcocks are capped and protected with a plastic bag for transport. The peritoneum is sutured with heavy-duty umbilical cotton tape and the abdominal wall repaired with 1.0 silk (Resorba Wundversorgung) using a continuous suture. The skin along the suture line is infiltrated with the long-acting local anesthetic bupivacaine (0.25 %, AstraZeneca Ltd., Auckland, NZ) and the flank wounds repaired with a purse string suture using 1.0 silk. The exit site on the maternal flank is placed sufficiently posterior to prevent the ewe from chewing the leads.

The maternal long saphenous vein is then catheterized to provide access for postoperative care and euthanasia. A small (4 cm) skin incision is made on the tarsus over the cleft formed by the tibia and fibula. A size-33 trochar and cannula (Surgical-access Pty Ltd) is inserted through the incision and proximally advanced under the skin, exiting at flank level through a stab wound. The trochar's cannula is removed and the distal end of a size 040 catheter tracked subcutaneously, exiting at the incision site. The catheter is then flushed with heparinized sterile saline solution (10 U/mL) and the vein catheterized as previously described. Ewes are then taken off anesthesia, extubated and when judged sufficiently awake, taken back to the laboratory in their cage. Ewes are constantly supervised until they can stand up in a stable manner and are seen to eat and drink. The anesthesia and surgery typically take approximately 2–4 h depending on the experience of the surgeon and the extent of instrumentation required for the particular study. These procedures are generally well tolerated by the ewe and fetus without complications.

3.4 Surgical Recovery

A postoperative recovery period of 4–5 days is allowed before experimentation during which time the health of the ewe and fetus is monitored continuously, including MAP, FHR and EEG activity. This period is important since in our experience it can take several days for fetal sleep state cycling to fully normalize and even longer for endocrine parameters such as cortisol levels to return to normal in term fetuses. Daily arterial blood samples (0.3 mL) are taken from the fetus to monitor pH and blood gases, blood glucose and lactate as measures of fetal wellbeing. Any animals not considered to have recovered by our laboratory criteria are assessed by the University veterinarian.

Antibiotics are administered intravenously to the ewe for a total duration of 4 days consisting of a daily administration of Crystapen (Benzylpenicillin sodium, 600 mg dissolved in 4 mL of saline, Biochemie GmbH., Vienna, Austria) and Gentamycin (80 mg) for a period of 2 days. The topical antibiotic Terramycin (Pfizer) is applied to surface wounds. Fetal catheters are maintained patent by continuous infusion of heparinized saline. The concentration varies depending on the gestational age: 10 U/mL at 0.2 mL/h at 0.6 ga and 20 U/mL at 0.2 mL/h at 0.7 and 0.85 ga. The fetal and maternal vascular lines are flushed periodically with heparinized saline (20 U/mL) to remove any vascular obstructions.

3.5 Data Acquisition and Hardware Processing

Clearly the type and intensity of fetal monitoring will vary depending on the scientific questions to be addressed. However, it is critical to appreciate that successful use of this paradigm is associated with fetal hypotension that at its nadir is very close to that associated with cardiac arrest [28, 29]. Thus, to avoid excess mortality it is essential to continuously monitor the FHR and MAP during and shortly after asphyxia. We describe one particular approach here, but any strategy that provides accurate, real-time information will be effective.

All electrical and monitoring equipment is housed in a separate experimental monitoring room adjacent to the animal holding rooms. Catheters and electrodes coming from the ewe are connected to the appropriate signal leads in a Perspex box bolted to the side of the cage at approximately the height of the fetal head when the ewe is standing. All signal leads are then passed to the experimental monitoring room and connected to monitoring equipment and a computer workstation.

Signals are acquired through a 16 bit analogue to digital card with circular buffering DMA (National Instruments, Texas, USA) and collected by customized software (Labview for Windows, National Instruments) on a computer workstation. Blood pressure, amniotic trace, ECG, EEG and blood flow signals are recorded continuously from 12 h before the experiment until termination of the preparation. Data are then transferred to a server for off-line analysis.

3.6 Occlusion of the Umbilical Cord

Complete compression of the umbilical cord is performed by inflating the occluder with 5 mL of sterile saline, a volume known to completely occlude the umbilical cord as determined in pilot experiments with a Transonic flow probe placed around an umbilical vein [30]. The total duration of occlusion required to induce hypotension and neural injury, is a function of maturity, and may need to be fine tuned in different settings or with different breeds of sheep. For reference, in our studies we found that the optimal durations of near-terminal single periods of occlusion were as follows: 30 min at 0.6 ga [20, 30], 25 min at 0.7 ga [26, 27] and 15 min at 0.85 ga [23, 28].

Successful occlusion is confirmed by rapid fetal bradycardia and arterial hypertension. FHR and MAP traces are monitored continuously throughout the occlusion and blood composition measurements are taken at standard time points to assess fetal condition. Upon completion of the occlusion protocol the occluder is deflated. It is unknown whether rapid deflation of the occluder has any material effect on fetal recovery compared with controlled deflation over 10 s. Successful recovery from occlusion is confirmed by a rapid, overshoot increase of FHR and MAP. If bradycardia persists for more than 30 s or blood pressure does not increase to over 50 % of baseline in the first 60 s after release of occlusion then a dose of epinephrine (0.1 mL/kg estimated weight, 1:10,000 epinephrine, Health Support Ltd.) should be given to the fetus via the brachial vein by slow i.v. push.

4 Experimental Design

4.1 Prolonged Partial Umbilical Cord Occlusion

Prolonged periods of partial asphyxia in utero can be produced by partial compression of the umbilical cord. Partial cord compression for 90 min induced severe asphyxial insults in near-term fetal sheep [31], similar in nature to those following uterine hypoperfusion [32] except for some relatively minor hemodynamic differences, followed by delayed development of seizures. Although this approach might seem to be less controllable than occlusion of the uterine blood supply, in practice it is easier to vary the degree of occlusion, and umbilical venous flow can be measured directly with an ultrasonic flow probe. Both methods produce similar levels of asphyxia and evidence of encephalopathy with typically relatively variable neural injury [31, 32].

4.2 Complete Umbilical Cord Occlusion

Complete occlusion of the cord, continued until hypotension develops, has been widely studied. Term and preterm fetuses respond to asphyxia in a qualitatively similar manner, albeit the preterm fetuses can survive for much longer without injury as discussed below [25, 33, 34]. As shown in Fig. 3 at all ages occlusion was initially associated with rapid onset of bradycardia and an increase in MAP, associated with intense peripheral vasoconstriction (the "compensation" phase in which cerebral perfusion is maintained). At this time carotid blood flow is maintained at around baseline values, with profound suppression of EEG activity [19, 33]. Microsphere studies have shown that although total brain flow does not change in this initial phase, within the brain, blood flow is diverted away from the cerebrum towards the brain stem [34]. As umbilical cord occlusion was continued MAP eventually falls. The key mediators include impaired cardiac function secondary to hypoxia, acidosis, depletion of myocardial glycogen and cardiomyocyte injury [35] and loss of the initial peripheral vasoconstriction [19, 28]. Once MAP falls below baseline (the so called "decompensation" phase [18, 19]), carotid blood flow falls in parallel, consistent with the known relatively narrow low range of autoregulation of cerebrovasculature in the fetus [36], and there is loss of redistribution of flow within the brain [34].

In the near-term fetus an isolated 10 min episode of umbilical cord occlusion has been consistently associated with selective neuronal loss, predominantly in the dorsal hippocampus, after 72 h recovery [14, 33, 37]. A more prolonged period of 15 min cord occlusion is associated with more extensive neuronal loss, but a correspondingly higher cardiac mortality [23, 40].

4.3 Repeated "Prolonged" Cord Occlusion

Intermittent exposure to asphyxia is much more common in labor than sustained insults. Four episodes of 5 min each of total umbilical cord occlusion in the near term fetal sheep repeated 30 min apart were associated with moderate to severe striatal damage but minimal neuronal loss in the cortex or hippocampus as shown in Fig. 4 [38, 39]. In turn, histological injury is associated with delayed development of intermittent intense post asphyxial seizures.

These two approaches to asphyxia: 10 min of total cord occlusion, and repeated cord occlusion, are perhaps the first

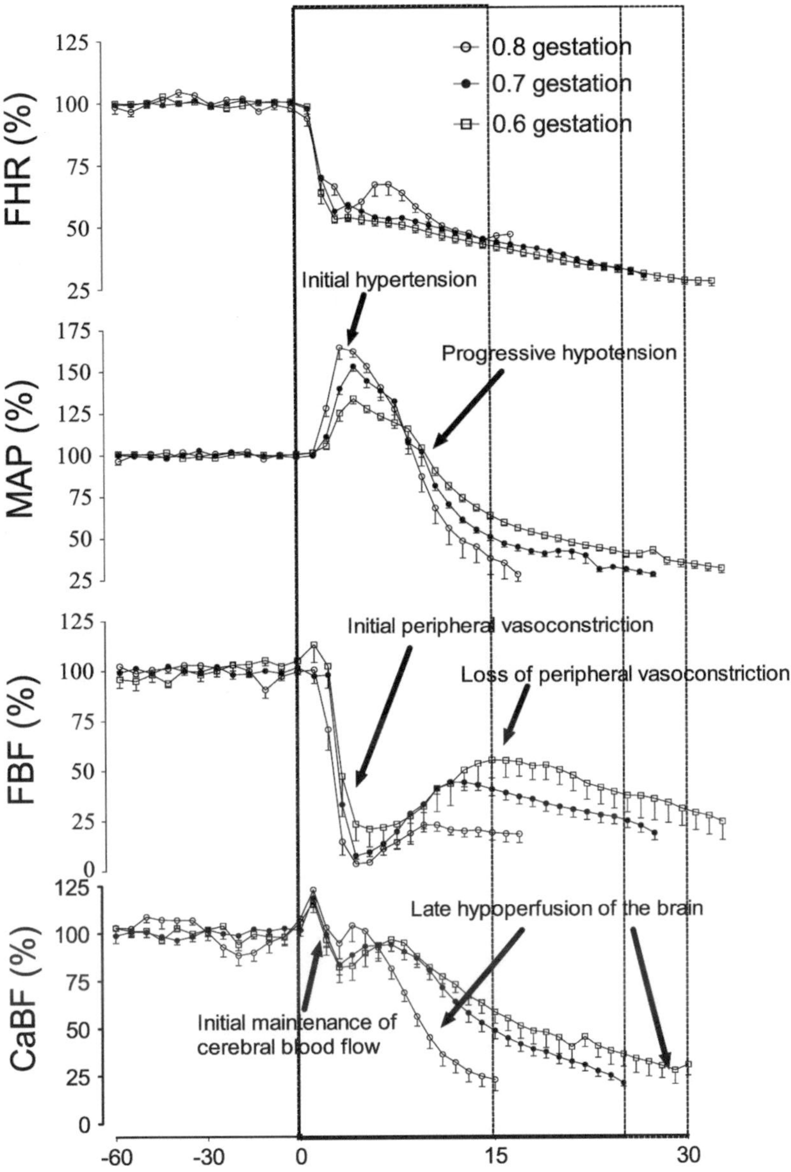

Fig. 3 Cardiovascular responses to prolonged umbilical cord occlusion in fetal sheep at 0.6, 0.7, or 0.85 gestation. Fetal heart rate (FHR, *top panel*), mean arterial pressure (MAP, *second panel*), femoral blood flow (FBF, *third panel*), and carotid blood flow (CaBF, *bottom panel*) data represent 1 min averages and are expressed as percentages of baseline. The period of umbilical cord occlusion for each group is indicated by the *rectangles*. Data are mean ± SE. Data modified from Wassink et al. [19]

pathophysiologic models to share the advantages of consistency, and limited mortality, with the functional models such as cerebral ischemia [23]. As such they are particularly suitable for evaluating potential therapeutic interventions in labor.

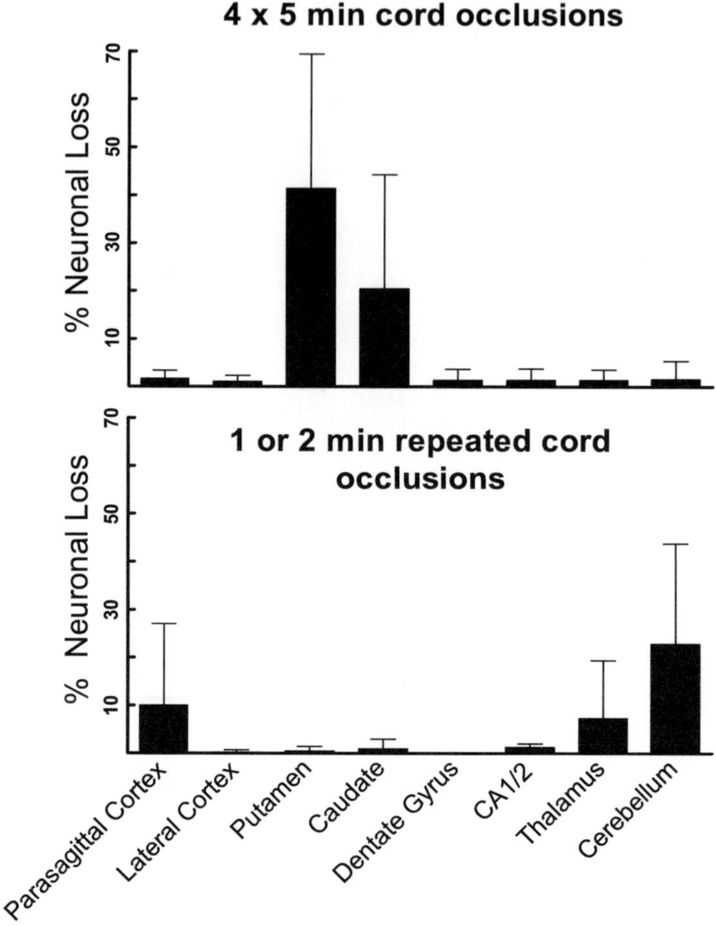

Fig. 4 The effects of altering the interval between periods of systemic asphyxia on the distribution of cerebral damage in near-term fetal sheep. Neuronal loss was assessed after 3 days recovery from asphyxia. The *top panel* shows the effect of 5 min episodes of umbilical cord occlusion, repeated four times, at intervals of 30 min. There is marked neuronal loss in the putamen and caudate nuclei of the striatum. The *bottom panel* shows the effects of brief umbilical cord occlusions repeated at frequencies consistent with established labor, either 1 min every 2.5 min or 2 min every 5 min. Occlusions were terminated after a variable time, when the fetal blood pressure fell below 20 mmHg for two successive occlusions. This insult led to damage in the watershed regions of the parasagittal cortex and cerebellum with sparing of the striatum. CA 1/2 and the dentate gyrus are regions of the hippocampus. Data are mean ± SD and derived from de Haan et al. [15, 42]

4.4 Repeated Brief Umbilical Cord Occlusions

During typical deliveries uterine contractions are relatively brief (approximately 1 min) compared with the models discussed above [18]. In one series of studies the effect of 1 min of umbilical cord occlusion repeated every 5 min (1:5 group) was compared with that of 1-min occlusions repeated every 2.5 min (1:2.5 group) in the near-term fetal sheep. The former frequency of decelerations every 5 min is consistent with early labor while, the latter with decelerations every 2.5 min is consistent with late first stage and second stage labor. The fetal heart rate and blood pressure changes were monitored continuously and occlusions were continued for 4 h or until fetal hypotension developed (defined as $MAP < 20$ mmHg) [15, 40–43]. This insult was associated with development of focal areas of infarction in the parasagittal cortex, thalamus, and cerebellum, in approximately half of surviving asphyxiated fetuses, and mild selective neuronal loss in these regions in nearly all fetuses (Fig. 4). Severe injury was associated with a longer period of blood pressure below baseline levels, with more epileptiform activity, and with slower normalization of the EEG [15].

5 Determinants of Outcome

In the near-term fetus neural injury has most often been reported in the parasagittal cortex, the dorsal horn of the hippocampus, and the cerebellar neocortex after a range of insults including prolonged single complete umbilical cord occlusion [37], prolonged partial asphyxia [12, 32, 48] and repeated brief cord occlusion (e.g., *see* bottom panel of Fig. 4) [15]. These areas are "watershed" zones within the borders between major cerebral arteries, where perfusion pressure is lowest. In both adults and children prolonged systemic hypotension is associated with lesions in these areas [44].

Although limited, or localized white or gray matter injury may occur even when significant hypotension is not seen [32, 48], the magnitude of damage is typically rather modest [45] and there is a strong correlation between either the depth or duration of hypotension and the amount of neuronal loss within individual studies of acute asphyxia [12, 15, 39, 46]. In fetal lambs exposed to prolonged severe partial asphyxia, induced by partial occlusion of the uterine artery, neuronal loss occurred only in fetuses in whom one or more episodes of acute hypotension occurred (Fig. 1) [12]. In contrast, in a similar study where an equally "severe" insult was induced gradually and titrated to maintain normal or elevated blood pressure throughout the insult, no neuronal loss was seen except in the cerebellum [47].

5.1 Basal Ganglia Injury

Although the watershed-type injuries described above are commonly recognized on clinical MRI, basal ganglia and thalamic damage is also common in clinical hypoxic ischemic encephalopathy (HIE). There is some evidence that this pattern is particularly associated with more severe acute events [7, 51]. This raises the possibility that it may be a function of more severe cardiovascular collapse, leading to loss of the initial redistribution of blood flow within the brain to the basal ganglia/thalamus and brainstem [34].

Another contributing factor is suggested by the experimental association between repeated but relatively prolonged episodes of asphyxia with selective neuronal damage to the striatal nuclei (putamen and caudate nucleus, Fig. 4, top) [15, 38, 48]. Consistent with this study, three episodes of 10 min of ischemia induced by occlusion of the carotid arteries repeated 1–5 h apart were associated with severe striatal injury with milder cortical neuronal loss whereas a single 30 min period of cerebral ischemia leads to predominantly parasagittal cortical neuronal loss, with only moderate injury to the dorsolateral striatum [49]. Intriguingly, significant striatal damage was also seen after prolonged partial asphyxia associated with intermittent hypotension [12].

The striatum is not in a watershed zone but rather within the territory of the middle cerebral artery. Thus it is likely that the pathogenesis of striatal involvement in the near-term fetus is related to a combination of the precise timing of the relatively prolonged episodes of asphyxia, and to the apparent vulnerability of the inhibitory neurons within the striatum to this type of insult [5], raising the possibility that this enhanced injury is related in part to abnormal excitatory inputs to these neurons.

5.2 Preexisting Metabolic Status and Chronic Hypoxia

The impact of metabolic status on the degree and distribution of brain injury remains controversial. There is evidence that hyperglycemia during hypoxia-ischemia reduces damage in the infant rat whereas it was associated with greater damage in adult rats [50, 51], but had no effect in the piglet [52]. These differences may reflect species specific maturational differences in the activity of cerebral glucose transporters [51].

The most common metabolic disturbance to the fetus is intrauterine growth retardation (IUGR) associated with placental dysfunction. Experimentally, normally grown singleton fetuses with spontaneous hypoxia showed more rapid centralization of circulation after umbilical cord occlusion and a delayed elevation of the ST waveform and slower fall, suggesting that exposure to hypoxia alters, and in some respects may even improve, myocardial dynamics during asphyxia [53]. In contrast, chronically hypoxic fetuses from multiple pregnancies developed much more severe, progressive metabolic acidosis than previously normoxic fetuses during brief (1 min) umbilical cord occlusions repeated every 5 min (pH 7.07 ± 0.14 vs. 7.34 ± 0.07) and hypotension (a nadir of 24 ± 2 mmHg vs. 45.5 ± 3 mmHg after 4 h of repeated occlusion) [54]. The fetuses with preexisting hypoxia were smaller on average than the normoxic fetuses, and had lower blood glucose values and higher $PaCO_2$ values. Similarly, in normally grown fetuses, 5 days of induced chronic hypoxemia was associated with increased striatal damage after acute exposure to repeated umbilical cord occlusion for 5 min every 30 min for a total of four occlusions [48].

Overall, these data support the clinical concept that growth retardation and placental insufficiency can increase the vulnerability of fetuses to even relatively infrequent periods of hypoxia in early labor. At the same time, they also highlight the complex effects of hypoxia, growth (and thus likely cardiac glycogen stores [55]) on fetal systemic and neural responses to asphyxia.

It is also important to appreciate that milder changes to the intrauterine environment can also modify fetal responses to asphyxia. There is considerable interest on the effects of stimuli such as maternal undernutrition and steroid exposure, particularly at critical times in pregnancy, not only on the fetal responses to challenges to its environment such as hypoxia, but also on risks for adverse health outcomes in adult life [56]. Intriguingly, mild maternal undernutrition that does not alter fetal growth may still affect development of the fetal hypothalamic-pituitary-adrenal function, with reduced pituitary and adrenal responsiveness to moderate hypoxia [57].

5.3 Brain Maturity

Maturation of the brain and body has dramatic and underappreciated effects on sensitivity to asphyxia. Experimental studies demonstrate that the premature fetus is actually much more tolerant to a given duration of asphyxia than at term, and further that tolerance to hypoxia-ischemia falls with postnatal age [58]. For example, the preterm sheep fetus at 90 days gestation (term is 147 days), prior to the onset of cortical myelination, can tolerate extended periods of up to 20 min of umbilical cord occlusion without neuronal loss [25, 60]. As asphyxia is continued after this point, there is progressive failure of combined ventricular output, with a fall in both central and peripheral perfusion, both associated with falling blood pressure. This phase is much less likely to be seen for any significant duration in the term fetus as cardiac glycogen stores are depleted more quickly at term [55]. Thus, at 0.6 gestation the majority of fetuses survived up to 30 min of complete umbilical cord occlusion [30].

In contrast, term fetuses are unable to survive such prolonged periods of sustained hypotension, and typically will recover spontaneously from up to a maximum of 10–12 min of cord occlusion, whereas after a 15 min period of complete occlusion the majority of fetuses either died or required active resuscitation with adrenaline after release of occlusion [19, 37, 59]. Thus, as a consequence of this extended survival during severe asphyxia, the preterm fetus is exposed to extremely prolonged and profound hypotension and hypoperfusion. At 0.6 gestation, for example, no injury occurs after 20 min of complete umbilical cord occlusion even though hypotension is already present, as shown in Fig. 1 [60], but severe subcortical injury occurs if the occlusion is continued for 30 min [20]. It may be speculated that during this final 10 min of asphyxia profound hypotension is associated with failure of redistribution of blood flow within the fetal brain, which places previously protected areas of the brain such as the brainstem at risk of injury [61].

6   Notes

Umbilical cord occlusion is a straightforward experimental technique that is logically linked with the pathogenesis of clinical hypoxic-ischemic encephalopathy. Moderate insults, such as 10 min of occlusion in the near-term fetal sheep, are associated with little risk of mortality in healthy fetuses, who then go on to develop selective neuronal loss as detailed above [33]. In contrast, more prolonged, complete occlusion is intrinsically associated with the development of severe hypotension [29, 53]. In such experiments fetal death may be associated either with asystole or with more rapid onset of hypotension as previously described [28]. Fortunately, because hypotension develops progressively, it is possible to identify warning signs such as a more rapid fall in blood pressure, or marked blood pressure or heart rate instability before it is terminal by continuously monitoring fetal arterial blood pressure. If the goal of the study is to test post-insult treatment it may be reasonable to stop occlusion a little earlier (1–2 min) in such cases to avoid cardiac death, provided that the experimenters use consistent criteria and robustly randomize all fetuses between groups.

7   Conclusions

The experimental models outlined in this chapter will continue to be developed and refined, aiming to mimic human pathophysiology as closely as possible at the whole body level, while allowing the factors contributing to the wide variability of outcomes observed after perinatal asphyxia to be dissected. Both systemic and cerebral factors are important in real life, and include respectively the determinants of cardiovascular decompensation and factors which modulate the intrinsic vulnerability of the brain, such as environmental temperature, metabolic status, and expression of neurotrophic factors, and are highly likely to affect responses to treatment. As the critical events which precipitate significant perinatal hypoxic-ischemic encephalopathy become better understood, our understanding and ability to intervene in clinical asphyxia will also improve.

Acknowledgments

The authors' work reported in this review has been supported by the Health Research Council of New Zealand, Lottery Health Board of New Zealand, the Auckland Medical Research Foundation, the National Institutes of Health, and the March of Dimes Birth Defects Trust.

References

1. Gunn AJ, Gunn TR (1997) Changes in risk factors for hypoxic-ischaemic seizures in term infants. Aust N Z J Obstet Gynaecol 37:36–39

2. Westgate JA, Gunn AJ, Gunn TR (1999) Antecedents of neonatal encephalopathy with fetal acidaemia at term. Br J Obstet Gynaecol 106:774–782

3. Wyatt JS, Gluckman PD, Liu PY, Azzopardi D, Ballard RA, Edwards AD, Ferriero DM, Polin RA, Robertson CM, Thoresen M, Whitelaw A, Gunn AJ, and on behalf of the CoolCap study group (2007) Determinants of outcomes after head cooling for neonatal encephalopathy. Pediatrics 119: 912–921

4. MacLennan A, The International Cerebral Palsy Task Force, Gunn AJ, Bennet L, Westgate JA (1999) A template for defining a causal relation between acute intrapartum events and cerebral palsy: international consensus statement. BMJ 319: 1054–1059

5. Gunn AJ, Bennet L (2009) Fetal hypoxia insults and patterns of brain injury: insights from animal models. Clin Perinatol 36:579–593

6. Steinman KJ, Gorno-Tempini ML, Glidden DV, Kramer JH, Miller SP, Barkovich AJ, Ferriero DM (2009) Neonatal watershed brain injury on magnetic resonance imaging correlates with verbal IQ at 4 years. Pediatrics 123: 1025–1030

7. Miller SP, Ramaswamy V, Michelson D, Barkovich AJ, Holshouser B, Wycliffe N, Glidden DV, Deming D, Partridge JC, Wu YW, Ashwal S, Ferriero DM (2005) Patterns of brain injury in term neonatal encephalopathy. J Pediatr 146:453–460

8. Bennet L, Westgate J, Gluckman PD, Gunn AJ (2003) Fetal responses to asphyxia. In: Stevenson DK, Sunshine P (eds) Fetal and neonatal brain injury: mechanisms, management, and the risks of practice, 2nd edn. Cambridge University Press, Cambridge, pp 83–110

9. Calvert JW, Zhang JH (2005) Pathophysiology of an hypoxic-ischemic insult during the perinatal period. Neurol Res 27:246–260

10. Johnston MV, Trescher WH, Ishida A, Nakajima W (2001) Neurobiology of hypoxic-ischemic injury in the developing brain. Pediatr Res 49:735–741

11. Dijkhuizen RM, Beekwilder JP, van der Worp HB, Berkelbach van der Sprenkel JW, Tulleken KA, Nicolay K (1999) Correlation between tissue depolarizations and damage in focal ischemic rat brain. Brain Res 840:194–205

12. Gunn AJ, Parer JT, Mallard EC, Williams CE, Gluckman PD (1992) Cerebral histologic and electrocorticographic changes after asphyxia in fetal sheep. Pediatr Res 31:486–491

13. Mallard EC, Williams CE, Johnston BM, Gluckman PD (1994) Increased vulnerability to neuronal damage after umbilical cord occlusion in fetal sheep with advancing gestation. Am J Obstet Gynecol 170:206–214

14. Fujii EY, Takahashi N, Kodama Y, Roman C, Ferriero DM, Parer JT (2003) Hemodynamic changes during complete umbilical cord occlusion in fetal sheep related to hippocampal neuronal damage. Am J Obstet Gynecol 188:413–418

15. de Haan HH, Gunn AJ, Williams CE, Gluckman PD (1997) Brief repeated umbilical cord occlusions cause sustained cytotoxic cerebral edema and focal infarcts in near-term fetal lambs. Pediatr Res 41:96–104

16. Giussani DA, Spencer JAD, Hanson MA (1994) Fetal and cardiovascular reflex responses to hypoxaemia. Fetal Matern Med Rev 6:17–37

17. Jensen A, Garnier Y, Berger R (1999) Dynamics of fetal circulatory responses to hypoxia and asphyxia. Eur J Obstet Gynecol Reprod Biol 84:155–172

18. Westgate JA, Wibbens B, Bennet L, Wassink G, Parer JT, Gunn AJ (2007) The intrapartum deceleration in center stage: a physiological approach to interpretation of fetal heart rate changes in labor. Am J Obstet Gynecol 197:e1–e11.236

19. Wassink G, Bennet L, Booth LC, Jensen EC, Wibbens B, Dean JM, Gunn AJ (2007) The ontogeny of hemodynamic responses to prolonged umbilical cord occlusion in fetal sheep. J Appl Physiol 103:1311–1317

20. George S, Gunn AJ, Westgate JA, Brabyn C, Guan J, Bennet L (2004) Fetal heart rate variability and brainstem injury after asphyxia in preterm fetal sheep. Am J Physiol Regul Integr Comp Physiol 287:R925–R933

21. Bennet L, Roelfsema V, Dean J, Wassink G, Power GG, Jensen EC, Gunn AJ (2007) Regulation of cytochrome oxidase redox state during umbilical cord occlusion in preterm fetal sheep. Am J Physiol Regul Integr Comp Physiol 292:R1569–R1576

22. Bennet L, Roelfsema V, Pathipati P, Quaedackers J, Gunn AJ (2006) Relationship between evolving epileptiform activity and delayed loss of mitochondrial activity after asphyxia measured by near-infrared spectroscopy in preterm fetal sheep. J Physiol 572:141–154

23. Gunn AJ, Gunn TR, de Haan HH, Williams CE, Gluckman PD (1997) Dramatic neuronal rescue with prolonged selective head cooling

after ischemia in fetal lambs. J Clin Invest 99:
248–256

24. Bennet L, Roelfsema V, George S, Dean JM,
Emerald BS, Gunn AJ (2007) The effect of
cerebral hypothermia on white and grey matter
injury induced by severe hypoxia in preterm
fetal sheep. J Physiol 578:491–506

25. Barlow RM (1969) The foetal sheep: morphogenesis of the nervous system and histochemical aspects of myelination. J Comp Neurol 135:
249–262

26. McIntosh GH, Baghurst KI, Potter BJ, Hetzel
BS (1979) Foetal brain development in the
sheep. Neuropathol Appl Neurobiol 5:103–114

27. Eshleman JR (1968) Methods used for sterilization or disinfection of instruments. J Dent
Educ 32:330–333

28. Bennet L, Booth LC, Ahmed-Nasef N, Dean
JM, Davidson J, Quaedackers JS, Gunn AJ
(2007) Male disadvantage? Fetal sex and cardiovascular responses to asphyxia in preterm
fetal sheep. Am J Physiol Regul Integr Comp
Physiol 293:R1280–R1286

29. Wibbens B, Westgate JA, Bennet L, Roelfsema
V, de Haan HH, Hunter CJ, Gunn AJ (2005)
Profound hypotension and associated ECG
changes during prolonged cord occlusion in
the near term fetal sheep. Am J Obstet Gynecol
193:803–810

30. Bennet L, Rossenrode S, Gunning MI,
Gluckman PD, Gunn AJ (1999) The cardiovascular and cerebrovascular responses of the
immature fetal sheep to acute umbilical cord
occlusion. J Physiol 517:247–257

31. Ball RH, Espinoza MI, Parer JT, Alon E,
Vertommen J, Johnson J (1994) Regional
blood flow in asphyxiated fetuses with seizures.
Am J Obstet Gynecol 170:156–161

32. Ikeda T, Murata Y, Quilligan EJ, Parer JT,
Theunissen IM, Cifuentes P, Doi S, Park SD
(1998) Fetal heart rate patterns in postasphyxi-
ated fetal lambs with brain damage. Am J
Obstet Gynecol 179:1329–1337

33. Hunter CJ, Bennet L, Power GG, Roelfsema
V, Blood AB, Quaedackers JS, George S, Guan
J, Gunn AJ (2003) Key neuroprotective role
for endogenous adenosine A1 receptor activation during asphyxia in the fetal sheep. Stroke
34:2240–2245

34. Jensen A, Hohmann M, Kunzel W (1987)
Dynamic changes in organ blood flow and oxygen consumption during acute asphyxia in fetal
sheep. J Dev Physiol 9:543–559

35. Gunn AJ, Maxwell L, de Haan HH, Bennet L,
Williams CE, Gluckman PD, Gunn TR (2000)
Delayed hypotension and subendocardial
injury after repeated umbilical cord occlusion

in near-term fetal lambs. Am J Obstet Gynecol
183:1564–1572

36. Parer JT (1998) Effects of fetal asphyxia on
brain cell structure and function: limits of tolerance. Comp Biochem Physiol A Mol Integr
Physiol 119:711–716

37. Mallard EC, Gunn AJ, Williams CE, Johnston
BM, Gluckman PD (1992) Transient umbilical
cord occlusion causes hippocampal damage in
the fetal sheep. Am J Obstet Gynecol 167:
1423–1430

38. Mallard EC, Williams CE, Johnston BM,
Gunning MI, Davis S, Gluckman PD (1995)
Repeated episodes of umbilical cord occlusion
in fetal sheep lead to preferential damage to the
striatum and sensitize the heart to further
insults. Pediatr Res 37:707–713

39. de Haan HH, Gunn AJ, Williams CE,
Heymann MA, Gluckman PD (1997)
Magnesium sulfate therapy during asphyxia in
near-term fetal lambs does not compromise the
fetus but does not reduce cerebral injury. Am J
Obstet Gynecol 176:18–27

40. de Haan HH, Gunn AJ, Gluckman PD (1997)
Fetal heart rate changes do not reflect cardiovascular deterioration during brief repeated
umbilical cord occlusions in near-term fetal
lambs. Am J Obstet Gynecol 176:8–17

41. Westgate JA, Gunn AJ, Bennet L, Gunning
MI, de Haan HH, Gluckman PD (1998) Do
fetal electrocardiogram PR-RR changes reflect
progressive asphyxia after repeated umbilical
cord occlusion in fetal sheep? Pediatr Res 44:
297–303

42. Westgate JA, Bennet L, de Haan HH, Gunn AJ
(2001) Fetal heart rate overshoot during
repeated umbilical cord occlusion in sheep.
Obstet Gynecol 97:454–459

43. Westgate JA, Bennet L, Brabyn C, Williams
CE, Gunn AJ (2001) ST waveform changes
during repeated umbilical cord occlusions in
near-term fetal sheep. Am J Obstet Gynecol
184:743–751

44. Torvik A (1984) The pathogenesis of watershed infarcts in the brain. Stroke 15:221–223

45. Rees S, Mallard C, Breen S, Stringer M, Cock
M, Harding R (1998) Fetal brain injury following prolonged hypoxemia and placental
insufficiency: a review. Comp Biochem Physiol
A Mol Integr Physiol 119:653–660

46. Ikeda T, Murata Y, Quilligan EJ, Choi BH,
Parer JT, Doi S, Park SD (1998) Physiologic
and histologic changes in near-term fetal lambs
exposed to asphyxia by partial umbilical cord
occlusion. Am J Obstet Gynecol 178:24–32

47. de Haan HH, van Reempts JL, Vles JS, de
Haan J, Hasaart TH (1993) Effects of asphyxia

on the fetal lamb brain. Am J Obstet Gynecol 169:1493–1501

48. Pulgar VM, Zhang J, Massmann GA, Figueroa JP (2007) Mild chronic hypoxia modifies the fetal sheep neural and cardiovascular responses to repeated umbilical cord occlusion. Brain Res 1176:18–26

49. Mallard EC, Williams CE, Gunn AJ, Gunning MI, Gluckman PD (1993) Frequent episodes of brief ischemia sensitize the fetal sheep brain to neuronal loss and induce striatal injury. Pediatr Res 33:61–65

50. Simpson IA, Carruthers A, Vannucci SJ (2007) Supply and demand in cerebral energy metabolism: the role of nutrient transporters. J Cereb Blood Flow Metab 27:1766–1791

51. Vannucci SJ, Maher F, Simpson IA (1997) Glucose transporter proteins in brain: delivery of glucose to neurons and glia. Glia 21:2–21

52. LeBlanc MH, Huang M, Vig V, Patel D, Smith EE (1993) Glucose affects the severity of hypoxic-ischemic brain injury in newborn pigs. Stroke 24:1055–1062

53. Wibbens B, Bennet L, Westgate JA, de Haan HH, Wassink G, Gunn AJ (2007) Pre-existing hypoxia is associated with a delayed but more sustained rise in T/QRS ratio during prolonged umbilical cord occlusion in near-term fetal sheep. Am J Physiol Regul Integr Comp Physiol 293:R1287–R1293

54. Westgate J, Wassink G, Bennet L, Gunn AJ (2005) Spontaneous hypoxia in multiple pregnancy is associated with early fetal decompensation and greater T wave elevation during

brief repeated cord occlusion in near-term fetal sheep. Am J Obstet Gynecol 193:1526–1533

55. Shelley HJ (1961) Glycogen reserves and their changes at birth and in anoxia. Br Med Bull 17:137–143

56. McMillen IC, Robinson JS (2005) Developmental origins of the metabolic syndrome: prediction, plasticity, and programming. Physiol Rev 85:571–633

57. Hawkins P, Steyn C, McGarrigle HH, Saito T, Ozaki T, Stratford LL, Noakes DE, Hanson MA (2000) Effect of maternal nutrient restriction in early gestation on responses of the hypothalamic-pituitary-adrenal axis to acute isocapnic hypoxaemia in late gestation fetal sheep. Exp Physiol 85:85–96

58. Gunn AJ, Quaedackers JS, Guan J, Heineman E, Bennet L (2001) The premature fetus: not as defenseless as we thought, but still paradoxically vulnerable? Dev Neurosci 23:175–179

59. Bennet L, Peebles DM, Edwards AD, Rios A, Hanson MA (1998) The cerebral hemodynamic response to asphyxia and hypoxia in the near- term fetal sheep as measured by near infrared spectroscopy. Pediatr Res 44:951–957

60. Keunen H, Blanco CE, van Reempts JL, Hasaart TH (1997) Absence of neuronal damage after umbilical cord occlusion of 10, 15, and 20 minutes in midgestation fetal sheep. Am J Obstet Gynecol 176:515–520

61. Barkovich AJ, Sargent SK (1995) Profound asphyxia in the premature infant: imaging findings. AJNR Am J Neuroradiol 16:1837–1846

Chapter 8

The Sheep as a Model of Brain Injury in the Premature Infant

Stephen A. Back, Art Riddle, and A. Roger Hohimer

Abstract

Preterm fetal sheep preparations provide unique experimental access to the complex pathophysiological processes that contribute to injury to the human brain during successive periods in development. Recent refinements have resulted in preparations that offer a number of significant advantages to model key aspects of human preterm cerebral injury. We describe a global cerebral ischemia preparation that replicates major features of acute and chronic human cerebral injury and which has provided access to complex clinically relevant studies of cerebral blood flow and neuro-imaging that are not feasible in smaller laboratory animals. Despite the higher costs and technical challenges of instrumented preterm fetal sheep models, they allow an integrated analysis of the spectrum of insults that appear to contribute to cerebral injury in human preterm infants.

Key words Hypoxia-ischemia, White matter injury, Oligodendrocyte, Cerebral blood flow, MRI, Necrosis, Apoptosis

1 Introduction

1.1 Overview of the Clinical Problems to Be Addressed in Preterm Fetal Sheep

The last decade has seen a resurgence of interest in instrumented preterm fetal sheep preparations to study the complex pathophysiological processes that contribute to brain injury in the preterm infant. The considerable merits of a large preclinical animal model have become increasingly recognized as it has become apparent that rodent models have significant limitations to study injury to the developing human brain [1]. Not only do preterm fetal sheep preparations closely replicate major features of acute and chronic human preterm brain injury, but they also provide access to complex clinically relevant studies of cerebral blood flow and neuro-imaging that are not feasible in smaller laboratory animals.

In recent years, studies of hypoxic-ischemic injury to the developing brain have yielded an increasingly complex and controversial set of observations related to the pathogenetic mechanisms that

Jerome Y. Yager (ed.), *Animal Models of Neurodevelopmental Disorders*, Neuromethods, vol. 104, DOI 10.1007/978-1-4939-2709-8_8, © Springer Science+Business Media New York 2015

result in injury to white and gray matter structures within the neuraxis. We focus on approaches to generate cerebral white matter injury (WMI), the major form of brain injury and the leading cause of chronic neurological disability in survivors of premature birth [2]. The broader questions of the pathogenetic mechanisms that relate WMI to cortical and subcortical gray matter injury have been recently reviewed [3].

Brain injury in preterm survivors has an unexplained predilection for cerebral white matter. The period of highest risk for WMI is ~23–32 weeks post-conceptional age. Although major advances in the care of premature infants have resulted in striking improvements in the survival of very low birth weight (VLBW) infants (<1.5 kg), improved survival has been accompanied by a significant increase in the number of preterm survivors with long-term neurological deficits [4]. The major consequences of this injury are permanent motor impairment (i.e., cerebral palsy; CP) ranging from mild to profound spastic motor deficits [5–9] as well as a broad spectrum of cognitive, social-behavioral, attentional, visual, and learning disabilities that manifest by school age in 25–50 % of children [10–14]. In preterm survivors, MRI-defined WMI but not gray matter injury manifests in the first months of life as abnormal movements that are predictive of CP [15–17]. The impact of WMI can be appreciated from a recent large population based study of children with CP. Perinatal WMI, including periventricular leukomalacia (PVL), was the most common finding, seen in almost half (42.5 %) of affected children [18]. WMI is not exclusively associated with prematurity and is increasingly appreciated in term infants [19–21]. Infants with complex congenital heart disease (CHD) are at particular risk for WMI and delayed brain maturation [22–24]. Since very low birth weight infants comprise about 1.5 % of the four million live births in the USA alone each year, the worldwide social and economic burden is considerable. The average lifetime costs per person with CP is estimated to be ~1 million dollars in the USA [25]. An understanding of the cellular and molecular basis of preterm WMI is thus urgently needed to develop effective interventions to prevent these lifelong neurological disabilities. Hence, there continues to be a critical need for suitable animal models of WMI in preterm survivors.

1.2 Advantages and Disadvantages of the Fetal Sheep to Model WMI

Fetal sheep preparations require considerable cost and infrastructure to support the surgical instrumentation and postoperative care of large laboratory animals. Fetal sheep studies require a highly skilled surgical team, a large animal operating facility suitable for sterile operations, specialized veterinary care and access to reliable breeders. Despite these challenges, many preparations have yielded very reproducible results with low morbidity and mortality, thereby limiting the number of animals required. Presently, the ovine genome has not been fully sequenced and the molecular tools available to study ovine brain injury are more limited than in rodents.

However, preterm (0.65 gestation or 95 days) fetal sheep models have multiple distinct advantages relative to small fetal and neonatal animals, which are limited by a lissencephalic brain that does not resemble the gyrencephalic human cerebrum. In terms of its neurodevelopment, the immature ovine brain is similar to preterm human between approximately 24–28 weeks in terms of the completion of neurogenesis, the onset of cerebral sulcation, and the detection of the cortical component of the auditory and somatosensory evoked potentials [26–29]. The long gestation of fetal sheep (145 days) allows selection of an appropriate developmental stage over which brain insults can be induced and evaluated.

The abundance of cerebral white matter and its anatomic similarities to that of the preterm infant make the fetal sheep ideal for neuropathological correlation with human [30, 31]. Rodents have a paucity of cerebral white matter that differs markedly from human, whereas the fetal sheep generates acute [32] and chronic [33] WMI that is very similar to human in histopathological features. White matter maturation in fetal sheep can be defined relative to human through assessment of oligodendrocyte lineage progression and myelination (Fig. 1). Oligodendrocyte development in the 0.65 gestation sheep fetus is similar to that of

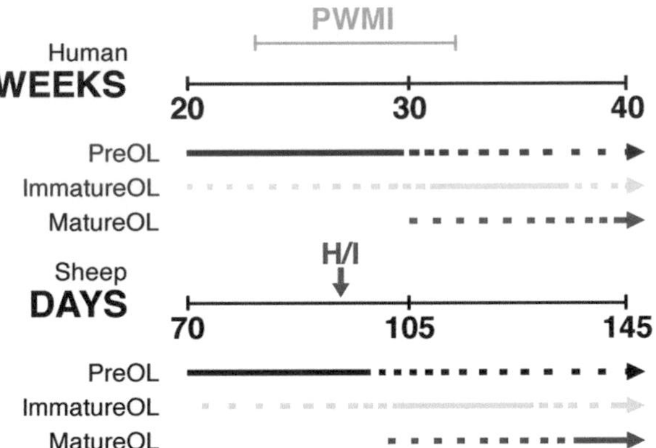

Fig. 1 Summary diagram that compares the timing of appearance of human versus ovine immature oligodendrocytes (O4+O1+). Human is depicted during the latter half of gestation (20–40 weeks) and is based upon data previously reported [105]. The progression of the oligodendrocyte lineage in fetal sheep white matter development has been previously described [32, 34]. The period shown (~70–145 days) roughly corresponds with fetal human during the latter half of gestation. *Solid lines* indicate the developmental period when each oligodendrocyte lineage stage predominates. The *dotted lines* indicate the period when these stages are a minor population. Note that in many studies hypoxia-ischemia (H/I; *arrow*) is administered at 90–95 days, which coincides with the high-risk period for periventricular white matter injury (PWMI; ~23–32 weeks)

24–28 week human [32]. The late gestation ovine fetus (0.9 gestation; 135 days) displays oligodendrocyte development similar to term human [34]. Thus, the investigator can choose the appropriate developmental timing for the insult to be given and evaluated.

In contrast to the fetal sheep, the cerebrovascular supply of the rodent white matter is also structurally and physiologically very dissimilar to human, which is an additional factor that is likely to contribute to the markedly different patterns of cerebral injury in rodents and fetal sheep. Whereas the fetal sheep cerebrum has a predilection for relatively selective WMI under conditions of moderate cerebral ischemia [32], rodents have a propensity for mixed cerebral injury such that substantial gray matter injury accompanies WMI [35–40]. This shortcoming, for example, limits the relevance of rodent hypoxia-ischemia models for the study of myelination disturbances associated with chronic human WMI. Necrotic injury to cerebral gray matter contributes substantially to neuro-axonal degeneration as a cause of dysmyelination, which is not a prominent feature of WMI in either fetal sheep [41] or contemporary human cases of WMI [42]. In contrast to fetal sheep preparations, the small size of rodents is also a major technical limitations for a wide range of invasive physiological measurements as well as for studies that seek to achieve high resolution neuroimaging by MRI [33, 43].

The size of the preterm sheep fetus allows for chronic instrumentation to enable hemodynamic measurements, repeated access to blood and CSF and chronic electrophysiological recording of the fetal electroencephalogram [44]. Thus, it is feasible to study well-defined brain insults with reliable measurements of blood pressure, oxygenation and cerebral blood flow. Chronic instrumentation also allows a wide range of practical and clinically pertinent cerebral insults to be administered. These include global cephalic ischemia [45], systemic hypotension or hypoxemia [46–48], single or repeated cord occlusion [49–51], increased intracranial pressure [52], and administration of infectious agents or exogenous inflammatory mediators [53–55]. These insults can be graded in intensity and duration to mimic the human situation. In each case, the stressor can be well described if not regulated by conventional monitoring measurements. Thus, a wide range of pathogenetic events can be evaluated physiologically and correlated neuro-pathologically with the distribution and extent of white matter damage. Moreover, the natural progression of various types of WMI can be evaluated in a time frame ranging from days-to-weeks-to-months after the insult.

From a practical standpoint, the availability, cost and ease of breeding of the sheep makes it a more practical large animal model than the nonhuman primate. In addition, the size and docile nature of the sheep supports the feasibility of both in utero and ex vivo neuro-imaging studies [33, 56]. Finally, fetal sheep preparations provide powerful access to large animal preclinical testing as exemplified by preclinical studies in near-term fetal sheep that lead to the head cooling trials for neonatal encephalopathy [57].

1.3 Hypoxia-Ischemia (H-I) in Fetal Sheep Generates Pathological Features of WMI

The ovine fetus offers significant advantages to analyze systemic hemodynamic disturbances that regulate cerebral blood flow and metabolism in preterm cerebral white matter. The sheep fetus displays cerebral hemodynamics similar to human and permits repeated physiological measurements in utero in the unanesthetized state. Measurements of blood pressure, electroencephalography, blood oxygenation and other vital variables can be correlated with acute changes in cerebral blood flow and metabolism. Importantly, like the human fetus [58–61], the fetal sheep displays a very limited range of cerebral autoregulation under normal conditions and a pressure-passive cerebral circulation when subjected to systemic hypoxia and associated hypotension [46, 62–65].

Multiple lines of evidence support a role for cerebral ischemia in the pathogenesis of WMI in very low birth weight human infants [2, 66, 67]. Given the limitations of human studies to directly link blood flow disturbances with WMI, studies in fetal sheep have greatly strengthened our understanding of the contribution of cerebral H-I to WMI. These experimental studies support that a complex interplay of factors related to cerebrovascular immaturity predispose preterm cerebral white matter to injury from H-I. A model of global cerebral hypoperfusion found that the midgestation animal displayed a predilection to subcortical WMI, whereas the near-term animal displayed predominantly parasagittal cortical neuronal injury [45, 68]. A variable degree of WMI was also detected after systemic hypotension arising from intermittent or partial umbilical cord occlusion [69, 70]. By contrast, in the near-term animal, repeated umbilical cord occlusion produced injury to both the periventricular white matter and the cerebral cortex [71]. Systemic hemorrhagic hypotension in the 0.75 gestation fetus resulted in mostly necrotic WMI with focal necrotic lesions or axonal swellings in the periventricular white matter [72]. Preterm ovine white matter lesions were detected after repeated systemic fetal endotoxin exposure that triggered both transient hypoxemia and hypotension [73, 74]. The importance of cerebral ischemia is supported by studies where WMI was detected only infrequently in models of hypoxemia in which a restriction in uteroplacental blood flow resulted in decreased oxygen delivery and mild acidemia to the fetus without systemic hypotension or cerebral hypoperfusion [47, 75, 76]. A model of fetal metabolic acidemia induced by maternal hypoxemia similarly produced mild-to-moderate injury in midgestation and near-term sheep [77]. Hence, cerebral hypoperfusion in conjunction with hypoxia appears to be a critical factor to generate significant WMI in the preterm fetal sheep.

Fetal sheep have also provided an important model to define mechanisms of oxidative injury from cerebral H-I. This is in part due to the large fetal cerebral hemispheres, which permit blood sampling from the venous sagittal sinus and placement of intracerebral probes for dialysis studies. Initial studies found that reperfusion after cerebral ischemia was an important source of free radical formation in the near-term fetal sheep brain [78]. Partial umbilical cord occlusion resulted in a delayed increase in lipid peroxidation in frontal and parietal white matter in near-term fetal sheep [79]. Dialysis studies in near-term [80] and preterm [81] sheep have both demonstrated the enhanced generation of stable adducts of reactive oxygen species.

2   Materials and Methods

2.1   Catheters and Other Materials

Each fetus has a non-occlusive V3 carotid catheter (0.023″ ID×0.039″ OD), which is inserted at the inguinal branch and advance about 4 cm centrally; and a V4 Amniotic catheter (0.030″ ID×0.048″ OD). A 4 mm hydraulic occluder is placed around the brachiocephalic artery or a 2 mm occluder is placed around each carotid artery. V3 and V4 catheters are 7′ in length and are made from medical grade micro vinyl manufactured by Scientific Commodities, Lake Havasu City, Arizona. The occluders are silastic and manufactured by In Vivo Metric, Healdsburg, California.

2.2   Animals

Time-bred sheep of mixed western breed (88–91 days gestation; term, 145 days) are raised on a farm from a reliable breeder before being transported and housed in the animal care facility for 3–7 days to acclimate the animals prior to surgery. The large animal facility is climate-controlled (ambient temperature 20±1 °C) with a 12 h light–dark cycle (light hours 06:00–18:00). Twin pregnancies are studied so that each experimental fetus has a twin control. Digital x-rays are performed at 87 days of gestation to confirm the ewe's pregnancy. Loose-fitting collars are routinely used as part of the preoperative procedure (to aid shearing surgical sites) and the ewes take no special notice of the collars.

2.3   Biohazard Precautions When Handling Sheep

Pregnant sheep may be infected with the human pathogen, *Coxiella burnetii* (Q-Fever), the organisms of which can be found in urine, feces, amniotic fluid, and placental tissue. Given that pregnant sheep are not tested for these organisms, we assume that all pregnant sheep may be infected. A vaccine for Q fever has been developed and used successfully in humans overseas, but the vaccine is not currently available for commercial use in the USA. Several procedures should be implemented to minimize infection with Q-Fever while working with pregnant sheep:

1. Animals are restricted to designated areas with Biosafety Level 2 practices for the containment of pregnant sheep. Specialized carts ("Q carts") are used for transport of animals outside the designated containment area. "Q-carts" are used for the safe transport of sheep between buildings; for example, between the loading dock at the time of delivery from the farm to their housing runs. The carts are fitted with air filters and are large enough for the sheep to stand or lie down. These carts are used for transport only, and the sheep are in the carts for the minimum amount of time possible.

2. All personnel are required to wear protective clothing when handling animals or their waste products. These include a disposible gown and shoe covers, surgical mask, and exam gloves.

3. Procedure rooms and other items in contact with the animals are routinely cleaned with a bleach solution (1 part 5 % hypochlorite to 30 parts water, changed daily). These include all equipment, carts, cages, and other items that need to be moved outside of the designated Q-fever containment area. Animal carcasses are "double bagged" before disposal.

2.4 Preoperative Care and Anesthesia

At 90 days of their 145 day gestation time-dated pregnant sheep with twin fetuses are prepared for sterile surgery. At ~15–18 h prior to surgery, food is withheld but the pregnant ewe has free access to water. The ewe's jugular vein is cannulated with a 20G intravenous (i.v.) cannula and injected with atropine (7.5 mg i.m.), ketamine (400 mg, i.v.), and diazepam (10 mg, i.v.) to allow tracheal intubation. If necessary, an additional half dose of each agent is given prior to anesthesia. The ewe is placed in a U-shaped trough on the surgery table in a supine position with its legs tied caudally and cranially to prevent rotation during surgery. Intubation is performed with a 8.5 mm Mallinckrodt inflatable endotracheal tube (Covidien-Nellcor, Boulder, Colorado) that is connected to the anesthesia mixer and ventilator. Under positive pressure ventilation (Norkovet II Anesthesia machine with a Hallowell Veterinary Ventilator, Hallowell EMC, Pittsfield, MA), isofluorane (1–2 % in a 70:30 mixture of oxygen and nitrous oxide) is initiated to maintain anesthesia during surgery with the isofluorane concentration adjusted to ensure a surgical level of anesthesia for the ewe and fetus. A continuous isotonic saline drip (500 ml/h) is administered to maintain maternal fluid balance.

2.5 Surgical Procedures

2.5.1 Maternal and Fetal Surgery for Fetal Global Cerebral Ischemia Protocols

Detailed surgical procedures are described in the accompanying article by Dr. Gunn and colleagues. Provided here are details pertinent to the generation of preterm fetal global cerebral ischemia. The maternal abdomen is shaved, disinfected with betadine and the operative site covered with sterile drapes. The abdomen is opened through a midline incision. A purse-string suture is placed around the uterine incision and is tightened to reduce amniotic fluid loss after the head and neck of the fetus are exposed.

We have employed two different approaches to achieve global cerebral ischemia. The first is to place an occluder on each of the carotid arteries in the neck. The second is to place a single occluder around the brachiocephalic artery in the chest. The relative advantages and disadvantages of each approach are discussed below. For placement of the carotid occluders, a small (less than 3 cm) midline incision is made in the fetal neck to expose the carotid arteries. To confine the cerebral blood supply to the carotid arteries, the vertebro-occipital arteries are ligated bilaterally. These anastomoses connect the vertebral arteries, supplied by the thoracic aorta, with the external carotid arteries that are fed by the brachiocephalic [82]. Silastic hydraulic occluders (2 mm diameter) are placed around each carotid artery. A non-occlusive indwelling polyvinyl catheter is placed in one carotid artery. A catheter is also attached to the fetal skin to allow subsequent measurement of amniotic fluid pressure. The fetal neck incision is closed with suture.

For placement of the brachiocephalic occluder (4 mm diameter), an incision is made at the second intercostal space on the left side. The brachiocephalic artery is isolated and a hydraulic occluder placed to allow subsequent controlled reduction of fetal cephalic blood pressure in order to cause global brain ischemia. The chest incision is closed in two layers.

After placement of either set of occluders, all catheters are anchored to the fetal skin on the neck and head or back to prevent them from pulling out or kinking off. The fetus is then gently returned to the uterus with special care taken to preserve and reseal the membranes. The uterine incision is closed and then oversewn to prevent leakage. The twin fetus is either left as an uninstrumented control or can be similarly instrumented. The maternal abdominal incision is closed in anatomical layers, and all catheters are tunneled subcutaneously to emerge on the flank of the ewe. Catheters are bundled and stored in a pouch stitched to the flank of the ewe, to prevent them from pulling out. One million units of penicillin-G are administered to the amniotic fluid (through a catheter) for the specific prevention of clostridia; no other antibiotics are routinely necessary (postoperative cultures have amply demonstrated the adequacy of this regime).

2.5.2 Placement of EEG Electrodes and EEG Recording Studies

EEG electrodes are placed via 2 cm scalp incisions on the fetal dura by making small holes in the fetal skull. One or two pairs of EEG electrodes (AS633-5SSF, Cooner Wire, Chatsworth, CA) are secured on the dura over the parasagittal parietal cortex (5 and 10 mm anterior to bregma and 5 mm lateral) with a reference electrode attached over the occiput. The electrodes are connected to high impedance amplifiers to provide a filtered input to a Stellate Systems (Montreal, Quebec) digital EEG recording and analysis system. We have recorded 2–5 channels of continuous EEG from ten animals for up to 16 h before and after global ischemia [83]. Seizure detection was done both by visual analysis and confirmed with a newborn seizure detection algorithm from Stellate Systems.

2.6 Postoperative Care

After surgery, the ewe is returned to a recovery pen for at least 24 h. The pen is covered with clear bedding material. The ewe is checked at least every half-hour until the animal is standing and feeding, usually within 0.5–1 h after surgery. Buprenex (buprenorphine, 0.3–0.6 mg, s.c.) is routinely administered after surgery and for the next 1.5 days, twice daily. The surgical incision site is assessed at least once daily and postsurgical infections are treated, as required in consultation with the veterinary staff. Fetal catheters are maintained patent by a daily infusion of 300 units/cm^3 of heparinized saline. Ewes are monitored daily for normal ruminal function and for signs of pain and distress including a hunched posture, reluctance to move around, unwillingness to lie down, or decreased appetite. Animals are assessed regularly for these signs after surgical procedures and at least once daily thereafter. A requirement for parenteral fluids or electrolytes is unusual in our experience, but their administration is at the discretion of veterinary staff, in consultation with investigators. In the rare situation that pain cannot be adequately managed, and treatments are proving to be ineffective, the animal is removed from the study and euthanized.

2.7 Study Protocols

2.7.1 Restraint of the Animals

After a minimum of 72 h of recovery, the ewe is placed in a mobile self-contained rolling cart and taken to the study room. In the cart, the ewe is loosely tethered to the cart via a collar and chain and has access to food and water. Typically, the maximum amount of time that the ewe is loosely restrained in the cart is 3 h. If the ewe is uneasy in the cart, a second sheep is put in a cart and brought into the study room for company. Alternatively, a mirror is placed in front of the study cart so that the ewe sees another apparent companion.

2.7.2 Cerebral Hypoperfusion Studies: Bilateral Carotid Occlusion

Animals are typically studied no earlier than the third postoperative day to allow fetal recovery from surgery. Pressure transducers are connected to appropriate amplifiers (TA 6000; Gould Instruments, Valley View, OH) to record mean arterial blood pressure (MABP) in the fetal artery relative to amniotic fluid pressure using ADI Powerlab software (Dagan Corp. Minneapolis, MN). Fetal heart rate (HR) is calculated from triplicate measurements of the arterial pressure pulse intervals over a continuous recording of >20 s. Fetuses are studied only if they demonstrated normal oxygenation (>6 ml O_2/100 ml blood) and blood indices [84]. Sustained cerebral hypoperfusion is initiated by bilateral carotid artery occlusion after inflation of the carotid occluders. Cerebral reperfusion is established by deflation of the occluders and was studied at either 15 or 60 min of restored flow. Verification of successful inflation of the occluders is verified by measurement of in an increase in pressure proximal to the occluders with an indwelling carotid or axillary artery catheter. Fetal heart rate and MAP traces are monitored continuously throughout the occlusion and arterial blood gas analysis is done at standard time points prior to, during and after the occlusion to assess the fetal response.

2.7.3 Cerebral Hypoperfusion and Hypoxia Studies: Brachiocephalic Artery Occlusion Coupled with Reduced Inspired Oxygen

To reduce fetal oxygen delivery, the ewe inspires a reduced content oxygen mixture. After a basal period where room air is inspired, a clear plastic bag is placed over the ewes head and the ewe inspires a 10–12 % oxygen mixture that flushes the bag at 20 l/min. Ewes respond by increasing their ventilation and generally tolerate the procedure well. After 4 min of lowered inspired oxygen, the fetus also becomes moderately hypoxic. At this time the occluder, which has been placed around the brachiocephalic artery, is inflated to reduce the perfusion pressure to the fetal head and arms for 25 min. Thereafter, the occluder is released to restore normal fetal cephalic blood pressure. The bag is removed and the ewe returns to breathing room air so that oxygenation of the ewe and fetuses oxygenation returns to normal.

2.7.4 Blood Analysis

One milliliter blood samples are taken anaerobically from the fetal carotid artery and analyzed for arterial p_aH, P_aO_2, P_aCO_2 (corrected to 39 °C), hemoglobin content, arterial oxygen content (CaO_2), arterial oxygen saturation, glucose and lactate (ABL 700 pH/Blood Gas Analyzer; Radiometer, Westlake, OH), and hematocrit (capillary microfuge). Fetuses are only studied if they demonstrate normal fetal oxygenation, defined as >6 ml O_2/100 ml blood, at a 24 h recovery from the operation.

2.7.5 Microsphere Injection Protocol

Fetal brain blood flow is measured spatially by the fluorescent microsphere distribution and reference sample method [85]. Fluorescent microspheres with four different colors (15 µm diameter; F-17047, F-17048, F-17048, F-17050; Molecular Probes, Eugene, OR) have the following peak excitation and emission: green (450/480) yellow (515/534), red (580/605) and scarlet (650/685). Approximately 3×10^6 microspheres suspended in 1 ml of saline with 0.05 % Tween are sonicated and then injected over 30 s into the fetal hindlimb vein followed by a 2 ml flush with saline. Starting just before and continuing 2 min after each injection, a reference blood sample is drawn at 0.75 ml/min into a syringe mounted in a syringe pump (Harvard Apparatus Co., Dover, MA).

2.7.6 Tissue Handling

The ewe and fetuses are killed by intravenous injection of the ewe with Euthasol (~10 ml/50 kg body weight; Virbac Inc., Ft. Worth, TX). After fetal brains are removed, we employ a variety of protocols for tissue preservation depending on the study criteria. These include immersion fixation at 4 °C in 4 % paraformaldehyde in 0.1 M phosphate buffer, pH 7.4 for 2 days followed by storage in PBS. Brains to be analyzed for microspheres or for MRI-histopathological correlation are subsequently immersed in 20 % sucrose until they sink and then rapidly frozen in OCT for acquisition of frozen sections with a cryostat.

3   Results and Discussion

3.1   Pathophysiological Mechanisms of WMI Related to the Brachiocephalic vs. Carotid Occlusion Models of Global Cerebral Ischemia

The dominant factors related to the generation of WMI in most if not all models are: (1) the content of oxygen and glucose in the blood; (2) the driving or perfusion pressure and hence cerebral blood flow (CBF); and (3) the duration of the insult. In preterm fetal sheep, it is often feasible for these important variables to be manipulated and measured either in groups of animals or in single individuals. The pathophysiological disturbances associated with these factors are variously weighted depending upon the model employed.

With models of maternal hypoxemia (e.g., high altitude or a low inspired oxygen fraction) fetal oxygenation falls while blood pressure (BP) rises transiently but then returns to normal and then falls below normal [47, 48, 77, 86–89]. Likewise systemic asphyxia models that utilize umbilical cord occlusion cause a reduced fetal arterial oxygen tension and content [50, 90, 91]. In cord occlusion models, BP initially increases to compensate for diminished oxygenation, but as systemic and cardiac hypoxia progressively intensify, hemodynamic and cardiac decompensation occurs with a resultant fall in BP. It appears that significant brain damage occurs only when the hypoxia and hypotension are allowed to progress until near death conditions occur. However, as cardiovascular compensations fail with time, substantial albeit variable reductions in BP occur. The residual brain blood flow has seldom been measured, but for brain injury to occur, pressures must fall below 1/3 normal for 10–15 min depending on the animal's age. The importance of the central autonomic system as well as adrenal stress hormones remains unclear as do systemic blood concentrations of metabolic substrates like glucose and products like lactate. The fetal sheep brain, in particular when immature and hypoxemic, has essentially no ability to autoregulate [62–64]. Hence, in asphyxia models, the fall in BP exacerbates the fetal brain hypoxia with a consequent reduction in CBF that leads to partial ischemia.

Occlusive cerebral ischemia models differ from asphyxia models in several significant ways. A major practical advantage of the occlusive models is that the heart is largely unaffected. Essentially, cerebral damage can be reliably generated without the significant fetal deaths and morbidity that are inherently associated with other models that require near death conditions to ensure brain damage. Injury in the umbilical cord occlusion models can be modulated by adjusting the duration of occlusion relative to the onset of systemic hypotension.

Unlike in utero systemic hypoxemia or asphyxia models, models that utilize carotid or brachiocephalic artery (BCA) occlusions cause an immediate fall in perfusion pressure and CBF, and thus have a well-defined onset of the insult [32, 45, 84]. Cerebral ischemia in these models is generally global and severe but not complete, especially when collateral circulation is left intact (discussed further below). Importantly, some CBF persists and the oxygenation and glucose levels of this residual flow have important influence on the extent of damage. The residual blood flow is the most difficult parameter to quantify. Even in fetal sheep models, CBF is only infrequently measured and BP is often used as a surrogate marker of CBF.

There are a number of considerations that guide the selection of a model that utilizes carotid versus common brachiocephalic artery (BCA) occlusion. However, the differences between bicarotid and BCA occlusion are not as great as the difference between either of those models and cord occlusion models. The carotid arteries in the 0.65 sheep fetus are small and there is a relatively greater risk that commercially available occluders can inadvertently obstruct flow chronically if placement and routing techniques are not optimal. By contrast, the BCA is a larger and much more stable vessel for the placement of a single occluder [65, 86]. In sheep, the BCA supplies the entire head including the carotids as well as both axillary arteries, which supply the forelegs. The BCA is the only major artery that supplies the upper body. Occlusion of the BCA is similar to bilateral carotid occlusion in terms of perfusion to the head and brain, but different in that forelimb perfusion is also reduced, which does not occur if flow to both carotids is completely restricted.

One important consequence of the BCA occlusion model is that proximal BP to the rest of the fetal body and the placental circulation is subject to an elevated pressure. This probably results in an increase in flow that causes a moderate but significant rise in arterial oxygenation. The same effect probably occurs with bicarotid occlusion, but would be expected to be smaller. Either preparation can be coupled with a lowered maternal inspired oxygen fraction to counteract elevations in arterial oxygenation or even generate a fetal hypoxemia in addition to cerebral ischemia.

The vertebral-occipital arteries (VCA) are small but potentially important in models where cephalic occlusions are studied [45, 82]. They provide an important anastomosis between the anterior circulation provided by the carotid arteries and the posterior circulation derived from the vertebral arteries. Vertebral-occipital arteries are likely to be variable in size from animal to animal, perhaps linked to variations in anatomy related to their supply at or near the Circle of Willis. They connect to the carotids near the lingual branch and are, thus, distal to the BCA occluder as well as to all but the most rostral placements of carotid occluders. Hence, they act to support brain blood flow when systemic pressure is normal but BCA or carotid occlusions are used.

Hence, in both the bi-carotid and BCA occlusion models, it is important to ligate these vessels to achieve near complete global cerebral ischemia. The exact amount of flow they provide, especially to the preterm fetus, has not been carefully determined. In normal sheep, it is not even clear whether there is a net flow and, if so, in what direction. During either carotid or BCA occlusions, flow is certainly from the vertebrals to the carotids distal to the site of placement of occluders. The amount of flow in the VCA, while likely to be variable from animal to animal, is clearly sufficient to cause variability in brain damage in the cephalic ischemia models unless they are ligated.

3.2 Role of Vascular End Zones in Cerebral WMI

The role of cerebral vascular immaturity in the pathogenesis of WMI has been difficult to study in preterm infants. Analysis of the vascular supply to the periventricular white matter has yielded controversial results. The periventricular white matter has two major blood supplies. Perforating arteries branch from leptomeningeal arteries, penetrate the cerebral cortex, and terminate as capillary beds adjacent to the ventricles. Branches of choroidal and striate arteries project toward the lateral ventricles and then deviate away from the ventricle toward their final termination in vascular capillary beds in the periventricular white matter. Although these vascular beds may collectively form vascular end zones and border zones that render the periventricular white matter particularly susceptible to ischemia, physiological studies in support of this concept are lacking. Hence, existence of these border zones remains controversial [92–94]. The presence of these vascular zones would provide a mechanism for WMI based on the notion that when periventricular white matter flow falls below a critical threshold, this region would display greater susceptibility to WMI relative to the better-perfused cerebral cortex.

To address this controversy, we developed methods to spatially quantify cerebral blood flow (CBF), since large differences in flow in small regions can be obscured when averaged with larger unaffected regions, as is the case with more global measures of cerebral blood flow [32]. In order to analyze disturbances in regions of periventricular white matter vulnerable to hypoxic-ischemic injury, we developed a novel method to achieve high resolution spatial blood flow measurements on the brains of immature fetal sheep. This was achieved by determining the location in the

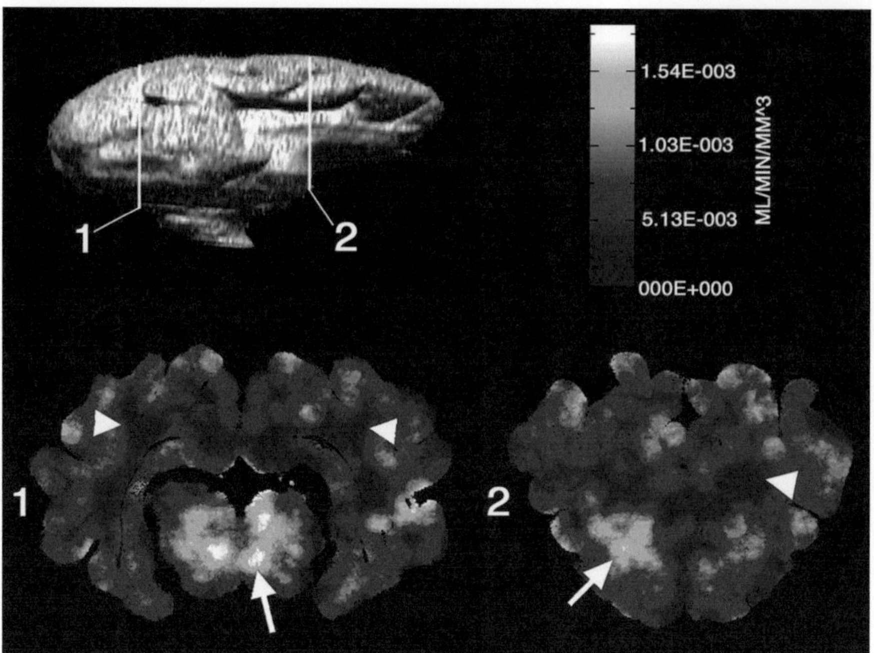

Fig. 2 Quantification of fetal cerebral blood flow in situ under conditions of basal flow. The top image represents a 3D surface reconstruction (Volocity, Improvision, Lexington, MA) of fluorescence images of a 0.65 gestation ovine control brain that indicates the frontal and parietal levels to which the lower blood flow images correspond. Representative gray scale basal flow images show higher blood flow (*arrows*) in the pons (image 1) and sub-cortical gray matter (image 2) and lower flow (*dark gray*) in the periventricular white matter (PVWM; *arrowheads*)

fetal brain of thousands of individual fluorescently labeled microspheres [32, 83]. This high-resolution blood flow technique is illustrated in Fig. 2. Basal or control blood flow in 2 of 16 "virtual" 2 mm thick coronal sections of a 0.65 gestation fetal sheep brain are shown. The image at upper left represents a 3D reconstruction of the approximately 1,000 Imaging CryoMicrotome™ images (Barlow Scientific, Olympia, WA) obtained to define the location of each microsphere. Two coronal 2D virtual sections, at the level of the parietal (1) frontal (2) white matter, where white matter damage was observed, show the 2D representation of blood flow under basal conditions. Blood flow is displayed as a "convolved" image of the density of microsphere distribution and quantified (ml/min) in the pseudocolor scale shown. Basal blood flow was particularly high, for example, in the thalamic nuclear groups in the midbrain (image 1, arrow) relative to the parietal cortex. By contrast, the basal blood flow of the parietal and frontal periventricular white matter was markedly lower (arrowheads). Cortical blood flow is much higher than periventricular white matter flow. This degree of spatial resolution can only be achieved in brains at least the size of the immature fetal sheep.

To seek evidence for vascular end and border zones in fetal cerebral white matter, we measured blood flow in histopathologically defined regions of injury in cerebral cortex and white matter in preterm fetal sheep [83]. While white matter blood flow was lower than cerebral gray matter, there was no evidence for pathologically significant gradients of fetal blood flow within the periventricular white matter under conditions of global partial ischemia or reperfusion. White matter lesions did not localize to regions susceptible to greater ischemia; nor did less vulnerable regions of cerebral white matter have greater flow during ischemia. An alternative explanation for the topography of cerebral white matter lesions is the distribution of susceptible cell types (see below), particularly late oligodendrocyte (OL) progenitors (pre-OLs), that are particularly susceptible to hypoxia-ischemia.

3.3 Relative Contributions of Hypoxia-Ischemia and OL Lineage Immaturity to Acute WMI

The preterm fetal sheep (0.65 gestation) displays heterogeneous OL lineage maturation in frontal periventricular white matter [32], which allowed us to define the relative contributions of oligodendroglial maturational factors and vascular factors to acute WMI. OL lineage maturation in medial periventricular white matter (PVWM) was similar to human (~23–28 weeks gestation) in that preOLs were the major OL stage present. By contrast, lateral PVWM was more differentiated and contained predominantly pre-myelinating and early myelinating immature OLs. Surprisingly, we found that moderate cerebral ischemia did not uniformly damage the PVWM. The medial and lateral PVWM sustained differing degrees of acute injury even though they sustained a similar degree of low flow during prolonged ischemia-reperfusion. Hence, while global ischemia was necessary for WMI, no regional differences in blood flow were found within the PVWM under basal or ischemic conditions to account for the differences in cell death between medial and lateral PVWM. Rather, differences in the topography of WMI were closely correlated with the distribution of vulnerable preOLs. Interestingly, in regions of preOL degeneration, other neural cell types (astrocytes, microglia, and axons) were markedly more resistant to injury.

Recently, in a fetal rabbit model of placental insufficiency, significant global fetal hypoxia-ischemia caused minimal WMI at fetal day 22, but a similar insult 3 days later in gestation caused pronounced WMI [95]. The relative susceptibility of the white matter at these two developmental ages coincided with the timing of appearance of susceptible preOLs. Taken together, these findings suggest that perturbations in cerebral blood flow are necessary but not sufficient to explain the distribution of WMI. The developmental predilection for WMI appears to be related to both the timing of appearance and regional distribution of susceptible preOLs.

3.4 Role of High Field MRI to Define Pathological Features of WMI

Although MRI is the optimal imaging modality to define WMI in preterm survivors [96–99], the histopathological features of MRI signal abnormalities have mostly been defined for WMI where PVL predominates [43, 100–104]. With the pronounced shift to milder forms of human WMI defined by quantitative and diffusion-weighted MRI, there is a need to define the cellular features of these lesions. We have recently analyzed diffuse lesions in a preterm fetal sheep model where animals survived for 1 or 2 weeks after global cerebral ischemia. This preparation generated a spectrum of WMI very similar to that observed from human autopsy studies, as well as a reduction in cerebral WM volume similar to that observed in preterm survivors [97–99]. We developed registration algorithms to analyze the histopathological features of three classes of MRI-defined lesions [33]. Each lesion type displayed unique astroglial and microglial responses that corresponded to distinct forms of necrotic or non-necrotic WMI. At 1 week after injury, for example, high field MRI (12 T) identified a novel hypointense signal abnormality on T2-weighted images with high sensitivity and specificity for lesions with astrogliosis. This unexpected finding suggests that current clinical MRI field strength may be a limiting factor to detect diffuse gliosis as well as microscopic necrosis (see below). Additional clinical-pathological studies are needed to determine whether high-field MRI can provide greater sensitivity to identify diffuse WMI than is currently feasible at lower field strengths.

3.5 MRI-Guided Ultrastructural Studies of Axonal Degeneration in WMI

Whether primary axonal injury occurs in cases of diffuse WMI that lack PVL has received limited study. During the acute phase of WMI, after global hypoxia-ischemia, fetal sheep that lack necrotic WMI do not show evidence of acute axonal degeneration [32]. We employed quantitative electron microscopy studies in our preterm fetal sheep model of diffuse WMI to define the extent of axonal degeneration in MRI-defined diffuse WMI [41]. During the chronic phase of WMI from this same fetal sheep preparation [33] and human [42], microscopic foci of necrosis were observed that are rich in microglia, but lack astrocytes or axons. However, no significant axonal degeneration, axonal loss or shift in the distribution of axon calibers was observed by quantitative electron microscopy studies in preterm fetal sheep [41]. Hence, the contribution of microscopic necrosis to axonal loss with secondary myelination failure appears to be low, but there is a need for further human neuropathological studies with sensitive markers of axonal injury that are applied to cases of WMI without necrosis.

3.6 Conclusions

The preterm human infant displays unique patterns of cerebral injury that now can be closely replicated in the preterm fetal sheep. Despite the higher costs and technical challenges of preterm fetal sheep models, they provide powerful access to preclinical questions related to the pathophysiology of WMI. Recent advances include spatially defined measurements of cerebral blood flow in utero, the definition of cellular-maturational factors that define the topography of WMI and the application of high field neuro-imaging to define MRI signatures for specific types of chronic WMI. There is a critical need to further define the cellular and molecular mechanisms that mediate the progression of cerebral white and gray matter injury. Such information is critical for the rationale design of therapies targeted to block the initial phase of injury or to promote regeneration and repair during the chronic phase. With few exceptions, most studies have focused on models of cerebral hypoxia-ischemia or maternal fetal infection. Future improved fetal sheep models are needed that more closely reproduce the spectrum of insults that are likely to converge to generate cerebral injury in human preterm infants.

Acknowledgments

Supported by the National Institutes of Neurological Diseases and Stroke: 1RO1NS054044, R37NS045737-06S1/06S2 to SAB and 1F30NS066704 to AR, the American Heart Association (SAB) and the March of Dimes Birth Defects Foundation (SAB).

References

1. Ferriero DM (2006) Can we define the pathogenesis of human periventricular white-matter injury using animal models? J Child Neurol 21:580–581

2. Volpe JJ (2008) Neurology of the newborn. W.B. Saunders, Philadelphia, PA

3. Back S (2012) Mechanisms of acute and chronic brain injury in the preterm infant. In: Miller S, Shevell M (eds) Acquired brain injury. Mac Keith Press, London

4. Wilson-Costello D, Fridedman H, Minich N, Fanaroff A, Hack M (2005) Improved survival rates with increased neurodevelopmental disability for extremely low birth weight infants in the 1990s. Pediatrics 115:997–1003

5. Hack M, Taylor H, Drotar D, Schluchter M, Cartar L, Andreias L, Wilson-Costello D, Klein N (2005) Chronic conditions, functional limitations, and special health care needs of school-aged children born with extremely low-birth-weight in the 1990's. JAMA 294:318–325

6. Miller SP, Ferriero DM, Leonard C, Piecuch R, Glidden D, Partridge JC, Perez M, Mukherjee P, Vigneron D, Barkovich AJ (2005) Early brain injury in premature new-borns detected with magnetic resonance imaging is associated with adverse neurodevelopmental outcome. J Pediatr 147:609–616

7. Beaino G, Khoshnood B, Kaminski M, Pierrat V, Marret S, Matis J, Ledesert B, Thiriez G, Fresson J, Roze JC, Zupan-Simunek V, Arnaud C, Burguet A, Larroque B, Breart G, Ancel PY (2010) Predictors of cerebral palsy in very preterm infants: the EPIPAGE prospective population-based cohort study. Dev Med Child Neurol 52:e119–e125

8. Mercier CE, Dunn MS, Ferrelli KR, Howard DB, Soll RF (2010) Neurodevelopmental outcome of extremely low birth weight infants from the Vermont Oxford network: 1998-2003. Neonatology 97:329–338

9. Liu J, Li J, Qin GL, Chen YH, Wang Q (2008) Periventricular leukomalacia in premature infants in mainland China. Am J Perinatol 25:535–540

10. Litt J, Taylor H, Klein N, Hack M (2005) Learning disabilities in children with very low birthweight:prevalence, neuropsychological correlates and educational interventions. J Learn Disabil 8:130–141

11. Jacobson LK, Dutton GN (2000) Periventricular leukomalacia: an important cause of visual and ocular motility dysfunction in children. Surv Ophthalmol 45:1–13

12. Glass HC, Fujimoto S, Ceppi-Cozzio C, Bartha AI, Vigneron DB, Barkovich AJ, Glidden DV, Ferriero DM, Miller SP (2008) White-matter injury is associated with impaired gaze in premature infants. Pediatr Neurol 38:10–15

13. Soria-Pastor S, Gimenez M, Narberhaus A, Falcon C, Botet F, Bargallo N, Mercader JM, Junque C (2008) Patterns of cerebral white matter damage and cognitive impairment in adolescents born very preterm. Int J Dev Neurosci 26:647–654

14. Anderson PJ, De Luca CR, Hutchinson E, Spencer-Smith MM, Roberts G, Doyle LW (2011) Attention problems in a representative sample of extremely preterm/extremely low birth weight children. Dev Neuropsychol 36:57–73

15. Constantinou JC, Adamson-Macedo EN, Mirmiran M, Fleisher BE (2007) Movement, imaging and neurobehavioral assessment as predictors of cerebral palsy in preterm infants. J Perinatol 27:225–229

16. Spittle AJ, Brown NC, Doyle LW, Boyd RN, Hunt RW, Bear M, Inder TE (2008) Quality of general movements is related to white matter pathology in very preterm infants. Pediatrics 121:e1184–e1189

17. Spittle AJ, Boyd RN, Inder TE, Doyle LW (2009) Predicting motor development in very preterm infants at 12 months' corrected age: the role of qualitative magnetic resonance imaging and general movements assessments. Pediatrics 123:512–517

18. Bax M, Tydeman C, Flodmark O (2006) Clinical and MRI correlates of cerebral palsy: the European Cerebral Palsy Study. JAMA 296:1602–1608

19. Pagliano E, Fedrizzi E, Erbetta A, Bulgheroni S, Solari A, Bono R, Fazzi E, Andreucci E, Riva D (2007) Cognitive profiles and visuoperceptual abilities in preterm and term spastic diplegic children with periventricular leukomalacia. J Child Neurol 22:282–288

20. Li AM, Chau V, Poskitt KJ, Sargent MA, Lupton BA, Hill A, Roland E, Miller SP (2009) White matter injury in term newborns with neonatal encephalopathy. Pediatr Res 65:85–89

21. Lasry O, Shevell MI, Dagenais L (2010) Cross-sectional comparison of periventricular leukomalacia in preterm and term children. Neurology 74:1386–1391

22. Wernovsky G, Shillingford A, Gaynor J (2005) Central nervous system outcomes in children with complex congenital heart disease. Curr Opin Cardiol 20:94–99

23. Miller S, McQuillen P, Hamrick S, Xu D, Glidden D, Charlton N, Karl T, Azakie A, Ferriero D, Barkovich A, Vigneron D (2007) Abnormal brain development in newborns with congenital heart disease. N Engl J Med 357:1971–1973

24. Licht DJ, Shera DM, Clancy RR, Wernovsky G, Montenegro LM, Nicolson SC, Zimmerman RA, Spray TL, Gaynor JW, Vossough A (2009) Brain maturation is delayed in infants with complex congenital heart defects. J Thorac Cardiovasc Surg 137:529–536, discussion 536–527

25. (CDC), C. f. D. C. a. P (2004) Economic costs associated with mental retardation, cerebral palsy, hearing loss and vision impairment – United States, 2003. MMWR Morb Mortal Wkly Rep 53:57–59

26. Barlow R (1969) The foetal sheep: morphogenesis of the nervous system and histochemical aspects of myelination. J Comp Neurol 135:249–262

27. Bernhared C, Kolmodin G, Meyerson B (1967) On the prenatal development of function and structure in the somesthetic cortex of the sheep. Prog Brain Res 2:60–77

28. Cook C, Gluckman P, Johnston B, Williams C (1987) The development of the somatosensory evoked potential in the unanaesthetized fetal lamb. J Dev Physiol 9:441–456

29. Cook C, Williams C, Gluckman P (1987) Brainstem auditory evoked potential in the fetal lamb, in utero. J Dev Physiol 9:429–440

30. Gluckman P, Parsons Y (1983) Stereotaxic method and atlas for the ovine fetal forebrain. J Dev Physiol 5:101–128

31. Vanderwolf C, Cooley R (1990) The sheep brain: a photographic series, 2nd edn. A. J. Kirby Co., London, ON

32. Riddle A, Luo N, Manese M, Beardsley D, Green L, Rorvik D, Kelly K, Barlow C, Kelly J, Hohimer A, Back S (2006) Spatial heterogeneity in oligodendrocyte lineage maturation and not cerebral blood flow predicts fetal ovine periventricular white matter injury. J Neurosci 26:3045–3055

33. Riddle A, Dean J, Buser JR, Gong X, Maire J, Chen K, Ahmad T, Cai V, Nguyen T, Kroenke C, Hohimer A, Back S (2011) Histopathological correlates of MRI-defined chronic perinatal white matter injury. Ann Neurol 70:493

34. Back SA, Riddle A, Hohimer AR (2006) Role of instrumented fetal sheep preparations in defining the pathogenesis of human periventricular white matter injury. J Child Neurol 21:582–589

35. Segovia K, Mcclure M, Moravec M, Luo N, Wang Y, Gong X, Riddle A, Craig A, Struve J, Sherman L, Back S (2008) Arrested oligodendrocyte lineage maturation in chronic perinatal white matter injury. Ann Neurol 63:517–526

36. Back SA, Han BH, Luo NL, Chrichton CA, Tam J, Xanthoudakis S, Arvin KL, Holtzman DM (2002) Selective vulnerability of late oligodendrocyte progenitors to hypoxia-ischemia. J Neurosci 22:455–463

37. Follet PL, Rosenberg PA, Volpe JJ, Jensen FE (2000) NBQX attenuates excitotoxic injury to the developing white matter. J Neurosci 20:9235–9241

38. Uehara H, Yoshioka H, Kawase S, Nagai H, Ohmae T, Hasegawa K, Sawada T (1999) A new model of white matter injury in neonatal rats with bilateral carotid artery occlusion. Brain Res 837:213–220

39. Olivier P, Baud O, Evrard P, Gressens P, Verney C (2005) Prenatal ischemia and white matter damage in rats. J Neuropathol Exp Neurol 64:998–1006

40. Marret S, Mukendi R, Gadisseux J-F, Gressens P, Evrard P (1995) Effect of ibotenate on brain development: an excitotoxic mouse model of microgyria and posthypoxic-like lesions. J Neuropathol Exp Neurol 54:358

41. Riddle A, Maire J, Gong X, Chen K, Kroenke CD, Hohimer AR, Back SA (2012) Differential susceptibility to axonopathy in necrotic and non-necrotic perinatal white matter injury. Stroke 43:178

42. Buser J, Maire J, Riddle A, Gong X, Nguyen T, Nelson K, Luo N, Ren J, Struve J, Sherman L, Miller S, Chau V, Hendson G, Ballabh P, Grafe M, Back S (2012) Arrested pre-oligodendrocyte maturation contributes to myelination failure in premature infants. Ann Neurol 71:93

43. Lodygensky G, West T, Moravec M, Back S, Dikranien K, Holtzman D, Neil J (2011) Diffusion characteristics associated with neuronal injury and glial activation following hypoxia-ischemia in the immature brain. Magn Reson Med 66:839–845

44. Gunn AJ, Bennet L (2009) Fetal hypoxia insults and patterns of brain injury: insights from animal models. Clin Perinatol 36:579–593

45. Reddy K, Mallard C, Guan J, Marks K, Bennet L, Gunning M, Gunn A, Gluckman P, Williams C (1998) Maturational change in the cortical response to hypoperfusion injury in the fetal sheep. Pediatr Res 43:674–682

46. Szymonowicz W, Walker A, Yu V, Stewart M, Cannata J, Cussen L (1990) Regional cerebral blood flow after hemorrhagic hypotension in the preterm, near-term, and newborn lamb. Pediatr Res 28:361–366

47. Rees S, Breen S, Loeliger M, McCrabb G, Harding R (1999) Hypoxemia near midgestation has long-term effects on fetal brain development. J Neuropathol Exp Neurol 58:932–945

48. Gleason CA, Hamm C, Jones MD Jr (1990) Effect of acute hypoxemia on brain blood flow and oxygen metabolism in immature fetal sheep. Am J Physiol 258:H1064–H1069

49. Falkowski A, Hammond R, Han V, Richardson B (2002) Apoptosis in the preterm and near term ovine fetal brain and the effect of intermittent umbilical cord occlusion. Dev Brain Res 136:165–173

50. Bennet L, Rossenrode S, Gunning MI, Gluckman PD, Gunn AJ (1999) The cardiovascular and cerebrovascular responses of the immature fetal sheep to acute umbilical cord occlusion. J Physiol 517(Pt 1):247–257

51. Welin AK, Svedin P, Lapatto R, Sultan B, Hagberg H, Gressens P, Kjellmer I, Mallard C (2007) Melatonin reduces inflammation and cell death in white matter in the mid-gestation fetal sheep following umbilical cord occlusion. Pediatr Res 61:153–158

52. Harris AP, Koehler RC, Gleason CA, Jones MD Jr, Traystman RJ (1989) Cerebral and peripheral circulatory responses to intracranial hypertension in fetal sheep. Circ Res 64:991–1000

53. Rees S, Hale N, De Matteo R, Cardamone L, Tolcos M, Loeliger M, Mackintosh A, Shields A, Probyn M, Greenwood D, Harding R (2010) Erythropoietin is neuroprotective in a preterm ovine model of endotoxin-induced brain injury. J Neuropathol Exp Neurol 69:306–319

54. Duncan J, Cock M, Scheerlinck J, Westcott K, McLean C, Harding R, Rees S (2006) White matter injury after repeated endotoxin exposure in the preterm ovine fetus. Pediatr Res 52:941–949

55. Dean J, van de Looij Y, Sizonenko S, Lodygensky G, Lazeyras F, Bolouri H, Kjellmer I, Huppi P, Hagberg H, Mallard C (2011) Delayed cortical impairment following lipopolysaccharide exposure in preterm fetal sheep. Ann Neurol 70:846

56. Sorensen A, Pedersen M, Tietze A, Ottosen L, Duus L, Uldbjerg N (2009) BOLD MRI in sheep fetuses: a non-invasive method for measuring changes in tissue oxygenation. Ultrasound Obstet Gynecol 34:687–692

57. Gunn A, Bennet L (2008) Brain cooling for preterm infants. Clin Perinatal 35:735–748

58. Pyrds O (1991) Control of cerebral circulation in the high-risk neonate. Ann Neurol 30:321–329

59. Menke J, Michel E, Hildebrand S (1997) Cross-spectral analysis of cerebral autoregulation dynamics in high risk preterm infants during the perinatal period. Pediatr Res 42:690–699

60. du Plessis A (2008) Cerebrovascular injury in premature infants: current understanding and challenges for future prevention. Clin Perinatol 35:609–641

61. Soul J, Hammer P, Tsuji M, Saul J, Bassan H, Limperopoulous C, Disalvo D, Moore M, Akins P, Ringer S, Volpe J, Trachtenberg F, du Plessis A (2007) Fluctuating pressure-passivity is common in the cerebral circulation of sick premature infants. Pediatr Res 61:467–473

62. Papile L, Rudolph AM, Heymann M (1985) Autoregulation of cerebral blood flow in the preterm fetal lamb. Pediatr Res 19:159–161

63. Tweed W, Cote J, Pash M, Lou H (1985) Arterial oxygenation determines autoregulation of cerebral blood flow in fetal lamb. Pediatr Res 17:246–249

64. Helou S, Koehler RC, Gleason CA, Jones MD, Traystman RJ (1994) Cerebrovascular autoregulation during fetal development in sheep. Am J Physiol 266:H1069–H1074

65. Hohimer AR, Bissonnette JM (1989) Effects of cephalic hypotension, hypertension, and barbiturates on fetal cerebral flood flow and metabolism. Am J Obstet Gynecol 161:1344–1351

66. Greisen G (2009) To autoregulate or not to autoregulate – that is no longer the question. Semin Pediatr Neurol 16:207–215

67. Tsuji M, Saul J, du Plessis A, Eichenwald E, Sobh J, Crocker R, Volpe J (2000) Cerebral intravascular oxygenation correlates with mean arterial pressure in critically ill premature infants. Pediatrics 106:625–632

68. Raad RA, Tan WK, Bennet L, Gunn AJ, Davis SL, Gluckman PD, Johnston BM, Williams CE (1999) Role of the cerebrovascular and metabolic responses in the delayed phases of injury after transient cerebral ischemia in fetal sheep. Stroke 30:2735–2741

69. Clapp J III, Peress N, Wesley M, Mann L (1988) Brain damage after intermittent partial cord occlusion in the chronically instrumented fetal lamb. Am J Obstet Gynecol 159:504–509

70. Ikeda T, Murata Y, Quuilligan E, Choi B, Parer J, Doi S, Park S-D (1998) Physiologic and histologic changes in near-term fetal lambs exposed to asphyxia by partial umbilical cord occlusion. Am J Obstet Gynecol 178:24–32

71. Ohyu J, Marumo G, Ozawa H, Takashima S, Nakajima K, Kohsaka S, Hamai Y, Machida Y, Kobayashi K, Ryo E, Baba K, Kozuma S, Okai T, Taketani Y (1999) Early axonal and glial pathology in fetal sheep brains with leukomalacia induced by repeated umbilical cord occlusion. Brain Dev 21:248–252

72. Matsuda T, Okuyama K, Cho K, Hoshi N, Matsumoto Y, Kobayashi Y, Fujimoto S (1999) Induction of antenatal periventricular leukomalacia by hemorrhagic hypotension in the chronically instrumented fetal sheep. Am J Obstet Gynecol 181:725–730

73. Duncan J, Cock M, Scheerlinck J, Westcott K, McLean C, Harding R, Rees S (2002) White matter injury after repeated endotoxin exposure in the preterm ovine fetus. Pediatr Res 52:941–949

74. Dalitz P, Harding R, Rees S, Cock M (2003) Prolonged reductions in placental blood flow and cerebral oxygen delivery in preterm fetal sheep exposed to endotoxin: possible factors in white matter injury after acute infection. J Soc Gynecol Investig 10:283–290

75. Rees S, Stringer M, Just Y, Hooper S, Harding R (1997) The vulnerability of the fetal sheep brain to hypoxemia at mid-gestation. Dev Brain Res 103:103–118

76. Mallard E, Rees S, Stringer M, Cock M, Harding R (1998) Effects of chronic placental insufficiency on brain development in fetal sheep. Pediatr Res 43:262–270

77. Penning D, Grafe J, Hammond R, Matsuda Y, Patrick J, Richardson B (1994) Neuropathology of the near-term and midgestation ovine fetal brain after sustained in utero hypoxemia. Am J Obstet Gynecol 170:1425–1432

78. Bagenholm R, Nilsson U, Gotborg C, Kjellmer I (1998) Free radicals are formed in the brain of the fetal sheep during reperfusion after cerebral ischemia. Pediatr Res 43:271–275

79. Ikeda K, Murata Y, Quilligan EJ, Parer J, Doi S, Park S-D (1998) Brain lipid peroxidation and antioxidant levels in fetal lambs 72 hours

after asphyxia from partial umbilical cord occlusion. Am J Obstet Gynecol 178:474–478

80. Castillo-Melendez M, Chow J, Walker D (2004) Lipid peroxidation, caspase-3 immunoreactivity, and pyknosis in late-gestation fetal sheep brain after umbilical cord occlusion. Pediatr Res 55:864–871

81. Welin A-K, Sandberg M, Lindblom A, Arvidsson P, Nilsson U, Kjellmer I, Mallard C (2005) White matter injury following prolonged free radical formation in the 0.65 gestation fetal sheep brain. Pediatr Res 58:100–105

82. Baldwin B, Bell F (1963) The anatomy of the cerebral circulation of the sheep and ox. The dynamic distribution of the blood supplied by the carotid and vertebral arteries to cranial regions. J Anat 97:203–215

83. McClure M, Riddle A, Manese M, Luo N, Rorvik D, Kelly K, Barlow C, Kelly JJ, Vinecore K, Roberts C, Hohimer A, Back S (2008) Cerebral blood flow heterogeneity in preterm sheep: lack of physiological support for vascular boundary zones in fetal cerebral white matter. J Cereb Blood Flow Metab 28:995–1008

84. Chao CR, Hohimer AR, Bissonnette JM (1991) Fetal cerebral blood flow and metabolism during oligemia and early postoligemic reperfusion. J Cereb Blood Flow Metab 11:416–423

85. Bernard SL, Ewen JR, Barlow CH, Kelly JJ, McKinney S, Frazer DA, Glenny RW (2000) High spatial resolution measurements of organ blood flow in small laboratory animals. Am J Physiol Heart Circ Physiol 279(5):H2043–H2052

86. Hohimer AR, Chao CR, Bissonnette JM (1991) The effect of combined hypoxemia and cephalic hypotension on fetal cerebral blood flow and metabolism. J Cereb Blood Flow Metab 11:99–105

87. Iwamoto HS, Kaufman T, Keil LC, Rudolph AM (1989) Responses to acute hypoxemia in fetal sheep at 0.6-0.7 gestation. Am J Physiol 256:H613–H620

88. Rurak DW, Richardson BS, Patrick JE, Carmichael L, Homan J (1990) Oxygen consumption in the fetal lamb during sustained hypoxemia with progressive acidemia. Am J Physiol 258:R1108–R1115

89. Richardson BS, Rurak D, Patrick JE, Homan J, Carmichael L (1989) Cerebral oxidative metabolism during sustained hypoxaemia in fetal sheep. J Dev Physiol 11:37–43

90. Yan EB, Baburamani AA, Walker AM, Walker DW (2009) Changes in cerebral blood flow, cerebral metabolites, and breathing movements in the sheep fetus following asphyxia produced by occlusion of the umbilical cord. Am J Physiol Regul Integr Comp Physiol 297:R60–R69

91. Richardson BS (1993) The fetal brain: metabolic and circulatory responses to asphyxia. Clin Invest Med 16:103–114

92. Nelson MD Jr, Gonzalez-Gomez I, Gilles FH (1991) Dyke Award. The search for human telencephalic ventriculofugal arteries. AJNR Am J Neuroradiol 12:215–222

93. Mayer PL, Kier EL (1991) The controversy of the periventricular white matter circulation: a review of the anatomic literature. AJNR Am J Neuroradiol 12:223–228

94. Volpe JJ (2001) The structure of blood vessels in the germinal matrix and the autoregulation of cerebral blood flow in premature infants - reply. Pediatrics 108:1050

95. Buser J, Segovia K, Dean J, Nelson K, Beardsley D, Gong X, Luo N, Ren J, Wan Y, Riddle A, McClure M, Ji X, Derrick M, Hohimer A, Back S, Tan S (2010) Timing of appearance of late oligodendrocyte progenitors coincides with enhanced susceptibility of preterm rabbit cerebral white matter to hypoxia-ischemia. J Cereb Blood Flow Metab 30:1053–1065

96. Miller S, Ferriero D (2009) From selective vulnerability to connectivity: insights from newborn brain imaging. Trends Neurosci 32:496–505

97. Ment LR, Hirtz D, Huppi PS (2009) Imaging biomarkers of outcome in the developing preterm brain. Lancet Neurol 8:1042–1055

98. Mathur AM, Neil JJ, Inder TE (2010) Understanding brain injury and neurodevelopmental disabilities in the preterm infant: the evolving role of advanced magnetic resonance imaging. Semin Perinatol 34:57–66

99. Rutherford MA, Supramaniam V, Ederies A, Chew A, Bassi L, Groppo M, Anjari M, Counsell S, Ramenghi LA (2010) Magnetic resonance imaging of white matter diseases of prematurity. Neuroradiology 52:505–521

100. Hope PL, Gould SJ, Howard S, Hamilton PA, Costello AM, Reynolds EO (1988) Precision of ultrasound diagnosis of pathologically verified lesions in the brains of very preterm infants. Dev Med Child Neurol 30:457–471

101. Schouman-Claeys E, Henry-Feugeas MC, Roset F, Larroche JC, Hassine D, Sadik JC, Frija G, Gabilan JC (1993) Periventricular leukomalacia: correlation between MR imaging and autopsy findings during the first 2 months of life. Radiology 189:59–64

102. Felderhoff-Mueser U, Rutherford MA, Squier WV, Cox P, Maalouf EF, Counsell SJ,

Bydder GM, Edwards AD (1999) Relationship between MR imaging and histopathologic findings of the brain in extremely sick preterm infants. AJNR Am J Neuroradiol 20: 1349–1357

103. Inder TE, Neil JJ, Kroenke CD, Dieni S, Yoder B, Rees S (2005) Investigation of cerebral development and injury in the prematurely born primate by magnetic resonance imaging and histopathology. Dev Neurosci 27:100–111

104. Childs AM, Cornette L, Ramenghi LA, Tanner SF, Arthur RJ, Martinez D, Levene MI (2001) Magnetic resonance and cranial ultrasound characteristics of periventricular white matter abnormalities in newborn infants. Clin Radiol 56:647–655

105. Back SA, Luo NL, Borenstein NS, Levine JM, Volpe JJ, Kinney HC (2001) Late oligodendrocyte progenitors coincide with the developmental window of vulnerability for human perinatal white matter injury. J Neurosci 21:1302–1312

Chapter 9

The Rabbit as a Model of Cerebral Palsy

Kehuan Luo, Jessica Baker, Matthew Derrick, and Sidhartha Tan

Abstract

Rabbits, like humans, are perinatal brain developers. We have developed a model of fetal hypoxia-ischemia that results in postnatal rabbit kits with a phenotype analogous to that observed in humans with cerebral palsy. This chapter gives a practical approach to performing the model and shows the range methods available for study using the model.

Key words Behavior, Brain, Dystonia, Fetus, Hypoxia, Infant, Newborn, Infant, Premature, Ischemia, Placental insufficiency, Uterus

1   Introduction

Cerebral palsy refers to a group of disorders that affect a person's ability to move and to maintain balance and posture. It is due to non-progressive brain abnormalities, though the exact symptoms can change over a person's lifetime [1]. The most common signs and symptoms are due to hypertonia. Cerebral palsy is primarily caused by perinatal brain injury and results in one of the highest burdens of disease due to the lifelong consequences. It is therefore of enormous cost to society. This makes it imperative to develop better animal models that mimic the human condition. To establish an animal model of cerebral palsy the combination of a clinically relevant causation and a hypertonia phenotype is necessary.

In mammalian perinatal brain injury models, the most common models currently used involve rodents. Rodents however, are postnatal brain developers, i.e., white matter and motor development occur primarily after birth. Other perinatal models use ungulates or nonhuman primates. Both ungulates and nonhuman primates are prenatal brain developers, i.e., most of the motor development have been completed before birth. Hypertonia is not the principal manifestation of perinatal brain injury in rodent or ungulate models. Nonhuman primates may develop hypertonia but long term evaluation can be cost-prohibitive. In selecting an animal

Jerome Y. Yager (ed.), *Animal Models of Neurodevelopmental Disorders*, Neuromethods, vol. 104,
DOI 10.1007/978-1-4939-2709-8_9, © Springer Science+Business Media New York 2015

model for human perinatal disease, one must take into account the normal brain development of that animal. Humans are unlike other primates in that humans are perinatal brain developers. Rabbits are similar in this regard [2]. The ideal model for studying perinatal injury and subsequent motor deficits may not necessarily be in the closest animal relative to man, such as the chimpanzee or other great apes. Animals that are perinatal brain developers such as the rabbit may be more suitable.

A second important requirement for animal models is to mimic human disease conditions. Cerebral palsy is believed to frequently be due to perinatal hypoxia-ischemia. Acute placental insufficiency that occurs in conditions such as placental abruption is a cause of fetal hypoxia-ischemia. Placental abruptio has been shown to result in a higher incidence of cerebral palsy (Odds ratio 28) compared to controls [3]. Modeling acute placental insufficiency states in animals is problematic as it usually involves a laparotomy [4]. We have established a model without entering the abdominal cavity that subjects the fetus to hypoxia-ischemia [5, 6].

The model subjects rabbit fetuses to hypoxia-ischemia by transiently occluding the descending aorta of a timed pregnant dam using a Fogarty catheter introduced into the femoral artery. After delivery a battery of neurobehavioral tests which confirm a hypertonia phenotype. The model can be refined to predict which animal will develop the phenotype by using diffusion-weighted magnetic resonance imaging (MRI) during the uterine ischemia [7]. We have validated the phenotype as one analogous to cerebral palsy, as determined by its static nature, by further analyzing non-weight-bearing movement [8].

2   Methods

2.1   Uterine Ischemia in Pregnant Rabbits

New Zealand White Rabbits, with a gestation of 31.5 days, are obtained from Myrtle's Tennessee. We have studied uterine ischemia at E22 (70 % gestation), E25 (79 % gestation), and E29 (92 % gestation). The rabbits are anesthetized with a combination of intravenous fentanyl and droperidol followed by spinal anesthesia. These were chosen over other anesthetic agents that have potential additional confounding effects. Recovery time is also relatively rapid. Spinal/epidural anesthesia allows the dam to breathe spontaneously and to avoid the use of a ventilator. In addition, the amount of medications used for sedation or general anesthesia is considerably reduced when combined with spinal/epidural anesthesia.

2.1.1   Anesthesia

Materials

22 GA (0.9 × 25 mm) 1.00 in. sterile shielded I.V. Catheter (BD, CE 0086).

Fentanyl.

Droperidol.

25 G spinal needle.

19 G epidural needle.

Fur clippers.

Programmable infusion pump.

Restraint device.

Topical anesthetic for the skin infiltration (1 % lidocaine or 0.125 % bupivacaine).

Spinal anesthetic injection (2 % lidocaine or 0.25 % bupivacaine).

Methods

Induction

1. The rabbit dam is placed in restraining device and the ear shaved.

2. Intravenous access is obtained in lateral ear vein.

3. The Dam is anesthetized using intravenous Fentanyl (0.075 mg/kg/h) and Droperidol (3.75 mg/kg/h) mixed with normal saline or Ringer's and administered at a rate of 100 ml/h IV through the ear vein. Induction takes 5–10 min.

Maintenance

The doses are decreased to sedative doses of Fentanyl (0.025–0.05 mg/kg/h) and Droperidol (1.25–2.5 mg/kg/h)

Spinal Anesthesia

1. The rabbit is removed from the restraining device when she is adequately sedated.

2. The fur over the lumbar spine is shaved.

3. Local anesthetic is injected into subarachnoid space in the midline between L4-5 or L5-6 intervertebral spaces.

4. The spinal needle is introduced in the midline of the selected space and angled cranially by 10–30° (approximately aiming at the umbilicus). The bevel of the needle should face upward to avoid spinal nerve injury. A characteristic "pop" is felt when the subarachnoid space is entered and clear CSF will flow after withdraw of the stylet. Bupivacaine is then injected. For a surgical procedure lasting less than 1.5 h, 0.3–0.5 ml of 2 % lidocaine or 0.25 % bupivacaine will produce satisfactory anesthesia; for a procedure between 1.5 and 2.5 h, we use 0.3–0.5 ml Marcaine (7.5 mg bupivacaine HCl and 82.5 mg dextrose per ml).

Maintenance

The intravenous infusion rates are decreased to sedative doses of Fentanyl (0.025–0.05 mg/kg/h) and Droperidol (1.25–2.5 mg/kg/h)

Epidural Anesthesia

1. An 19G epidural needle is employed. A loss of resistance (LOR) syringe is prepared prior to the procedure by filling the syringe with saline or simply a small amount of air. The bevel of the needle should face cranially when bilateral blockage is desired. This is different from spinal anesthesia in which the

bevel tip is facing up. When the needle tip enters into epidural space, the resistance suddenly disappears and saline or air suddenly starts moving into epidural space. After gentle aspiration in several directions to ensure absence of CSF, a test dose of 0.5–0.75 ml 0.25 % bupivacaine or 2 % lidocaine is given. When a response is confirmed 5 min later, 2 ml 2 % lidocaine or 0.25 % bupivacaine is injected. The effect usually lasts 60–90 min.

2.1.2 Uterine Ischemia Surgery

Materials

Fur Clippers.

Surgical scalpels (#10 and #11).

Rectal temperature probe.

Water-filled heating blanket.

4 French EMB40 Fogarty embolectomy catheter (usable length 40 cm—single or double lumen).

Single use micro blood vessel clips.

Micro vascular introducer.

1.0, 5, 10 ml syringes one of each, one 3-way stopcock.

Surgical instruments: scissors, tissue forceps, and clamps.

Methods

1. After spinal or epidural anesthesia, the dam is laid supine on an operating board with the heating blanket (Gaymar, Orchard Park, NY), and all four extremities are restrained in the extended position to expose the abdomen and groin. The dam is allowed to breathe spontaneously.

2. The left femoral artery has been traditionally chosen as the site of arterial catheterization by our laboratory (but either could be used). The groin area is shaved with an electric clipper.

3. The femoral artery is palpated and a 3–5 cm longitudinal incision is made down to the level of the fascia over the course of the artery.

4. The femoral artery is located right below the inner part of the inguinal ligament with its course running downwards in a longitudinal shallow grove between two muscle groups. The femoral neurovascular bundle underneath the deep fascia is identified. At the lateral middle point of the neurovascular bundle, a small (3–5 mm) incision is made in the fascia overlying fascia, the fascia is picked up with a tissue forceps and the fascial incision is extended up and down with tissue scissors along the fiber direction of the bundle to the length of the skin incision.

5. The femoral artery is bluntly dissected to expose at least 2 cm of artery. Small branches of the artery should be tied with 4-0 silk sutures and cut. A poorly exposed artery is the most common reason for failed catheterization.

6. The isolated artery is held with a micro-clip at its proximal end and looped with 2-0 silk sutures both proximally and distally. The two loop sutures are secured to control unexpected hemorrhage and help facilitate the incision of the artery and introduction of the catheter into the femoral artery.

7. A 1–2 mm longitudinal incision is made in the front wall of the artery along the longitudinal axis of the artery. The L-shaped microvascular introducer is inserted into the incision and the 4 French EMB40 Fogarty embolectomy balloon catheter introduced and advanced 10 cm from the incision. This distance will place the balloon in the descending aorta just above the uterine arteries and below the renal arteries.

8. The balloon is inflated with 0.2–0.3 ml of saline. The looped sutures are used to secure the catheter to the artery. This is important to prevent accidental removal of the catheter which could result in uncontrolled bleeding.

9. *Closure of vessel and incision:* After the end of uterine ischemia the balloon is deflated and the catheter withdrawn. An artery clip is placed proximal to the incision. The opening in the femoral artery is then closed with four to six interrupted sutures using 7-0 silk suture. Attention should be taken not to narrow the vessel lumen and prevent blood flow. The fascia and skin are closed in separate layers.

This protocol can be modified to give potential neuroprotective agents to the dam either before uterine ischemia to pretreat the fetuses or after uterine ischemia as a post-treatment. The agents can be given intravenously or intra-arterially which avoids first pass metabolism by the dam's liver.

2.1.3   MRI

We also use MRI to monitor the degree of fetal brain injury during uterine ischemia and the immediate reperfusion period. The MRI response predicts which fetuses will develop hypertonia [7].

Equipment

3 T GE clinical magnet using an 8-channel extremity phase-array coil.

Animal restraint.

Minibore tubing extension set (MPS Acacia) for infusion of anesthetics during MRI scanning.

Portable infusion pump (Harvard Apparatus PHD2000 programmable, Holliston, MA).

Airbag for ventilation in emergencies.

Tegaderm Film (3M Health Care, St. Paul, MN).

Methods

Precautions prior to transport: The femoral artery catheter is secured with an additional 4-0 suture to the skin and the incision covered with sterile gauze. The dam is then transferred to MRI imaging center in a restrainer.

MR imaging: Using 3 T GE clinical magnet and an 8-channel extremity phase-array coil. Single shot fast spin echo (SSFSE) images are used for anatomical reference. 25-32 axial slices covering all fetuses in litter are obtained with these parameters: slice 4 mm, matrix 256×192. The time course of H-I is monitored by diffusion-weighted echo-planar images (DWI EPI) with $b = 0$ and 0.8 ms/μm^2, TR/TE = 7,400/70 ms, 1 NEX. DWI EPI is performed during 10 min baseline before H-I, 40 min of fetal H-I and 20 min of reperfusion. Three more imaging sessions are performed at 4, 24, and 72 h after the H-I insult.

For each imaging session the dams are sedated with an intravenous infusion of fentanyl (75 μg/kg/h) and droperidol (3.75 mg/kg/h).

ADC maps are calculated using in-house software, written in Matlab (Natick, MA).

Surviving kits are imaged on a 4.7 T Bruker scanner, using surface 28 mm coil and multi-slice T2-weighted RARE sequence (TE/TR 80/4,000 ms) to determine the presence of gross anatomical abnormalities. Diffusion tensor imaging is performed using diffusion-weighted spin-echo sequence in 6 non-collinear directions and directionally invariant ADC maps are generated using in-house software [7].

2.1.4 Laparotomy

This procedure is provided for the investigator if there is need for direct manipulation of the fetus or direct administration of drugs to the fetus.

Equipment

A sterile laparotomy package (gauzes, hemostats, needle holder, micro-needle holder, scalpel, drapes, retractors, steel bowls).

Different size sutures (2-0, 4-0, 7-0).

Different size syringes (1.0, 5, 10 ml, one of each).

Sterile gloves, tray, 2 % lidocaine or 0.25 % bupivacaine and normal saline.

Methods

1. The abdominal fur is shaved.

2. The skin is cleaned three times using 70 % alcohol and povidone–iodine-soaked gauzes.

3. The disinfected area is then covered with sterile drapes, leaving the midline incision area in the center exposed. 1 % lidocaine or 0.125 % bupivacaine can be given along the incision if local anesthesia is needed.

4. An incision is made in the lower midline area, starting 2–3 cm above the umbilicus and extending in a cephalocaudal direction downward for 10–15 cm. Attention should be paid not to damage the mammary tissue. The midline of the abdominal fascia is then exposed.

5. Two toothed forceps, one by an assistant and one by the operator, are used to pick up the abdominal muscle and fascia layer making a small tent and ensuring no abdominal contents are caught in the tent. A small longitudinal incision is made in the middle with a scalpel; fine tissue scissors is then used to extend the incision and cut the peritoneum. A fat-covered bladder will appear underneath. The peritoneum can be picked up to avoid the possible intestinal damage and incised with tissue scissors. The intestines are pushed aside carefully and packed away with warm, moist gauze pads. The uterus filled with fetuses is usually inspected inside abdominal cavity, and can only be taken outside for short periods.

6. After the number and position of the fetuses are determined, an experimental drug solution can be injected with a 1-ml TB syringe into the fetal peritoneal cavity, subarachnoid space, or nasal cavity through the wall of the uterus. To avoid the placenta, inject through other side of the uterus.

2.1.5 Hysterotomy

In dams that have undergone an MRI assessment the uterine position of each kit needs to be verified and so they are delivered by hysterotomy. In addition, severely affected kits sometimes cannot tolerate vaginal delivery and die during the birth process, and need to be delivered by hysterotomy.

Methods

1. Following laparotomy the uterus is isolated and an incision is made in the avascular part of the uterus in a longitudinal plane. Fetuses can be safely brought out of uterus, the umbilical cord ligated, and the wound closed with 3-0 sutures.

2. After finishing all the steps of operation, use warm saline to wash the area. The peritoneum is closed by a running over-and-over suture with 2-0 silk in a cephalocaudal direction. Be sure not to suture the internal organs with peritoneum, which could lead to postoperative ileus. Check the leakage and use interrupted over-and-over suture with 2-0 silk to close fascia. Close subcutaneous soft tissue with 4–0 silk interrupted over-and-over suture. An interrupted vertical mattress suture is used to close the skin.

2.2 Neurobehavioral Assessment

We have developed a comprehensive examination and motor performance scale for neurobehavioral assessment. This has been used to evaluate kits from day 1 (P1) to 18.

2.2.1 Materials

Digital video camera.

Sterile cotton-tipped swabs.

Peppermint extract.

Amyl acetate.

Ethanol.

Pasteur pipettes.

Infant formula without animal milk.

2.2.2 Methods

Our neurological examination includes the following parameters: tests for ability to smell (peppermint extract, alcohol, and amyl acetate); locomotion, righting reflex, muscle tone, and possible dystonia. Scoring sheets are provided.

Factor	Criteria
Locomotion score	This is an assessment of NORMAL activity
0	No activity
0.5	Occasional activity (<25 % of the time)
1	Occasional activity (25 % of the time)
1.5	Intermittent activity (25–50 % of the time)
3	Intermittent activity (50 % of the time)
2.5	Frequent activity (50–75 % of the time)
3	Frequent activity (75 % of the time)
3.5	Constant activity (75–100 % of the time)
4	Constant activity (100 % of the time)
Dystonia duration	This is an assessment of ABNORMAL activity that is suggestive of dystonia. Definition: "Dystonia" is defined as a movement disorder in which involuntary sustained or intermittent muscle contractions cause twisting and repetitive movements, abnormal postures, or both
0	No dystonia
0.5	Occasional (<25 % of the time); predominantly submaximal
1	Occasional (<25 % of the time); predominantly maximal
1.5	Intermittent (25–50 % of the time); predominantly submaximal
3	Intermittent (25–50 % of the time); predominantly maximal
2.5	Frequent (50–75 % of the time); predominantly submaximal
3	Frequent (50–75 % of the time); predominantly maximal
3.5	Constant (>75 % of the time); predominantly submaximal
4	Constant (>75 % of the time); predominantly maximal
Eyes and upper face	Dystonia score
0	None
1	Mild: increased blinking or slight forehead wrinkling

(continued)

Factor	Criteria
2	Moderate: jaws open with force, can suck and swallow, constant dribbling, no tongue thrusting
3	Moderate: jaws open with moderate force, suck with swallow, dribbling, mild tongue thrusting
4	Extreme: jaws cannot open
Tone	This is a modification of the Ashworth scale that we use to assess a range from hypotonia to hypertonia.
0	No resistance to induced movement
1	Barely perceptible resistance to induced movement
2	Perceptible resistance to induced movement
2.5	Moderate resistance to induced movement
3	Severe resistance to induced movement
4	Unable to move the muscle group with force applied

A range of neurobehavioral deficits is observed in the P1 newborn kits following fetal hypoxia-ischemia at E22 [9]. There are increasing neurobehavioral deficits following greater than 30 min of sustained fetal hypoxia-ischemia at E22, with increasing hypoxia-ischemia there are increased stillbirths. What is striking is the observation of hypertonia and resulting postural abnormalities in ~75 % of the survivors. Some of the stillbirths also have postural abnormalities. We are able to keep alive some of the severely affected kits, by careful nursing and orogastric feeding of rabbit milk. In all cases with hypertonia and postural deficits, the hypertonia persists and the motor deficits at P11 are a stark contrast to the motor capabilities of normal controls [8].

2.2.3 Forced Swim Test

The test was developed to elicit motor movements in young rabbits by subjecting them to an environment that stimulates movement. Submersing the lower part of the body in water reliably induces a reflex swimming motion. Swimming provides the advantage the limb motion can be assessed without the need to bear weight removing muscle weakness as a confounding variable in the assessment of motor function [8].

2.2.4 Materials

Glass fish tank filled with 34–36 °C water.

Clear Perspex sheet.

Digital video camera.

Kits are marked with a water resistant marker on their toes, ankle, knee, hip, shoulder, elbow, wrist, and finger tips. They are then placed into fish tank filled in a lane bordered by the clear Perspex sheet to limit movement away from the observer. The kit is kept at right angles to the video camera and movement in a plane parallel to the camera is ensured. Kits are supported with simple harness to prevent them from sinking and keep them stationary. Swimming is recorded on video for approximately 60 s. Kits are then dried and returned to their dam or a neonatal incubator. Kits are swum on P1, P5, and P11. 80–100 frames of the video are subsequently digitized. The digitized video is assessed using a custom program written using Matlab software. Each frame is assessed and the joint positions marked manually. The joint angles (degrees) for each frame are then calculated from the joint positions. The distribution is assessed and the start and end of each stroke marked manually to exclude regions where the kit was not swimming. The average of maximal, minimal, range, mean, and median angle (degrees) and angular velocity (degrees/second) of each joint are calculated from at least 5–10 complete strokes and the stroke frequency (strokes/second) is assessed for both upper and lower limbs. The maximum and minimum angles are chosen to define maximal extension. The mean and median angles are chosen as measures of the mid joint motion

3 Summary

We have introduced major steps in this cerebral palsy model in pregnant rabbits. Some critical moments in the procedure have been photographed in Figs. 1, 2, 3, 4, 5, 6, 7, and 8. This model has produced consistent neurobehavioral deficits in newborn kits. It has been shown to be a reliable animal model capable of reproducing many aspects of human cerebral palsy. Investigational interventions based on this model are currently under way in our lab, which may provide promising therapies for cerebral palsy and neonatal encephalopathy.

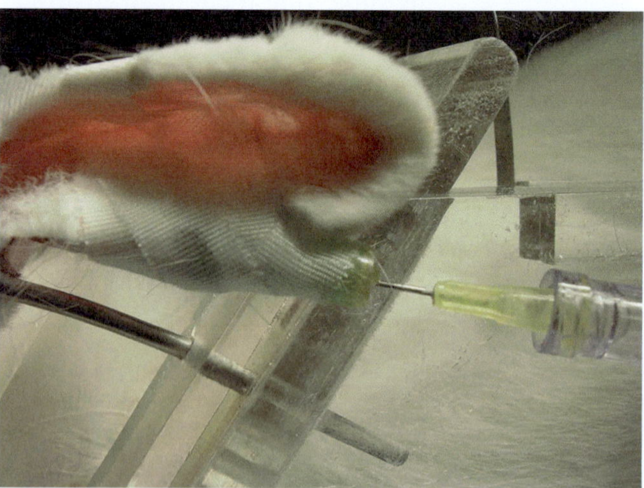

Fig. 1 Cannulation of ear vessels

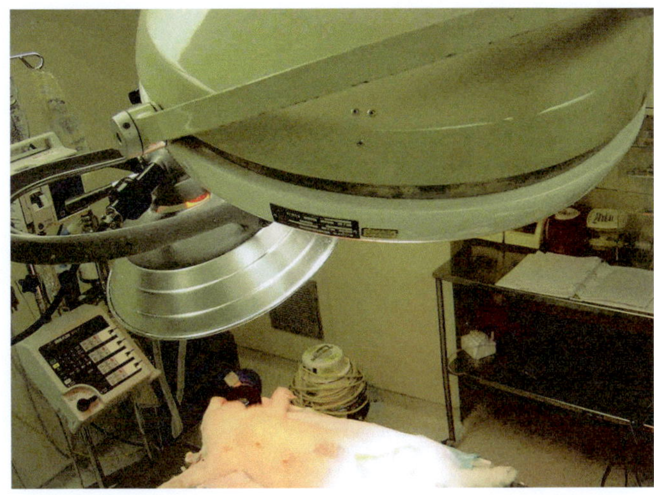

Fig. 2 Operating suite showing multiple channel I.V. pump, heat lamp, and operating lamps over pregnant dam

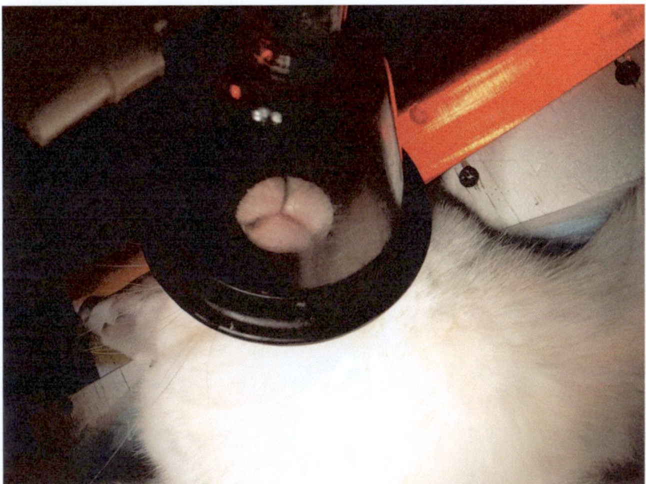

Fig. 3 Bag and mask ventilation given to the pregnant dam

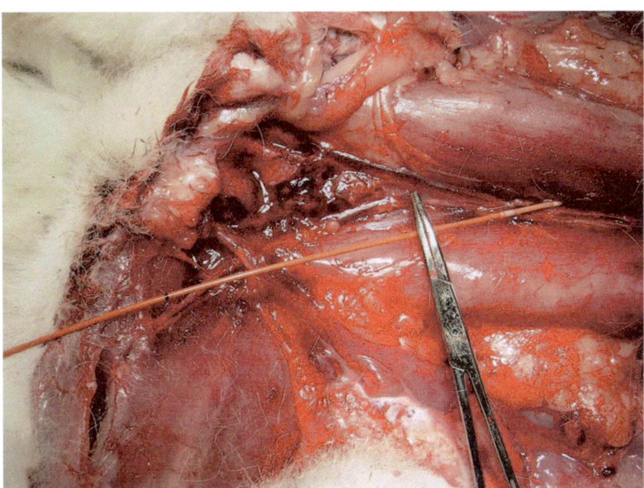

Fig. 4 Hemostat points to the femoral artery and overlaps the Fogarty balloon catheter with the *black color* mark being 10 cm from tip

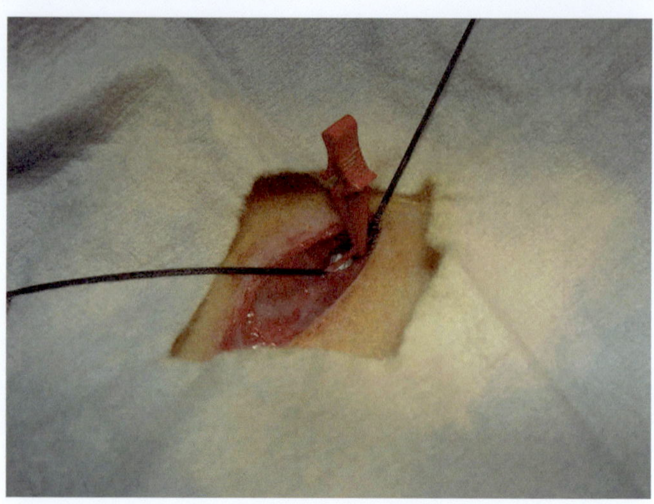

Fig. 5 Exposed femoral artery with two looped sutures and a micro-clip

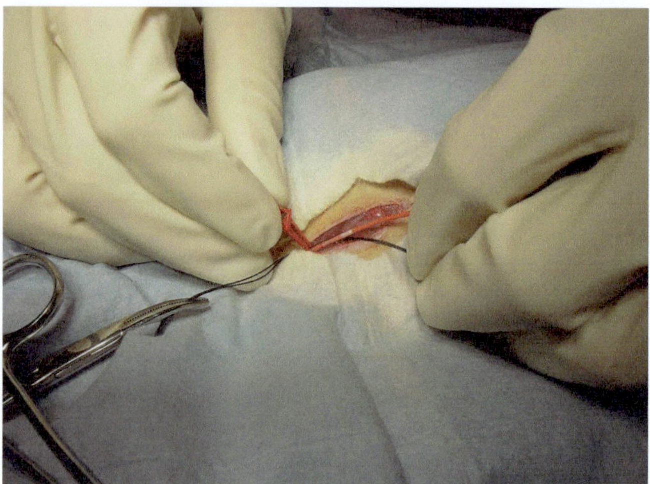

Fig. 6 The tip of the catheter is sliding into the femoral artery

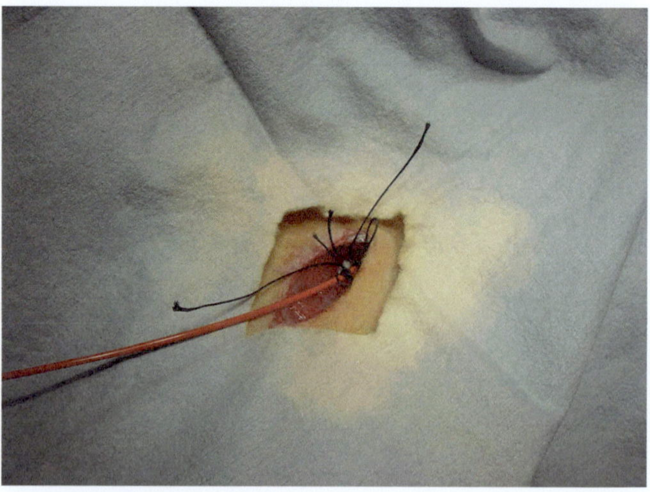

Fig. 7 The inserted catheter is secured to the artery with two looped sutures

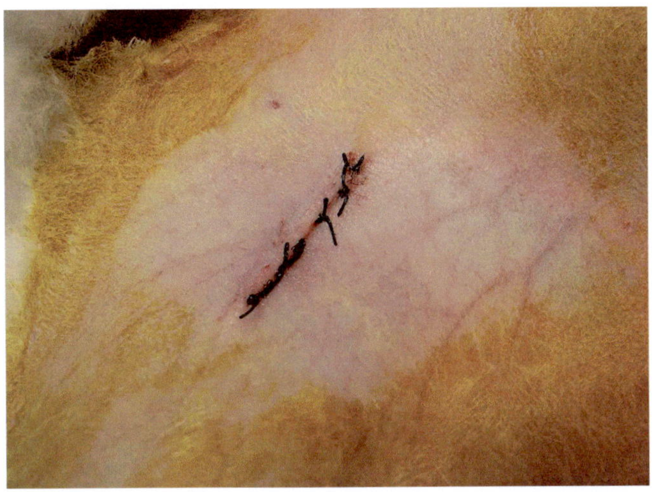

Fig. 8 Closure of skin by double mattress sutures

Acknowledgements

Supported by grants from NIH, HD01138, NS41476, NS43285, NS051402 (S.T.).

References

1. CDC (2015) Facts about Cerebral Palsy. http://www.cdc.gov/ncbddd/cp/facts.html. Accessed 12 Jan 2015.
2. Harel S, Shapira Y et al (1978) Neuromotor development in relation to birth weight in rabbits. Biol Neonate 33(1–2):1–7
3. Matsuda Y, Maeda T et al (2003) Comparison of neonatal outcome including cerebral palsy between abruptio placentae and placenta previa. Eur J Obstet Gynecol Reprod Biol 106(2): 125–129
4. Iwasa H, Aono T et al (1990) Protective effect of vitamin E on fetal distress induced by ischemia of the uteroplacental system in pregnant rats. Free Radic Biol Med 8(4):393–400
5. Derrick M, Luo NL et al (2004) Preterm fetal hypoxia-ischemia causes hypertonia and motor deficits in the neonatal rabbit: a model for human cerebral palsy? J Neurosci 24(1):24–34
6. Tan S, Drobyshevsky A et al (2005) Model of cerebral palsy in the perinatal rabbit. J Child Neurol 20(12):972–979
7. Drobyshevsky A, Derrick M et al (2007) Fetal brain magnetic resonance imaging response acutely to hypoxia-ischemia predicts postnatal outcome. Ann Neurol 61(4):307–314
8. Derrick M, Drobyshevsky A et al (2009) Hypoxia-ischemia causes persistent movement deficits in a perinatal rabbit model of cerebral palsy: assessed by a new swim test. Int J Dev Neurosci 27(6):549–557
9. Derrick M, Drobyshevsky A et al (2007) A model of cerebral palsy from fetal hypoxia-ischemia. Stroke 38(2):731–735

Chapter 10

A Newborn Piglet Survival Model of Post-hemorrhagic Ventricular Dilatation (PHVD)

Kristian Aquilina and Marianne Thoresen

Abstract

Intra-ventricular hemorrhage (IVH) and post-hemorrhagic ventricular dilatation (PHVD) are important issues in neonatal care and continue to contribute to significant motor and cognitive morbidity. Several questions about its pathophysiology remain unanswered and animal models have been useful in identifying relevant risk factors and potential mechanisms. In this chapter, we describe a neonatal piglet model of IVH and PHVD involving the injection of homologous blood with an elevated hematocrit into the ventricular system. The animals are capable of long-term survival and, through a ventricular access device inserted in the second week, allow repeated aspiration of cerebrospinal fluid and measurement of intraventricular pressure.

Key words Pig, Newborn, Neonatal intraventricular hemorrhage, Post-hemorrhagic ventricular dilatation, Hydrocephalus, Ventricular access device, Intraventricular pressure monitoring

1 Introduction

Intraventricular hemorrhage (IVH) remains an important problem in the management of the preterm newborn. Continuing progress in neonatal care has led to an increase in the survival rates of extremely premature neonates and the incidence of IVH has not decreased in recent years. Up to 50 % of neonates with large hemorrhages develop post-hemorrhagic ventricular dilatation (PHVD); a substantial proportion of these infants require permanent cerebro-spinal fluid (CSF) diversion, usually through a ventriculo-peritoneal shunt. These shunts are associated with significant complication rates, including infection and malfunction [1]. Clearly, the management of PHVD, together with the extent of periventricular infarction and white matter disease, determines motor and cognitive prognosis [2] for these children.

The development of PHVD is related to basal adhesive arachnoiditis, obstructing the flow of CSF in the basal cisterns [3]. Although several studies have confirmed the role of upregulated fibroblasts and the excessive production of transforming growth

Jerome Y. Yager (ed.), *Animal Models of Neurodevelopmental Disorders*, Neuromethods, vol. 104, DOI 10.1007/978-1-4939-2709-8_10, © Springer Science+Business Media New York 2015

factor beta in this process, the precise mechanism is poorly understood [4]. Multiple interventions have been attempted, including serial lumbar punctures, insertion of a ventricular access device (VAD) with repeated aspiration of CSF, the use of acetazolamide to reduce CSF production, and more recently, early CSF drainage with fibrinolysis and irrigation (DRIFT) [5–7]. No single technique has, however, been shown to reduce the requirement for permanent CSF diversion.

A large number of animal models of hydrocephalus have been described. Most of these are based on the injection of aluminum silicate (kaolin) into the cisterna magna of rodents or large animals [8–10]. Only a few studies have evaluated hydrocephalus induced by IVH, and fewer still have used neonatal animals in models that permit long-term survival [11–13]. The pathophysiology of PHVD is probably different from that following kaolin administration. Kaolin replicates the inflammatory reaction in the basal cisterns, but does not reflect the neuronal and ependymal injury mediated by free unbound iron and the reactive radicals it generates that occurs in PHVD [14]. This model involves the intraventricular injection of homologous blood in the anesthetized neonatal piglet, with the resulting ventricular dilatation followed up by ultrasonography through a surgical fontanelle [15]. A VAD was implanted into the dilated ventricles and ventricular CSF sampled and intraventricular pressure measured over an eight-week survival period.

2　Materials

2.1　Animal Preparation

1. Open incubator with accurate temperature control (e.g., Giraffe, Datex-Ohmeda, GE Healthcare, Finland).

2. Neonatal mechanical ventilator (e.g., SLE, Croydon, UK).

3. Side-stream multigas analyzer (e.g., Capnomac Ultima, Datex-Ohmeda, Finland).

4. Transcutaneous oxygen saturation probe (e.g., Masimo, Irvine CA, USA).

5. Sterile solution of 3.75 % dextrose in 0.45 % saline, as maintenance intravenous fluid.

6. Blood glucose electrodes (e.g., Precision Plus, Abbott Laboratories, Bedford, Mass, USA).

7. Umbilical arterial and venous catheters (e.g., Vygon, Ecouen, France).

8. Rectal thermometer (e.g., YSI 400 series thermistor probe, Yellow Springs Instruments, Yellow Springs, OH, USA).

9. Amplitude integrated EEG monitor (e.g., Olympic 6000 or BrainZ BRM3 brain monitor, Natus Medical Inc., San Carlos CA, USA).

10. Intra-parenchymal ICP monitor (e.g., Codman MicroSensor ICP transducer, Codman Johnson and Johnson, Raynham, MA, USA).

2.2 Injection of Intra-ventricular Blood

1. Harvard syringe driver (e.g., Harvard Apparatus, Kent, UK).

2.3 Recovery and Survival

1. Piglet milk (Faramate, Volac Feeds, Devon, UK).

2. Intramuscular iron supplement (Leodex 20 % iron injection (1 mL), Leo Pharmaceuticals, Bucks, UK).

2.4 Insertion of Ventricular Access Device

1. Ventricular access device, Ommaya reservoir (Codman Johnson and Johnson, Raynham, MA, USA).

2. Prepare intrathecal vancomycin by mixing intravenous vancomycin as recommended by the manufacturer and then passing the solution through a micro-filter. Intrathecal gentamicin is commercially available.

3 Methods

3.1 Animal Preparation

3.1.1 Anesthesia and Ventilation

1. Run a mixture of 2 % halothane and N_2O through a closed Perspex box.

2. Place the weighed neonatal piglet into the box and observe. As the animal slowly becomes anesthetized, observe the respiratory rate, and, as it starts to fall, transfer the animal to the open incubator for endotracheal intubation.

3. Place the piglet in the supine position and intubate using a laryngoscope and a neonatal cuffed endotracheal tube. Inflate the cuff to prevent an air leak around the tube. Secure the tube to the snout with tape. Auscultate the chest to confirm satisfactory and symmetrical air entry.

4. Connect the endotracheal tube to a mechanical ventilator and ventilate with 0.8 % halothane in a mixture of 30–40 % O_2 and 60–70 % N_2O. Adjust the inspired halothane concentration to keep its end-tidal concentration at 1 %, as monitored by a sidestream gas analyzer. Adjust the ventilation to keep the end tidal CO_2 at 4.5–5.5 kPa.

5. Measure transcutaneous SaO_2 by placing a pulse oximeter probe on a hind leg hoof; maintain SaO_2 at 95–98 % by adjusting the fraction of inspired O_2 as necessary (*see* **Note 1**).

3.1.2 Vascular Access and Fluid Administration

1. Under aseptic conditions, cannulate an ear vein using a neonatal venous cannula. Run an intravenous fluid solution of 3.75 % dextrose/0.45 % saline at 10 mL/kg/h. Maintain blood glucose between 3 and 10 m mol/L throughout the procedure.

2. Position the piglet on its side. Under sterile conditions, identify the umbilical vein and both umbilical arteries. Insert one umbilical artery and one umbilical vein catheter, confirming patency on withdrawal of arterial and venous blood from each, respectively. Take an arterial specimen to determine arterial blood gases and confirm adequacy of ventilation parameters. Connect the arterial catheter to a primed fluid-filled pressure transducer zeroed to the level of the right atrium; connect this to a blood pressure monitor. Confirm a satisfactory arterial wave form; the catheter may need repositioning if this is not obtained. Secure both catheters to the skin with a single 3/0 silk suture.

3. Transfer the intravenous fluid line to the umbilical venous catheter. Maintain patency of the ear cannula by running heparinized saline solution at 1 mL/h through a syringe driver.

4. Administer antibiotic prophylaxis (gentamicin 2.5 mg/kg and cephalothin 20 mg/kg) intravenously twice daily until all vascular lines are removed.

3.1.3 Temperature and Cardiovascular Monitoring

1. Record deep rectal temperature at 6 cm. Maintain between 38 and 39 °C, the normal body temperature for piglets.

2. Monitor heart rhythm with a three-lead electrocardiogram.

3. Monitor mean arterial blood pressure (MABP) continuously through the umbilical artery catheter. Treat any episodes of hypotension, defined as MABP below 40 mmHg, with boluses of 0.9 % saline, at 10 mL/kg. Administration of inotropic support has never been required in this model.

3.1.4 Monitoring of Electroencephalographic (EEG) Activity and Management of Seizures

1. Record two-channel amplitude-integrated EEG (aEEG) continuously. The needle electrodes of the BrainZ aEEG monitor are inserted aseptically in the parietal region of the head.

2. In this model, electroconvulsive activity was defined as spike or sharp wave activity with amplitude more than double the background activity, occurring at regular frequency and lasting longer than 20 s. Clinical seizures were defined as rhythmic pathological movements accompanied by electroconvulsive activity. If these EEG abnormalities last more than 10 min, with or without clinical evidence of a seizure, treat initially with Phenobarbital at 20 mg/kg. If necessary, follow with an additional dose. If seizure activity persists, give a 100 µg/kg bolus of midazolam intravenously.

3.1.5 Formation of Surgical Anterior Fontanelle and Insertion of Intracranial Pressure Monitor

1. Position the piglet prone with slight extension of the neck. Place a soft roll between the trunk and each limb to support the animal in a stable position without compromise of the umbilical catheters.

2. Shave the cranium with clippers and clean with alcoholic chlorhexidine solution. Drape the cranium with a sterile drape and secure a sterile area.

3. Under strict aseptic conditions, make a curved midline incision at the coronal suture. Strip the periosteum from the cranium at this point. Using a hand held burr, make 1.5 cm diameter burr hole exposing the dura. Enlarge the burr hole, if necessary, using bone or Kerrison rongeurs. This represents an artificial surgical fontanelle, and, apart from allowing some expansion of the cranium, also serves as an acoustic window for serial ultrasound scanning (Fig. 1a).

4. At the anterior end of the incision, over the frontal bone, perform twist-drill craniostomy. Perforate the dura gently with a small hypodermic needle and insert a Codman intraparenchymal intracranial pressure (ICP) monitor into the brain. Follow the manufacturer's instructions for zeroing and connecting to the express box and monitor. Secure the lead to the skin with a single 3/0 silk suture.

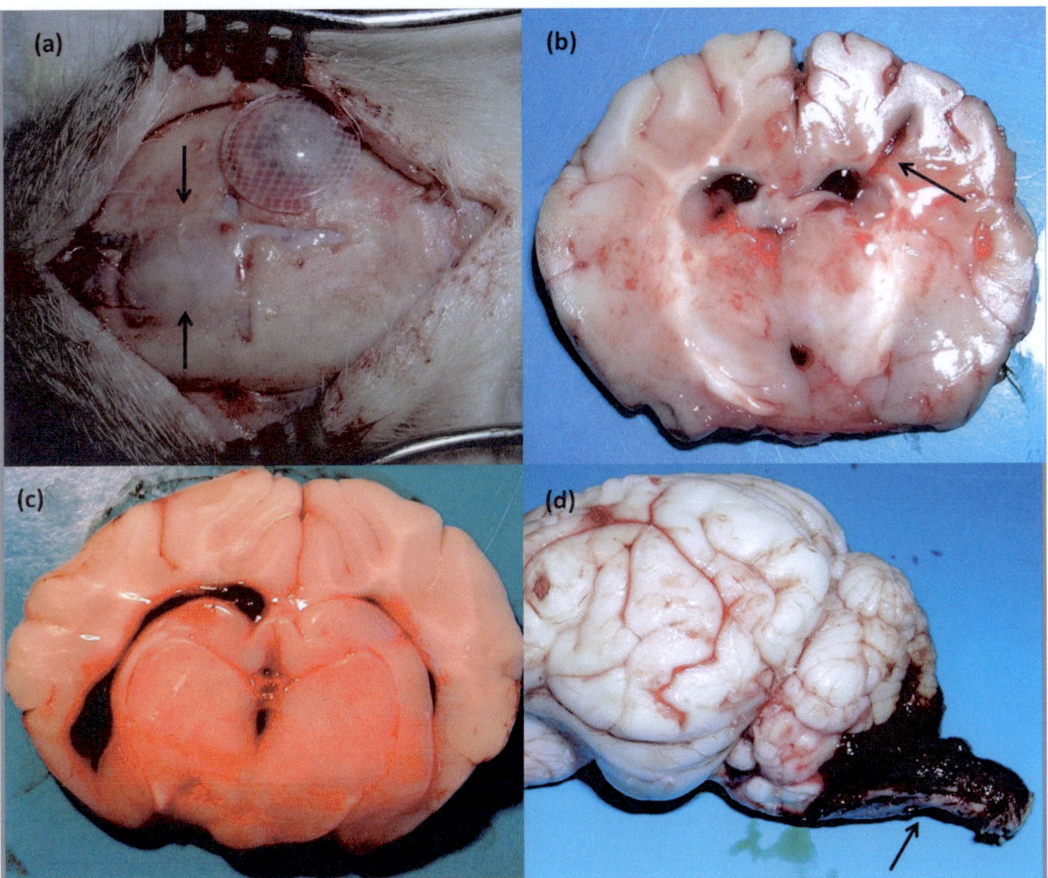

Fig. 1 (**a**) Reopening of the original incision at postmortem evaluation, demonstrating the surgical fontanelle anterior to the coronal suture, between *arrows*. A ventricular access device is implanted posterolateral to it; (**b, c**) Coronal sections of fresh brain, at the level of the foramen of Monro (**b**) and at the atrium of the lateral ventricle (**c**), obtained 24 h after injection of intraventricular blood, showing blood within the ventricular system; *arrow* in (**b**) Shows the needle track through which the blood was injected; (**d**) Fresh brain, also obtained 24 h after injection, demonstrating accumulation of injected blood within the basal cisterns around the brainstem (*arrow*)

3.2 Injection
of Intraventricular
Blood

1. Obtain approximately 15 mL of homologous blood from the umbilical artery just before injection. Prepare for injection by mixing with buffered sodium citrate and placing in a centrifuge for 10 min at 6,000 rpm. Remove four-fifths of the plasma and resuspend the red cells in the residual plasma and buffy coat. This process elevates the hematocrit of the blood to be injected to a level similar to that in preterm human neonates, where the hemoglobin level is 16–18 g/dL.

2. Drill a further twist drill craniostomy 8 mm lateral to the midline and 5 mm posterior to the bregma. Insert a 25G cannula into the brain parenchyma. A gentle pop is felt when the ependymal surface of the ventricle is breached. At this stage, clear CSF can usually be seen tracking up the cannula. Secure the cannula with bone wax to the surrounding bone (*see* **Note 2**).

3. Draw the blood for intraventricular injection into a 5 mL syringe connected to soft tubing. Prime the tubing and connect to the cannula. Insert the syringe into a syringe driver and start the intraventricular blood injection at a rate of 2.0 mL/h. Inject a total of 4 mL (Fig. 1b–d).

4. Monitor ICP throughout the injection period. If ICP rises over 15–20 mmHg, stop the injection until ICP falls below 10 mmHg and then re-start. This is a rare problem, and ICP is usually maintained below 20 mmHg throughout injection in most cases (Fig. 2).

5. Use intermittent ultrasound scanning through the fontanelle to confirm that blood is entering and filling the ventricular system (Fig. 3a).

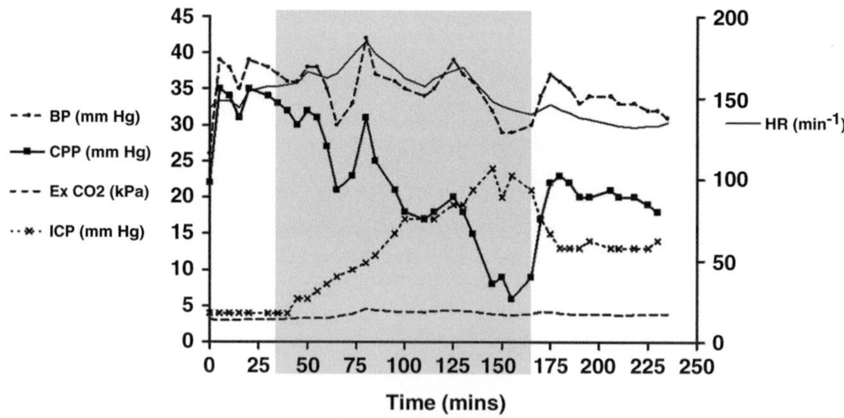

Fig. 2 Graph showing variation of ICP, expired CO_2 (Ex CO_2), cerebral perfusion pressure (CPP), BP and heart rate (HR) before during and after the 2-h intraventricular blood injection, delineated by the *shaded box*. Note how ICP rises progressively during injection; this was accompanied by a brief transient fall in CPP

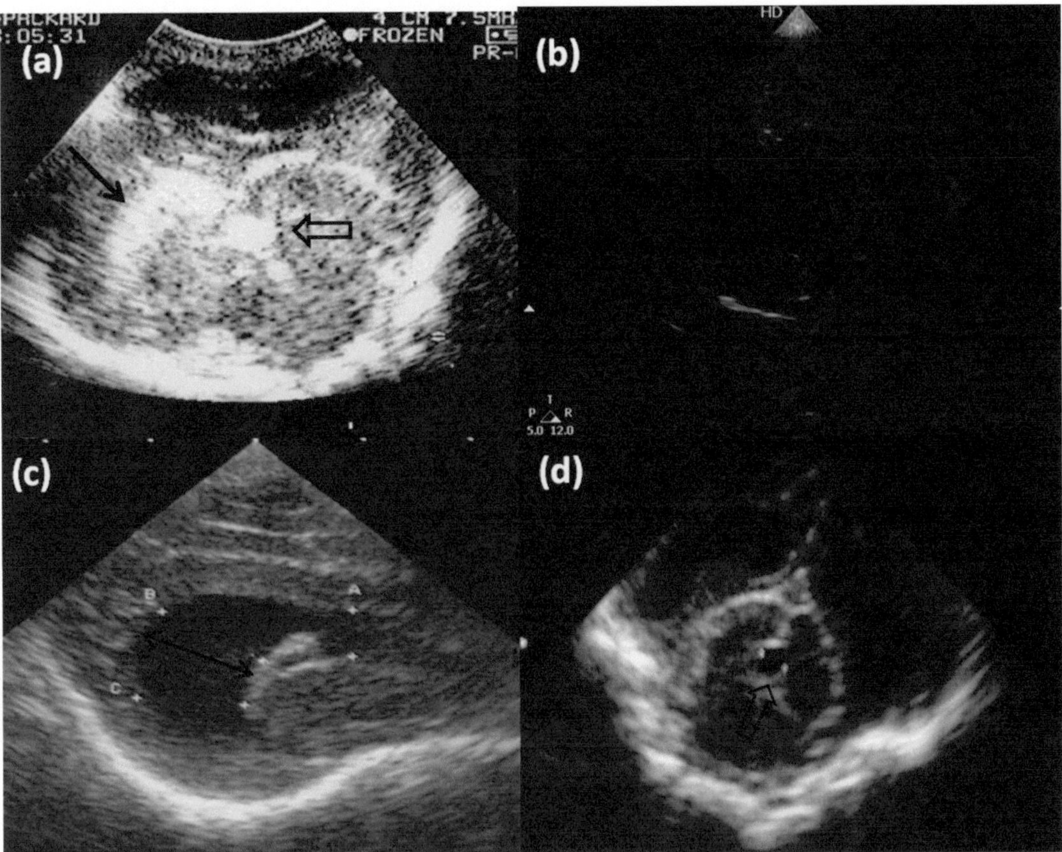

Fig. 3 (**a**) Coronal ultrasonographic view obtained through the surgical fontanelle towards the end of intraventricular blood injection, showing hyperechoic blood within the lateral and third ventricles (*solid* and *open arrows* respectively); (**b**) Coronal view showing dilated frontal horns—the *black arrow* measures the maximal frontal diameter; (**c**) Coronal view showing dilated occipital horns—the *black arrow* measures the thalamo-occipital distance; (**d**) Coronal view, angled posteriorly, identifies dilated occipital horns of the lateral ventricle and a dilated fourth ventricle (*open arrow*)

6. On completion of injection, withdraw the cannula slowly, over 20 min, to minimize back-tracking of blood through the parenchymal tract.

7. Close the cranial incision with a single layer of 4/0 nylon interrupted sutures. Remove the ICP monitor.

3.3 Recovery and Survival

1. On completion of intraventricular injection, administer intramuscular analgesia (buprenorphine, 20 µg/kg).

2. Discontinue halothane and N_2O anesthesia after skin closure. Run the ventilator on the mixture of O_2 and N_2. Needed to acquire SaO_2 in the range 95–98 % (usually 25 % O_2).

3. Slowly wean off ventilatory support and wait for the piglet to wake up. Remove the endotracheal tube when awake and

Table 1
Scheme for neurological examination of the neonatal piglet after injection of intraventricular blood

	12 h	24 h	Day 2	Day 3	Day 4
Walking					
Foreleg function					
Hind leg function					
Left/right tone					
Fore/hind tone					
Activity level					
Sucking					
Orientation					
Vocalization					
Respiration					
Consciousness					
Pupils					
Pain response					
Abnormal movements					

spontaneous respiratory movements occur at an appropriate rate. Remove the aEEG electrodes (*see* **Note 3**).

4. After extubation, continue intravenous fluids for about 6 h. Most animals will be able to bottle feed within 1 h of extubation. Continue to feed at 3 h intervals.

5. A neurological examination checklist, as in Table 1, facilitates serial evaluation of the animals during recovery. Although we had initially developed this examination routine for piglets undergoing hypoxic ischemic studies, we have also found the same criteria useful for IVH studies [16].

6. All lines, including the umbilical artery catheters, are removed 12 h after recovery from anesthesia. This allows establishment of a regular feeding pattern, confirmation of a satisfactory fluid intake and administration of a final dose of intravenous prophylactic antibiotic.

7. 12–24 h after recovery, the animals are taken to their pen where they can feed ad lib from bottles attached to the wall.

8. Administer intramuscular iron supplements during the first week.

9. Wean the animals at 21 days and phase over to dry chow.

10. Weigh the piglets and examine the wound daily. Remove the nylon sutures at 6 days. Serial ultrasonography through the surgical fontanelle allows monitoring of ventricular enlargement. This can be done by wrapping the piglet in a blanket and does not require general anesthesia or even sedation. Perform ultrasound in both the coronal and sagittal planes, measuring the maximal frontal horn diameter, thalamo-occipital distance and the width of the third ventricle, as described for human neonates (Fig. 3b–d) [17].

3.4 Insertion of Ventricular Access Device

1. Moderate ventricular dilatation facilitates safe insertion of a VAD. This procedure is usually performed approximately 1 week after intraventricular injection of blood.

2. Anesthetize the piglet as described above, using a closed box initially and then proceeding to endotracheal intubation under halothane and N_2O. A single peripheral cannula, usually inserted into one of the ear veins, is sufficient for administration of maintenance intravenous fluids and prophylactic antibiotics, as described above. The piglet is positioned prone, with the neck extended and the trunk supported on four soft rolls, as for intraventricular blood injection. A pulse oximeter and electrocardiogram are sufficient for anesthetic monitoring.

3. Use ultrasonography to identify the larger lateral ventricle, and the optimal position for insertion of the VAD. Select a location in the frontal lobe s with a thin overlying cortex, as far from eloquent cortex and subcortical tracts as possible. The tip of the catheter should lie within free CSF space, ideally close to the foramen of Monro or the atrium of the lateral ventricle (Fig. 4a) (*see* **Note 4**).

4. Shave the skin on the cranium with clippers. Prepare the skin with alcoholic chlorhexidine and drape a sterile field.

5. Make a curvilinear incision in the frontal region, over the selected optimal point for VAD insertion. Be careful to leave sufficient vascular supply to the skin between the previous and new incision to prevent flap necrosis. The flap needs to be large enough to allow the dome of a neonatal type 10 mm VAD to sit under the skin without tension, with the suture line clearly peripheral to the margin of the reservoir.

6. Pre-assemble a standard 10 mm neonatal VAD, also known as an Ommaya reservoir, by securing a 2 cm ventricular catheter to the bottom outlet, and securing it with a single 3/0 silk tie. Fill the reservoir with saline solution (Fig. 4b).

7. Using a handheld burr, make a burr hole to expose the underlying dura. Then make a cruciate incision in the dura to expose the cerebral cortex. Make a small cortical incision using a number 15 scalpel or, if available, bipolar diathermy. Cannulate the body of the lateral ventricle using either a neonatal venous cannula, or, if available, a Dandy brain cannula.

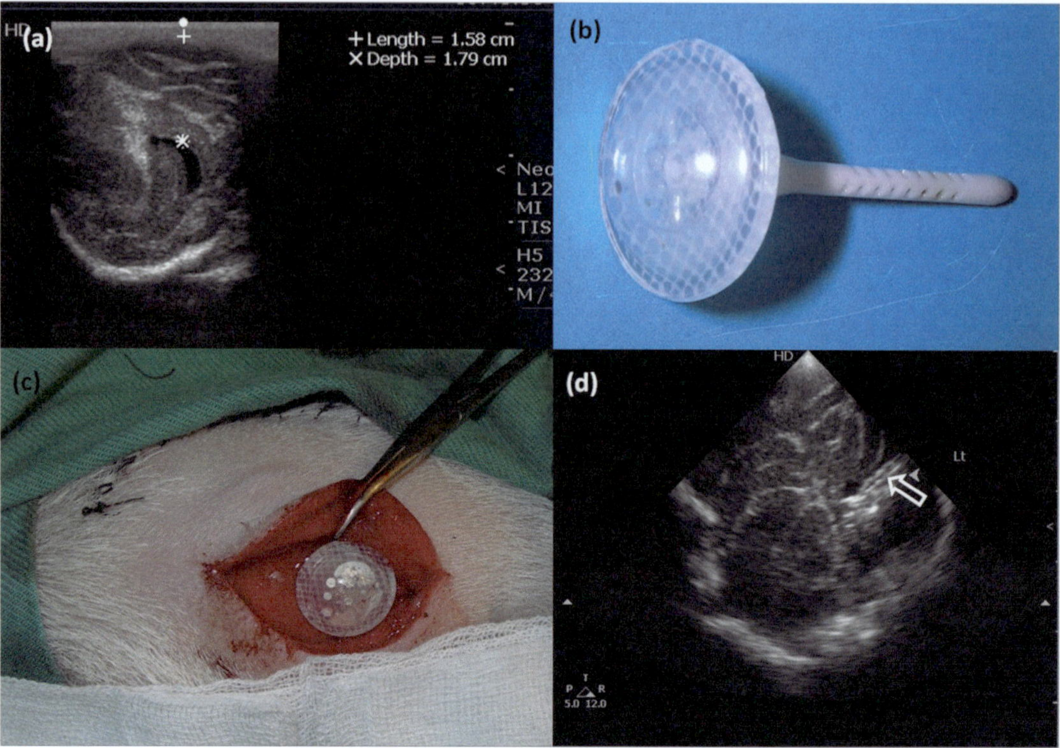

Fig. 4 (**a**) Sagittal ultrasound image used to identify the optimal position for insertion of a VAD. The measured length is used to guide the length of catheter required; (**b**) An assembled ommaya reservoir, 10 mm in diameter, with a flat base, prior to insertion; (**c**) Insertion of the device, under general anesthetic, through a new curved incision; (**d**) Coronal ultrasound image, demonstrating the catheter of the VAD within the body of the lateral ventricle (*open arrow*)

8. Remove the cannula and, at the same time, advance the pre-assembled VAD along the same tract. The bottom of the reservoir, flat in the neonatal version of the device, should sit easily on the surrounding cranial bone. Gentle aspiration of 1 mL of CSF with a 25G needle attached to a 1 mL syringe confirms satisfactory location of the catheter tip within the ventricle (Fig. 4c, d).

9. Administer 5 mg (0.5 mL) of intrathecal vancomycin and 1 mg (0.2 mL) of intrathecal gentamicin into the reservoir.

10. Pull the skin flap back over the reservoir. It is sometimes possible, depending on the size of the animal, to close the skin in two layers, using 4/0 vicryl for the galeal/subcutaneous layer and separate 4/0 nylon interrupted sutures for the skin. If not possible, a single skin 4/0 nylon suture layer is acceptable.

11. Administer intramuscular analgesia and discontinue anesthesia as described above. Proceed to extubate and recover the animal.

12. Carefully check the condition of the skin and the healing incision at least once daily. Remove sutures after 6 days if the wound is healing well (*see* **Note 5**).

3.5 Cannulation of the Ventricular Access Device

1. This procedure is preferably performed under sedation. Intramuscular injection of a combination of ketamine (5–10 mg/kg) and medetomidine (100 µg/kg) is usually sufficient to provide approximately 20 min of satisfactory sedation.

2. Position the piglet prone. Clean the skin over the VAD with alcoholic chlorhexidine.

3. Use a 25G (yellow) butterfly needle, primed with sterile normal saline, to penetrate the skin over the reservoir and subsequently the dome of the reservoir. Gentle aspiration with a 2 mL syringe confirms patency of the reservoir.

4. To measure intraventricular pressure, connect the tubing of the butterfly needle to a fluid-filled pressure transducer, zeroed to the middle of the piglet's head. An appropriate ICP waveform on the pressure monitor confirms patency of the fluid-filled system and reflects intra-ventricular pressure.

5. If a specimen of ventricular CSF is required, do not prime the tubing of the butterfly needle with saline. Aspirate gently, drawing up to 1 mL of CSF.

6. Before removing the needle, administer intrathecal vancomycin and gentamicin, as described above.

3.6 Perfusion-Fixation of the Brain

1. After final weighing anesthetize the animal as previously described, proceeding to endotracheal intubation and mechanical ventilation. Monitor the animal's cardiovascular status with pulse oximetry and electrocardiography. Secure a peripheral venous line and run maintenance fluid at 20 mL/h.

2. Position the piglet supine on the incubator or perfusion station.

3. Ensure deep halothane anesthesia; at this stage intravenous propofol and fentanyl may also be used to ensure adequate depth of anesthesia.

4. Make a midline longitudinal cervical incision and place a self-retaining retractor to expose the trachea and sternocleidomastoid muscles bilaterally. Identify the common carotid artery on either side, and dissect it free from the internal jugular vein and vagus nerve within the carotid sheath (Fig. 5a).

5. Pass a vascular sling around the carotid artery to isolate it. Insert a large bore cannula into the carotid artery and confirm brisk arterial backflow. Secure the cannula to the artery distally and proximally (Fig. 5b).

6. Once a cannula is in place in both carotid arteries, divide the internal jugular veins on each side. Connect the cannulae to 20 mL syringes filled with normal saline and flush slowly, manually, repeatedly until the venous effluent becomes clear.

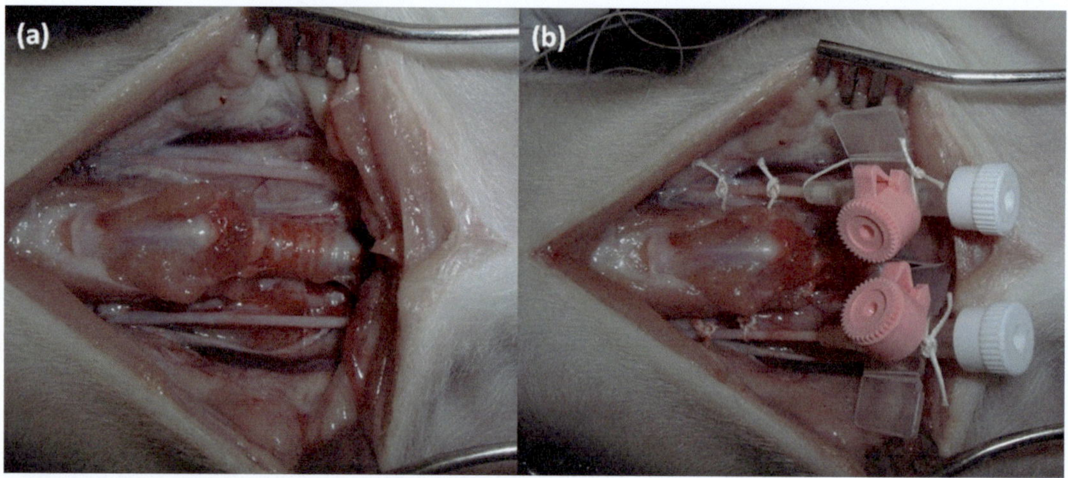

Fig. 5 (a) Midline cervical exposure of the carotid arteries prior to perfusion-fixation; (b) An intravascular cannula has been inserted in each carotid artery in preparation for injection of saline and formaldehyde

7. Connect the cannulae to 20 mL syringes containing 4 % neutral phosphate buffered formaldehyde and flush several times until the resistance to flow increases significantly. This ensures satisfactory fixation of the brain.

8. Once complete, the cranium is opened and the brain can be removed for further analysis and storage (Figs. 6 and 7).

4   Notes

1. Ensure that the neonatal piglet is well hydrated prior to anesthesia. Even mild dehydration causes significant hypotension on exposure to anesthetic agents. An intravenous fluid bolus of 10 mL/kg is often required immediately after endotracheal intubation to maintain a satisfactory mean arterial pressure. Use a laryngoscope with a long blade to intubate the trachea, maintaining hyper-extension of the neck. Keep the animal warm at all times.

2. The ventricular system of the normal piglet is slit-like. Placing an ultrasound probe over the surgical fontanelle facilitates cannulation of the ventricle for blood injection.

3. As it wakes up, the piglet usually starts to move the snout and neck, and subsequently the limbs. A brief period of SIMV may be helpful to confirm that the piglet is breathing spontaneously. A period of CPAP, with 30–40 % O_2, usually for about 30 min, is often required after extubation. Occasionally, the piglet may develop a focal or generalized seizure on waking up. If this is suspected, administer anticonvulsants, delay extubation

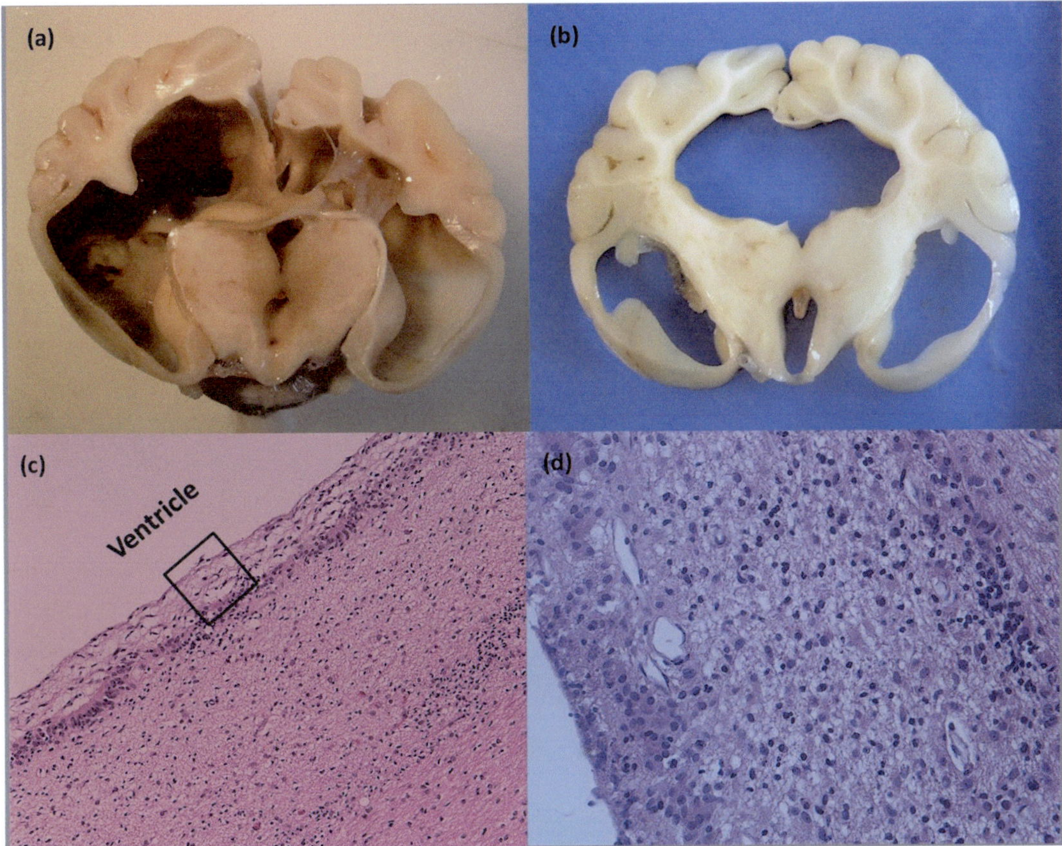

Fig. 6 (**a, b**) Sectioned brain, at 8 weeks post-injection, demonstrating enlarged occipital and temporal horns, thin cortical mantle and dilated cerebral aqueduct. Note extensive hemosiderin deposits on the ependymal surface of the ventricle in (**a**); (**c**) Hematoxylin and eosin stained section of the periventricular region, demonstrating disruption of the ependymal surface (magnification ×400); (**d**) Enlargement of the *boxed section* in (**c**), showing loss of neurones and vacuolation, in addition to the loss of ependymal cells at the ventricular surface (magnification ×630)

and continue anesthesia for at least an additional hour until all seizure activity is controlled.

4. Even slightly enlarged piglet ventricles contain only a small volume of CSF. When inserting the VAD [18], anything more than a very gentle aspiration through the reservoir will cause the ventricular ependyma to be sucked against the holes of the catheter, with consequent likely blockage.

5. Inspect the surgical wounds daily. Any sign of wound breakdown or infection requires urgent exploration under anesthesia. Removal of a VAD, either because it has become blocked or because of breakdown of the overlying skin, is associated with a significant risk of intraventricular hemorrhage. For this reason it is advised to leave the original VAD in situ and insert a second one on the contralateral side.

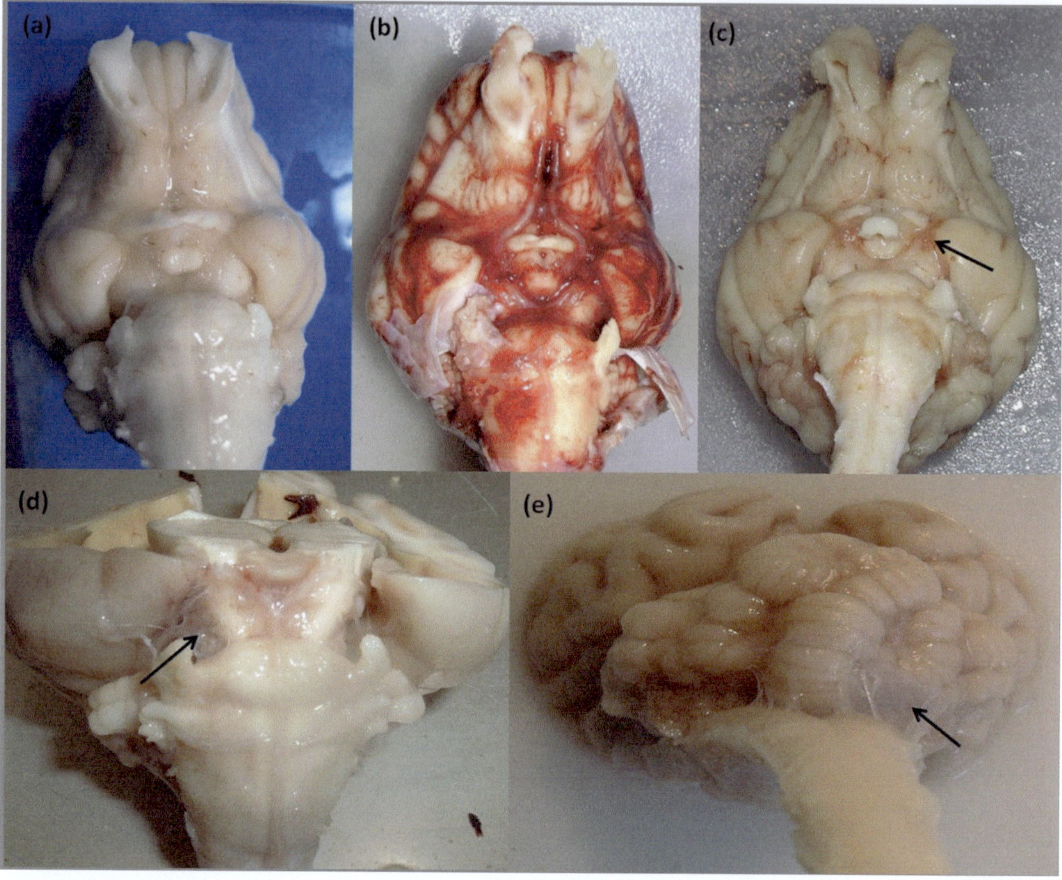

Fig. 7 (a, b, c) Ventral views of piglet brains in control animal (**a**), and in animals sacrificed at 2 and 8 weeks (**b** and **c**); note the progression of arachnoidal adhesions in the basal cisterns (*arrow* in **c**), more clearly visible in (**d**) and (**e**). This is the cause of the disturbance of CSF flow and consequent ventricular dilatation

References

1. Pople IK, Bayston R, Hayward RD (1992) Infection of cerebrospinal fluid shunts in infants: a study of etiological factors. J Neurosurg 77(1): 29–36

2. Volpe JJ (1989) Intraventricular hemorrhage and brain injury in the premature infant. Neuropathology and pathogenesis. Clin Perinatol 16(2):361–386

3. Larroche JC (1972) Post-haemorrhagic hydrocephalus in infancy. Anatomical study. Biol Neonate 20(3):287–299

4. Cherian S et al (2004) The pathogenesis of neonatal post-hemorrhagic hydrocephalus. Brain Pathol 14(3):305–311

5. Whitelaw A, Kennedy CR, Brion LP (2001) Diuretic therapy for newborn infants with posthemorrhagic ventricular dilatation. Cochrane Database Syst Rev 2, CD002270

6. Whitelaw A (2000) Repeated lumbar or ventricular punctures for preventing disability or shunt dependence in newborn infants with intraventricular hemorrhage. Cochrane Database Syst Rev 2, CD000216

7. Whitelaw A et al (2007) Randomized clinical trial of prevention of hydrocephalus after intraventricular hemorrhage in preterm infants: brain-washing versus tapping fluid. Pediatrics 119(5):e1071–e1078

8. Balasubramaniam J, Del Bigio MR (2006) Animal models of germinal matrix hemorrhage. J Child Neurol 21(5):365–371

9. Del Bigio MR, Zhang YW (1998) Cell death, axonal damage, and cell birth in the immature rat brain following induction of hydrocephalus. Exp Neurol 154(1): 157–169

10. Khan OH, Enno TL, Del Bigio MR (2006) Brain damage in neonatal rats following kaolin induction of hydrocephalus. Exp Neurol 200(2):311–320

11. Xue M et al (2003) Periventricular/intraventricular hemorrhage in neonatal mouse cerebrum. J Neuropathol Exp Neurol 62(11):1154–1165

12. Goddard J et al (1980) Intraventricular hemorrhage–an animal model. Biol Neonate 37(1–2):39–52

13. Mayfrank L et al (2000) Morphological changes following experimental intraventricular haemorrhage and intraventricular fibrinolytic treatment with recombinant tissue plasminogen activator. Acta Neuropathol 100(5):561–567

14. Wagner KR et al (2003) Heme and iron metabolism: role in cerebral hemorrhage. J Cereb Blood Flow Metab 23(6):629–652

15. Aquilina K et al (2007) A neonatal piglet model of intraventricular hemorrhage and posthemorrhagic ventricular dilation. J Neurosurg 107(2 Suppl):126–136

16. Thoresen M et al (1996) A piglet survival model of posthypoxic encephalopathy. Pediatr Res 40(5):738–748

17. Davies MW et al (2000) Reference ranges for the linear dimensions of the intracranial ventricles in preterm neonates. Arch Dis Child Fetal Neonatal Ed 82(3):F218–F223

18. Aquilina A et al (2011) Preliminary evaluation of a novel intraparenchymal capacitive intracranial pressure monitor. J Neurosurg 115(3):561–569 (manuscript number JNS10 - 1920)

Chapter 11

Physiologic Aspects of the Piglet as a Model of Neonatal Hypoxia and Reoxygenation

Richdeep S. Gill, David L. Bigam, and Po-Yin Cheung

Abstract

Every year, there are approximately one million newborns die because of birth asphyxia. For those who survive, they have an increased risk for adverse neurodevelopmental outcomes due to the insult. Newborn piglets are a useful model for hypoxia-ischemic encephalopathy injury in human newborns. We here describe a swine model of neonatal hypoxia and reoxygenation, the instrumentation procedures and hypoxia-reoxygenation protocol. The model allows observation of physiological changes of systemic and regional circulations including the cerebral perfusion, as well as biochemical and histological studies of hypoxia-reoxygenation injury. Further, translational evaluation of therapeutic interventions for hypoxia-ischemic encephalopathy can be conducted.

Key words Neonate, Asphyxia, Model, HIE

1 Introduction

1.1 Neonatal Asphyxia

Asphyxia is defined as a condition of impaired gas exchange leading to progressive hypoxemia and hypercapnia with a significant metabolic acidosis. Of the 130 million newborn infants born each year globally, 4 million die within the first 4 weeks [1, 2]. Birth asphyxia is estimated to be responsible for 25 % of newborn deaths worldwide [2].

1.2 Neonatal Piglet Model

The neonatal piglet is similar in size to a human infant at birth, weighing approximately 1.5–2.0 kg and has similar development to a 36–38 week gestation human fetus [3]. The size of piglets also allows instrumentation for accurate monitoring of hemodynamic changes. Further, swine are good models that are commonly used for experiments because their uniform and predictable size for each breed and short cycle of reproduction (114 days).

Piglets have similar physiological characteristics to humans including the responses to hypoxia and reoxygenation (H-R) [4].

Jerome Y. Yager (ed.), *Animal Models of Neurodevelopmental Disorders*, Neuromethods, vol. 104, DOI 10.1007/978-1-4939-2709-8_11, © Springer Science+Business Media New York 2015

In the cardiovascular system, coronary blood flow in pigs is almost analogous to humans, with 90 % similarities in both anatomy and function [5]. This allows for swine to be used as models for coronary blood flow, infarct production, and hemodynamic studies. Neonates undergo the transition from fetal to neonatal circulation with dramatic changes in the pulmonary vascular bed. The pulmonary vascular reactivity in gas exchange response to global and regional hypoxia in newborn piglets is similar to the human neonates [6]. They both share similar morphological and functional characteristics in the pulmonary vasculature. Pulmonary vasoconstriction occurs in response to alveolar hypoxia. The hypoxia pulmonary vasoconstriction serves as a mechanism to match perfusion with ventilation, to optimize gas exchange.

Newborn piglets are a useful model for hypoxia-ischemic encephalopathy injury in human newborns. The newborn piglets have a neurodevelopment similar to 36–38 week human fetus [5]. Cerebral metabolic data and cerebrovascular systems in the piglet are comparable to their human counterparts [5]. Cerebral blood flow is primarily via the carotid arteries. Furthermore, the relative similar anatomic location of the carotid arteries to the trachea, allows for excellent and convenient exposure for placement of flow probes with minimal surgery.

The gastrointestinal tract of swine has anatomic differences from humans but analogous physiologic function [5]. There are vascular differences in the small intestine, with vascular arcades forming in the muscularis mucosa rather than in the mesentery. The swine colon is also spiral and located in the upper left quadrant. The splanchnic blood flow characteristics are analogous to the human thereby serving as an excellent model for neonatal necrotizing enterocolitis.

Induction of anesthesia in the newborn piglet often requires inhalational administration of agents such as isoflurane, which can be replaced with intravenous medications once the intravenous access is established. Inhaled anesthetics may have hemodynamic effects including cerebral vasodilatation and shifts in myocardial, renal and splanchnic flow [7]. Thus, inhaled anesthetics should be minimized and limited for the induction of anesthesia.

1.3 Hypoxia–Reoxygenation

Oxygen serves as the basic fuel for cellular function. The lack of oxygen or hypoxia subsequently leads to cells switching to anaerobic metabolism. Reoxygenation restores the oxygen supply to deprived cells. However, this may salvage one population of previously hypoxic cells, while injuring other viable hypoxic cells. The reoxygenation or reperfusion injury is believed to be a consequence of cytotoxic and highly reactive oxygen-derived free radicals following the establishment of oxygen or blood supply.

To create a clinically relevant asphyxiating insult, we utilize alveolar hypoxia. This allows the creation of global hypoxia-ischemia as that commonly occurs in asphyxiated neonates, which leads to multiorgan injury and a systemic inflammatory response phenomenon, in contrast to the regional hypoxic-ischemic insult in some rodent experiments. The duration and severity of hypoxia is an important aspect of the protocol. In our experimental model, systemic hypoxemia is steadily maintained for 2 h to simulate the clinical setting. It is approximated that in the clinical situation of non-hemorrhagic fetal asphyxia, the time from onset of hypoxia, diagnosis and emergent resuscitation is about 2 h [8]. The aim for the severity of the hypoxic insult is an arterial oxygen saturation of 30–40 % [8]. Normocapnic hypoxemia is adjusted frequently to create progressive myocardial depression leading to cardiogenic shock (cardiac output 40–50 % of baseline, mean arterial pressure 30–35 mmHg) with a significant metabolic acidosis (pH = 6.95–7.05). Features similar to clinical asphyxia are created, leading to multiorgan dysfunction and failure. Although the experimental design of alveolar normocapnic hypoxia simulates the features of clinical asphyxia, the addition of hypercapnia in the protocol will further increase the translational value of the model but may complicate the study of responses to H-R.

Previously reoxygenation was initially started with 100 % oxygen lasting 0.5–1 h. The length of time was to mimic the clinical resuscitative situation if the event occurs in a community hospital at a distance away from the neonatal intensive unit [8]. Following initial reoxygenation with 100 % oxygen, room air (FiO_2 0.21) was used for the remaining period of the experiment. With the 2010 guidelines from the American Heart Association and the International Liaison Committee on Resuscitation, which recommend the initial use of room air in neonatal resuscitation, reoxygenation with 21 % oxygen can be administered first. In our experimental study, periodic arterial blood gas measurements are utilized to monitor hypoxic stress and recovery.

2 Materials

2.1 Major Monitoring and Maintenance Equipment

- Sechrist Infant Ventilator IV-100B (Sechrist Industries, Anaheim, California, USA).
- Ohmeda 5100 oxygen monitor (Arleta, California, USA).
- Fluotec 3 anesthesia machine (Yorkshire, England).
- Argyle™ catheters, 3.5 or 5 French, single- or double-lumen (Covidien, Mansfield, MA).
- Pulse oximeter (Nellcor, Hayward, California, USA).
- HP 78834A multichannel blood pressure system (Hewlett Packard Co, Palo Alto, California, USA).
- T206 Transonic flow meter (Transonic Systems Inc., Ithaca, NY).

- Transonic flow probes (various sizes)
- Overhead thermal lamp (Ohio NCI, Madison, Wisconsin, USA).
- Airworks heater (China).
- ABL700 blood gas machine (Radiometer, Copenhagen, Denmark).
- OSM3 blood oximeter (Radiometer, Copenhagen, Denmark).
- Variable pressure volumetric pump (IVAC Co., San Diego, California, USA).
- Medfusion 2001 syringe pumps (Medex Inc., Dulutch, GA, USA).
- Data recording computer (Data Translation, Ontario, Canada).

2.2 Surgical Equipment

- Electrosurgical cautery unit (Birtcher Model 771).
- Baby mixture (Sklar, West Chester, PA, USA).
- Toothed Adson pickups (Prestige Surgical Instruments Inc., Sialkot, PK).
- Non-toothed Adson pickups (Keir Surgical, Vancouver, BC, Canada).
- Long non-toothed pickups (Sklar, West Chester, PA, USA).
- Fine needle driver (Classic Plus, USA).
- 11′ blade scalpel (Bard-Parker, New Jersey, USA).
- Hemostat ×4 (Down, England).
- Small self-retaining retractor (Sklar, West Chester, PA, USA).
- Large self-retaining retractor (WPI Inc., Sarasota, FL).
- Size 3.0 or 3.5 pediatric endotracheal tube (non-cuffed) (Mallinckrodt, Washington DC, USA).
- Sutures: (0-Silk, 3-0 Silk, 3-0 prolene, 5-0 prolene) (Ethicon, North Ryde, NSW).

2.3 Drugs and Fluids

- Isoflurane.
- Propofol.
- Midazolam.
- Acepromazine.
- Fentanyl.
- Pancuronium.
- Dextrose 5 % water (D5W)—maintenance fluid.
- Normal saline—maintenance fluid.
- Ringer's Lactate—bolus fluid.

3   Methods

3.1   Animals

1. Newborn piglets 1–3 days of age weighing 1.5–2.5 kg are obtained from a local farm on the day of experimentation.

2. The newborn piglet is kept warm using an overhead thermal lamp prior to anesthesia (**Note 4.1**).

3. Check oxygen and nitrogen tanks for adequate supply

4. Prime intravenous lines with Dextrose 5 % water (D5W) and run at 10 ml/kg/h.

5. Turn on heating pad, which is underneath the surgical instrumentation bed. Turn on overhead fluorescent and heating lights overtop to the surgical bed. Maintain piglet temperature at 38.5–39.5 °C throughout the experimental period.

6. Place the electrocautery plate on operating table with wet gauze spread overtops.

3.2   Anesthesia and Femoral Venous/Arterial Access

1. Induce anesthesia with isoflurane 5 % with oxygen using a facemask for the newborn piglet.

2. Decrease isoflurane to 2–3 % once piglet is anesthetized (**Note 4.2**).

3. Place piglet on the operating table in supine position and tie all four limbs exposing the femoral regions on both sides.

4. Place pulse oximetry on the right forward foot for continuous oxygen saturation monitoring (**Note 4.3**).

5. Make an incision in femoral region; dissect until both the femoral artery and vein are well exposed. Gain proximal and distal control using a 3-0 silk suture (Fig. 1).

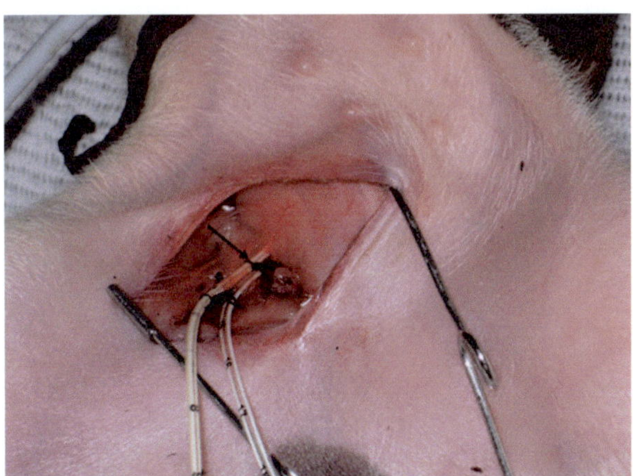

Fig. 1 Femoral region is exposed. Two 5-French Argyle™ catheters are in place. The single-lumen Argyle™ catheter is placed in the femoral artery (*black arrow*). The double-lumen Argyle™ catheter is placed in the femoral vein

6. Make a small nick in the femoral vein with an 11-scalpel blade and insert a 5-French Argyle™ double-lumen catheter and advance to the level of right atrium (15 cm). This catheter will be used for the administration of medications and fluids using the infusion pumps.

7. Similarly, make a small nick in the femoral artery and insert a single-lumen 5-French Argyle™ catheter and advance to the level of infra-renal, distal aorta (5 cm). Attach the catheter to a pressure transducer and monitor for continuous systemic arterial blood pressure measurements.

3.3 Tracheostomy and Mechanical Ventilation

1. Incise the skin overlying the trachea using electrocautery. Dissect and isolate the trachea. Gain proximal and distal control with 0-silk sutures.

2. Make a transverse incision across the trachea with an 11-blade scalpel and insert a size 3.0 or 3.5 endotracheal tube.

3. Secure in place with the 0-silk sutures and attach to mechanical ventilator.

4. Commence pressure-controlled assisted ventilation with pressures of 19/4 cm H_2O at a rate of 20 breaths/min.

5. Once mechanical ventilation is commenced, discontinue isoflurane and continue anesthesia with intravenous fentanyl 5–50 µg/kg/h, midazolam 0.2–0.5 µg/kg/h, and pancuronium 0.05–0.1 mg/kg/h, with additional boluses as needed (**Note 4.4**).

6. Inspired oxygen concentration (FiO_2) is measured (Ohmeda 5100 oxygen monitor) and maintained at 0.21–0.24 for percutaneous oxygen saturation between 90 and 100 %.

3.4 Surgical Procedures

1. With the neck region exposed, isolate the common carotid artery.

2. Encircle the common carotid artery with a 2-mm Transonic flow probe (2SS, Transonic Systems Inc., Ithaca, NY) (Fig. 2).

3. Place ultrasonic gel between the flow probe and artery to allow for optimal signal transduction.

4. Close the neck incision with 3-0 silk in a running fashion to minimize movement of the flow probe and leakage of ultrasonic gel.

5. Open retroperitoneum through a left flank incision (**Note 4.5**).

6. Isolate the superior mesenteric artery (SMA) and left renal artery (Fig. 3).

7. Place a 3-mm Transonic flow probe (3SB) around the SMA and a 2-mm flow probe (2SB) around the renal artery.

8. Make a left anterior thoracotomy incision at the fourth intercostal space (**Note 4.6**).

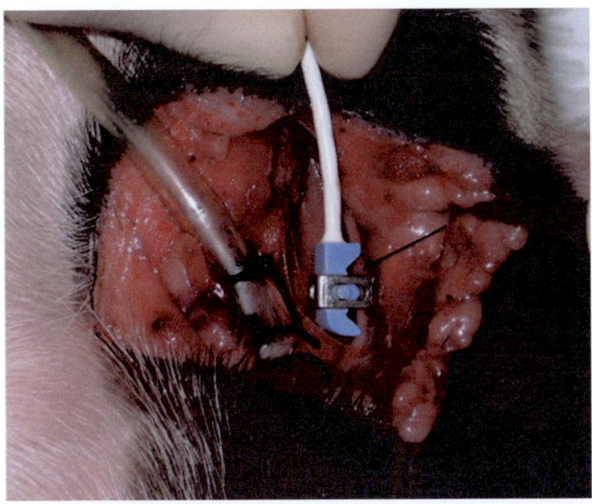

Fig. 2 Anterior neck region is exposed. An endotracheal tube is within the trachea via tracheostomy. The Transonic® flow probe is encircled around the common carotid artery (*black arrow*)

9. Open the pericardium and isolate the main pulmonary artery; place a 6-mm probe (6SB) Transonic flow probe at the main pulmonary artery (Fig. 4).

10. Place ultrasound gel between all flow probes and arteries and attach to a Transonic flow meter for continuous blood flow measurements.

11. Place a purse-string near the base of the pulmonary artery with 5-0 prolene suture.

12. Insert 14-G angiocatheter in middle of purse-string and tie in place (Fig. 4).

13. Attach angiocatheter to pressure transducer.

14. Close or cover all incisions

3.5 Stabilization

1. Allow piglets to recover, which usually requires 1 h, from surgical instrumentation until baseline hemodynamics are stabilized (change of ±10 % over 30 min).

2. Give a 10 ml/kg intravenous bolus of Ringer's lactate solution.

3. Perform arterial blood gas analysis and adjust ventilator rate accordingly (**Note 4.7**).

3.6 Hypoxia Protocol

1. Lowering the FiO_2 of inspired oxygen to 0.10 by increasing the concentration of inhaled nitrogen gas to induce hypoxemia.

2. Can adjust the FiO_2 0.10–0.15 to obtain a P_aO_2 of 20–40 mmHg for 2 h (**Note 4.8**).

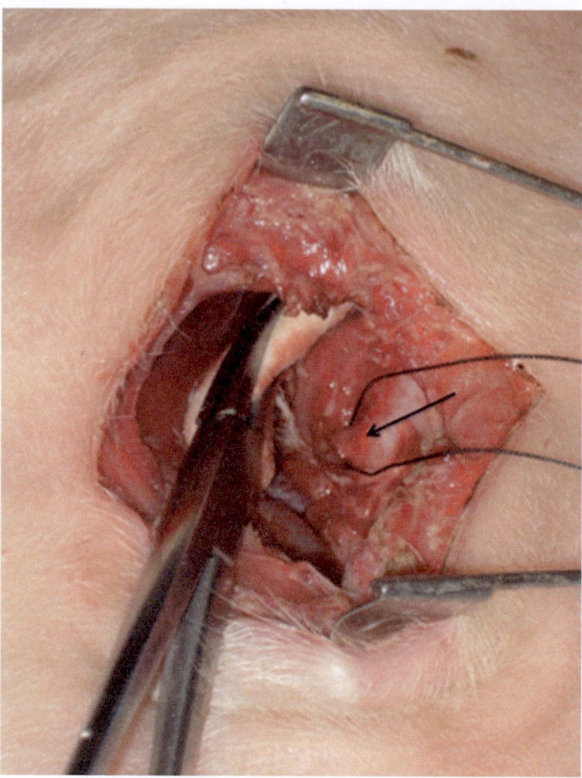

Fig. 3 The flank region is exposed. The kidney has been retracted anteriorly to expose the abdominal aorta. The superior mesenteric artery (*black arrow*) is encircled with a 0-silk suture at its take-off point from the abdominal aorta

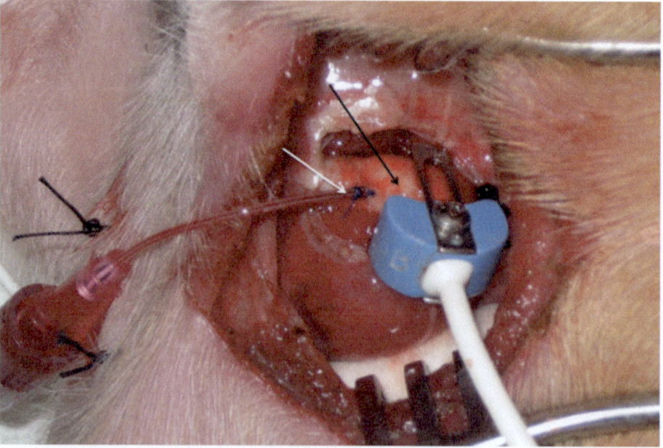

Fig. 4 The mediastinum is exposed via an anterior thoracotomy incision. The heart is visualized and pulmonary artery (*black arrow*) is isolated. A Transonic flow probe is encircled around the pulmonary artery. This allows continuous measurement of pulmonary artery flow, a surrogate for cardiac output. An angiocatheter (*white arrow*) is place in the pulmonary artery and secured in place with 5-0 prolene suture

3. Perform arterial blood analysis to assess P_aCO_2 and adjust ventilator rate accordingly.

4. With the induction of hypoxemia, the first hour is dedicated to steadily inducing a tachycardic (and cardiac output) response (**Note 4.9**).

5. Continue to monitor for changes in carotid artery, SMA and renal flow (**Note 4.10**).

6. During the second hour of hypoxia, the hypoxic stress is increased to steadily lower cardiac output to 30–40 % of baseline (**Note 4.11**), mean arterial pressure to 30–35 mmHg (**Note 4.12**) and pH 6.95–7.05 (**Note 4.13**).

7. Hypoxic stress may be prematurely terminated or extended by 15 min as appropriate (**Note 4.14**).

3.7 Reoxygenation Protocol

1. Increase FiO_2 to 21 % abruptly by discontinuing nitrogen gas (**Note 4.15**).

2. Monitor cardiac output and mean arterial pressure for rapid recovery (**Note 4.16**).

3. Continue reoxygenation with 21 % oxygen for the remaining period of experiment (**Note 4.17**). The FiO_2 can be titrated to 0.25 if needed.

---

4   Notes

4.1 Animals

Once the newborn piglet has arrived, allow 15–20 min for piglet to accommodate under the overhead thermal lamp. The piglet should be close to normal baseline temperature and physiologic state, prior to starting surgical instrumentation. However, during surgical instrumentation, care must be taken to prevent hyperthermia. Hyperthermia induces a tachycardic response.

4.2 Isoflurane Anesthesia

It has been noted that piglets required varying degrees of isoflurane to not respond to femoral incision with movements. Adjust the isoflurane by increasing it until the piglet does not move to pitching of the skin overlying the femoral region. It is also cautioned against isoflurane overdose.

4.3 Pulse Oximetry

We noted that oxygen saturation should be between 90 and 100 %. Important to make sure adequate oxygen is being supplied along with isoflurane.

4.4 Intravenous Sedation and Analgesia

We noted that giving intravenous sedation and analgesia with midazolam and fentanyl respectively maintains adequate anesthesia. The heart rate steadily decreases, usually between 150 and 200 beats/min. It is important to provide adequate anesthesia prior to giving pancuronium to limit piglet movements during instrumentation in thoracic and retroperitoneal surgeries. Boluses of fentanyl, acepromazine, and propofol will be required accordingly.

4.5 Flank Incision

Once the skin incision overlying the left flank is made, dissection is carried out until the kidney is visible. Mobilization of the left kidney can be completed with gentle blunt dissection using two peanuts (peanut in a snap/hemostat), starting inferiorly and progressive upwards. Care must be taken to minimize renal bruising during retraction and prevent lymph leak from large lymph nodes surrounding the aorta and renal artery. An uncontrolled lymph fluid leak can lead to significant hypovolemia.

4.6 Ductus Arteriosus

Prior to placing the Transonic flow probe around the pulmonary artery, the patent ductus arteriosus should be isolated and tied using a silk suture to make sure that it is closed. This will prevent shunting during H-R secondary to pulmonary hypertension thus making the assessment of cardiac output inaccurate.

4.7 Normocapnia

To maintain normocapnia ($PaCO_2$ between 35 and 45 mmHg) arterial blood gases are repeated between every 15 and 30 min during H-R. Adjusting the rate of mechanical breaths via the ventilator typically allows this. However, inspiratory and expiratory pressure may be adjusted to reach normocapnia and hypoxia.

4.8 Hypoxia Protocol

When decreasing the inspired oxygen FiO_2, monitor the oxygen saturation closely. An initial goal is to lower the oxygen saturation (pulse oximetry) to approximately 50 % by steadily lowering the FiO_2. Severity of normocapnic hypoxemia can be increased if there is no tachycardic response when decreasing the inspired oxygen FiO_2. Careful monitoring of the pulmonary arterial flow (a surrogate marker for cardiac output) will reflect an initial increase in flow to compensate for the reduced oxygen supply.

4.9 Cardiac Compensation

The induction of hypoxemia in the newborn piglet over the first hour of hypoxia should increase the cardiac output (pulmonary arterial flow) to 130–140 % of baseline. Between the first 0.5 and 1 h of hypoxia, cardiac output should reach its peak compensation.

4.10 Centralization of Blood Flow

As the newborn piglet is stressed with hypoxia, blood flow should become centralized. Three major organs are protected during hypoxemia by physiologic centralization, which include the heart, brain, and adrenals. Thus, the carotid arterial flow should increase with increasing cardiac output. The flow through the renal artery is typically decreased as this physiologic centralization occurs. The flow through the SMA usually does not change during the first hour of hypoxia but gradually decreases as the hypoxia progresses.

4.11 Cardiogenic Shock

During the second hour of hypoxia, the FiO_2 should be slowly reduced to allow for a steady decrease of cardiac output (pulmonary arterial flow). If cardiac output begins to fall quickly, the FiO_2 may be increased to maintain cardiac output. Changing the FiO_2 may not allow steady decrease of cardiac output. In this case, reducing the ventilator rate or decreasing the inspiratory pressure may allow more steady reduction of cardiac output.

4.12 Hypotension

As hypoxic stress leads to steady cardiac failure, the MAP is monitored closely. MAP should also steady decrease along with cardiac output. A general goal at the end of 2 h of hypoxia is a MAP between 30 and 35 mmHg. A MAP < 25 mmHg should be avoided in most circumstances, because we observe that it is correlated with extreme hypoxic stress, which the newborn piglet will likely not recover from, regardless of the proposed intervention.

4.13 Acidosis

Clinically relevant asphyxia includes progressive acidosis. A general rule is to lower the arterial pH by 0.1 every 30 min. For example, if the pH is 7.40 at the start of hypoxia, it will be 7.00 by the end of 2 h.

4.14 End of Hypoxia

The goal is to maintain cardiac output at 40–50 % of baseline in the piglet during the last 15–30 min of hypoxic stress. The piglet may "crash" quickly during this time, at which point hypoxia must be terminated. However, hypoxic stress can be extended by 15 min, if cardiac output, MAP and pH goals have not been reached.

4.15 Reoxygenation with 21 % O_2

Once nitrogen as is discontinued, and only 21 % inspired oxygen is being used, the ventilator will need readjustment of inspired and expiratory pressures. Also, the piglet may need a few manual breaths as part of the initial resuscitation.

4.16 Recovery of Cardiac Output and MAP

Cardiac output and MAP should recover to close to baseline values. If there is no recovery of cardiac output to at least 70–80 % of baseline, it is unlikely that the piglet will be able to complete the reoxygenation phase of the experiment.

4.17 Duration of Reoxygenation

The duration of reoxygenation can range from 4 h in an acute experiment, 72 h in a sub-acute experiment and multiple days (5–7 days) in a chronic experiment.

Acknowledgment

The project was supported by an operating grant from the Canadian Institutes of Health Research (P.Y.C.) and an establishment fund from the Alberta Heritage Foundation for Medical Research (P.Y.C.).

References

1. Lawn J, Shibuya K, Stein C (2005) No cry at birth: global estimates of intrapartum stillbirths and intrapartum-related neonatal deaths. Bull World Health Organ 83(6):409–417
2. Lawn JE, Cousens S, Zupan J, Lancet Neonatal Survival Steering Team (2005) 4 million neonatal deaths: when? Where? Why? Lancet 365(9462): 891–900
3. Chapados I, Cheung PY (2008) Not all models are created equal: animal models to study hypoxic-ischemic encephalopathy of the newborn. Commentary on gelfand SL et al.: a new model of oxidative stress in rat pups (neonatology 2008;94:293–299). Neonatology 94(4):300–303
4. Swindle MM, Smith AC, Hepburn BJ (1988) Swine as models in experimental surgery. J Invest Surg 1(1):65–79
5. Swindle MM, Smith AC (1998) Comparative anatomy and physiology of the pig. Scand J Lab Anim Sci 25(Supp 1):11–22
6. Hill DE (1985) Swine in perinatal research: an overview. In: Tumbleson M (ed) Swine in biomedical research. Plenum, New York, pp 1155–1159
7. Riebold TW, Thurmon JC (1985) Anesthesia in swine. In: Tumbleson M (ed) Swine in biomedical research. Plenum, New York, pp 243–253
8. Cheung PY, Stevens JP, Haase E, Stang L, Bigam DL, Etches W et al (2006) Platelet dysfunction in asphyxiated newborn piglets resuscitated with 21 % and 100 % oxygen. Pediatr Res 59(5): 636–640

Chapter 12

The Newborn Pig Global Hypoxic-Ischemic Model of Perinatal Brain and Organ Injury

Elavazhagan Chakkarapani and Marianne Thoresen

Abstract

Hypoxic-ischemic encephalopathy (HIE) remains an important cause of neonatal death and is a major contributor of disability. Animal models have contributed greatly to the understanding of the pathophysiology of HIE such as the mode of cell death, secondary energy failure, and the development of therapeutic interventions such as therapeutic hypothermia. Further research into the understanding of pathophysiology of HIE and neuroprotective interventions is required. The newborn pig global hypoxic-ischemic model of perinatal brain and organ injury is a valuable model simulating neonatal encephalopathy secondary to perinatal asphyxia in newborn term babies. This model involves induction of global hypoxia-ischemia to the whole body for 45 min while monitoring brain activity. In this model, reduction of inhaled oxygen to 6–7 % induces hypoxia and significant hypotension is induced by halothane anesthesia leading to ischemia. The pigs manifest clinical signs of encephalopathy including seizures and multiorgan dysfunction including hypotension and histopathological injury in the lungs, heart, kidneys, and the liver. This is a survival model; the pigs survive after receiving full intensive care in the acute period at the same level as that of asphyxiated newborns. The model is suitable to assess the effect of neuroprotective interventions on neurological recovery and on all organs in particular the brain.

Key words Pig, Hypoxia-ischemia, Encephalopathy, Seizures, Neuropathology

1 Introduction

Nearly 6/1,000 live births are affected by hypoxic-ischemic encephalopathy (HIE) in developed countries [1] and contribute to nearly 814,000 deaths per year worldwide [2]. Despite therapeutic hypothermia, nearly 45–50 % of cooled infants suffer from death or disability [3]. Hypoxic-ischemic encephalopathy is characterized by disturbance of brain function secondary to perinatal asphyxia including multiorgan failure [4]. Asphyxia denotes impaired blood gas exchange leading to progressive hypoxemia and hypercapnia with a significant metabolic acidosis [5]. The encephalopathy that develops secondary to the hypoxia and ischemia involved in the asphyxial process presents with seizures, secondary energy

Jerome Y. Yager (ed.), *Animal Models of Neurodevelopmental Disorders*, Neuromethods, vol. 104, DOI 10.1007/978-1-4939-2709-8_12, © Springer Science+Business Media New York 2015

failure in the neonatal period followed by abnormal neurodevelopment and disability.

Several animal models involving rats [6], pigs [7], and fetal lamb [8] have contributed to the understanding of the pathophysiology of the HIE and the investigation of therapeutic hypothermia. The newborn pig brain at term gestation (115 days) is more myelinated compared to newborn term infants [9]. However, newborn pigs (large white) within 48 h of age are of reasonable size (1–1.8 kg), which allows measurement of biochemical variables, monitoring cardiovascular and cerebrovascular hemodynamics and perform clinical examination. After induction of hypoxia by reducing the inhaled oxygen to 6–7 % and ischemia with hypotension caused by halothane inhalation, newborn pigs develop encephalopathy with seizures similar to the human infants [10]. Following this global hypoxia-ischemic insult, pigs develop multiorgan dysfunction characterized by hypotension, need for mechanical ventilation, necrotizing enterocolitis and biochemical abnormalities. The neuropathology and organ pathology is similar to the newborn infants following asphyxia. Brain injury is seen in cortical watershed regions, basal ganglia, hippocampus and cerebellum similar to infants after perinatal asphyxia [10]. Organ injury include infarction in the heart; necrosis, tubular and ductal dilatation, interstitial hemorrhage in kidneys; necrotic lesions in the intestine; subcapsular hematoma, necrosis, thrombus, and focal infarcts in the liver; hemorrhagic necrosis and hemorrhage in lungs and adrenal glands [11, 12].

2   Materials

2.1 Animal Preparation

2.1.1 Anesthetic Induction

1. Pen of suitable size (2 × 2 m) with straw bedding and heat source provided by infrared bulb placed 20–30 cm above the floor of the pen.

2. Thermometer to monitor temperature of the Pen (e.g., ATP Dual sensory memory indoor outdoor thermometer-MTH 968, digital meters, www.digital-meters.com, UK).

3. Weight scale (e.g., SECA, mod727, capacity 20 kg; graduation 0.1–0.2 oz; Birmingham, UK).

4. Perspex box (100 × 50 × 50 cm) with lid and inlet and outlet for connecting gas flow.

5. Veterinary anesthetic machine (e.g., Ohmeda).

6. Rectal thermometry (e.g., Datatherm® II continuous temperature monitor, Woodley Equipment Company LTD, UK).

2.1.2 Neurological Examination

1. Non-slippery mat for neurological examination.

2.1.3 Intubation, Ventilation, and Cardiorespiratory Monitoring

1. Open incubator with accurate temperature control (e.g., Giraffe, Datex-Ohmeda, GE Healthcare, Finland).

2. Bag valve mask with custom made mask to fit the snout.

3. Neonatal mechanical ventilator (e.g., SLE, Croydon, UK).

4. Laryngoscope with long blade (Miller blade) size 153 mm (e.g., Mercury Medical, Florida, USA).

5. Cuffed endotracheal tube (ETT) (size 3.0 or 3.5 mm) (e.g., Haylard Health Ltd, Reigate, Surrey, UK).

6. Main stream end-tidal CO_2 monitor (e.g., Tidal Wave™, Respironics, USA) or side stream end-tidal CO_2 monitor (e.g., Criticool, MTRE, UK).

7. Oxygen and anesthetic concentration analyzer (e.g., Criticool, MTRE, UK).

8. Transcutaneous oxygen saturation probe (e.g., Tidal Wave™, Respironics, USA or Criticool, MTRE, UK).

9. Tape to secure the Endo tracheal tube (e.g., Elastoplast).

10. 2.5 ml syringe to inflate the cuff.

11. Three cylindrical support rolls (10 cm long and 4 cm diameter).

12. Equipment to maintain temperature:
    - Radiant warmer and the temperature probe in the incubator (or).
    - Servo-controlled cooling equipment with thermometry (e.g., Criticool, MTRE, UK).

2.1.4 Lines, Fluids, and Biochemistry

1. 24G venous cannula.

2. Sterile 0.9 % saline.

3. Sterile solution of 5 % dextrose in 0.45 % saline.

4. Umbilical arterial and venous catheters size 4.0 Fr (Vygon (UK) Ltd, Gloucestershire, GL7 1PT).

5. Autoclaved umbilical catheter instrument set (two clamps, surgical drapes, two curved artery forceps, one needle holder, one toothed forceps, one non-toothed forceps, one blunt dilator, one sharp dilator, one scalpel blade holder, and one silver pot).

6. Blood gas analyzer (e.g., istat, Abbott point of care, USA).

7. Blood glucose analyzer (e.g., Precision Plus, Abbott Laboratories, Bedford, Mass, USA).

8. Blood lactate analyzer (e.g., Lactate Pro, HaB International Ltd, Warwickshire, UK).

9. Invasive blood pressure monitor (e.g., Hewlett Packard invasive blood pressure unit).

10. Single use blood pressure transducers (Becton & Dickinson, Oxford science park, Oxford).

2.1.5 Head Preparation

1. Razor.

2. Autoclaved cranial instrument set for preparation of fontanelle (two clamps, surgical drapes, one curved artery forceps, one straight artery forceps, one scalp tissue retractor, one rongeur forceps, one burr hole drill, two small twist drills, one toothed forceps, one silver pot, one scalpel blade holder).

3. Amplitude integrated electroencephalography to run at 1 mm/s (Lectromed UK Ltd, Hertfordshire, UK) for hypoxic-ischemic insult and at 6 cm/h for recovery (e.g., Olympic 6000 or BrainZ BRM3 brain monitor, Natus Medical Inc., San Carlos CA, USA).

4. Intra-parenchymal ICP monitor (e.g., Codman MicroSensor ICP transducer, Codman Johnson and Johnson, Raynham, MA, USA).

2.1.6 Recovery and Survival

1. Multimilk for pigs (Homestead Farm supplies, Oxfordshire, UK).

2. Buprenorphine for pain control.

2.1.7 Medications

1. Anesthetics (nitrous oxide, isoflurane, halothane).

2. Inotropes (dopamine, adrenaline).

3. Anticonvulsants (phenobarbital, phenytoin, clonazepam).

4. Heparinized saline (1 unit/ml).

5. Antibiotics (Amoxicillin and Gentamicin).

3 Methods

3.1 Animal Preparation

3.1.1 Clinical Examination

1. Newborn pigs (Large white/Landrace) aged less than 48 h and has had colostrum are received and held in the Penn. Monitor the rectal temperature every 30 min.

2. Weigh the pig.

3. Perform clinical neurological examination as in Table 1 along with video recording.

3.1.2 Anesthesia Induction, Intubation, and Ventilation

1. Prime the Perspex box with 2.0 % isoflurane, 65 % nitrous oxide and 33 % oxygen for 3 min using anesthetic machine and connect the outlet of the box to appropriate scavenging system. Place the radiant warmer above the Perspex box.

2. Place the pig in the box and cover the box with lid and dark colored sheet to keep the pig calm.

3. Once the pig is adequately anesthetized confirmed by the presence of regular respirations and lack of response to pain (see note 1), pig is taken to the open incubator.

4. In the open incubator, place the pig with the head on the edge of the incubator. Hold the pig in the supine position and keep

Table 1
Clinical neurology scoring system. Normal, mild, moderate and severe are scored as 2.0, 1.0, 0.5, and 0.0 respectively

Score	Item description
1. *Respiration*	
2	Normal respiratory pattern
1	Tachypnea, recessions with normal saturation in room air
0.5	Apnea or CPAP or need for supplemental oxygen
0	Artificially ventilated
2. *Consciousness*	
2	Alert with spontaneous eye opening
1	Response to sound, noise, or verbal stimuli
0.5	Response to painful stimuli
0	Unresponsive
3. *Orientation*	
2	Actively looking and investigating the surroundings
1	Some interaction
0.5	Looking at the surroundings but not interacting or investigating
0	No interaction
4. *Ability to walk*	
2	Ability to demonstrate coordinated gait with all four limbs
1	in coordinate gait
0.5	Falls during walking
0	Unable to walk
5. *Forelimb control*	
2	Ability to rise quickly from supine position using fore limbs
1	Ability to lift slowly from supine using forelimbs
0.5	Brief upright posture with forelimbs, but unable to sustain and falls down
0	Unable to use fore limbs
6. *Limb tone*	
2	Normal tone in forelimbs and hind limbs
1	Slightly increased/decreased tone in either fore or hind limbs
0.5	Slightly increased/decreased tone in both fore and hind limbs
0	Increased /decreased tone in both fore and hind limbs
7. *Activity*	
2	Almost continuous activity in the awake state
1	Some periods of activity in the awake state
0.5	Awake and moving in the lying down position but not walking around
0	No activity
8. *Suck*	
2	Presence of sucking reflex
1	Dysconjugate sucking reflex
0.5	Biting on the teat but not sucking
0	Absence of sucking reflex

(continued)

Table 1
(continued)

Score	Item description
9. *Vocalization*	
2	Spontaneous continuous vocalization
1	Spontaneous intermittent and sparse vocalization
0.5	Vocalization in response to stimuli
0	No vocalization
10. *Pathological movements*	
2	Absence of pathological movements including tonic posturing and clonus
1	Infrequent tonic posturing and clonus
0.5	Infrequent pathological movements, tonic posturing and clonus
0	Frequent or continuous pathological movements
11. *Sniff test*	
2	Ability to actively sniff an apple or marmite placed 30 cm from the pig and nibble them
1	Sniffs apple or marmite but does not nibble
0.5	Intermittently sniffs apple or marmite
0	No sniffing

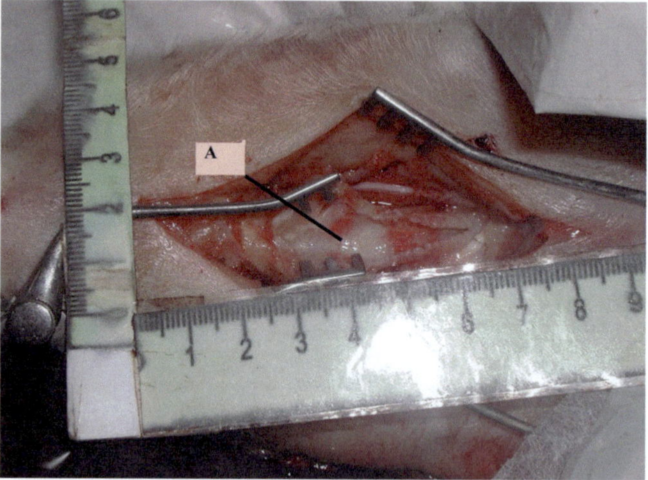

Fig. 1 Section of newborn pig trachea. (**A**) curvature in the trachea, due to which a resistance is felt when intubating the trachea

the mouth open by the assistant, while other person intubates the pig with a cuffed endotracheal tube to a length of 11 cm. The pig trachea has a curvature as shown in (Fig. 1) and hence the intubation can be difficult (see note 2).

5. Connect the end-tidal CO_2 monitor to ETT and confirm air entry on both hemithorax to ensure the correct position of the ETT. Connect the saturation probe to the hind limb hoof.

6. Connect the ETT to the ventilator set at a rate of 15–20 per minute, Peak inspiratory pressure of 18–20 cm H_2O, peak end expiratory pressure of 4–6 cm H_2O, inspiratory time of 0.45–0.55 s and fraction of inspired oxygen to 21 %. Secure the ETT with tape to the jaws. The time cycled pressure limited ventilation is adjusted to maintain the end-tidal CO_2 between 4.7 and 6.0 kPa (see note 3).

7. A rectal temperature probe from the servo-controlled cooling machine is inserted to 6 cm and the surface probe is placed on the forehead of the pig. The core temperature is set at 38.5 °C, which is the core temperature of the pig [13].

8. The chest is shaved and three electrocardiogram electrodes are attached to the chest [2] and left thigh [1].

3.1.3 Lines

1. The ear vein is cannulated; venous blood gas performed along with blood sugar. Connect the maintenance infusion, 5 % dextrose in 0.45 % saline at the rate of 60 ml/kg/day to maintain blood glucose between 3 and 10 mmol/L.

2. Give an initial dose of amoxicillin 30 mg/kg and gentamicin 5 mg/kg through the cannula.

3. Place the pig in the supine position, with support rolls on either sides of the forelimb.

4. Under sterile precautions, umbilical arterial (UAC) and venous catheters (UVC) are inserted to 13 and 7 cm respectively (see note 4).

5. The UAC is connected to the blood pressure transducer and the invasive blood pressure monitor with the appropriate module. The blood pressure is zeroed at the level of heart in the mid-axillary line before commencing measurement. The UAC is infused with heparinized saline (1 unit/ml) at 0.5 ml/h through the transducer to prevent thrombus formation.

6. Connect the maintenance infusion to the UVC and lock the ear cannula with heparinized saline of 1 ml.

3.1.4 Head Preparation

1. Place the pig in the prone position with one support roll under each axilla and another support roll under the inguinal region to allow chest movement.

2. Shave the head and clean the surface of any shaving foam with saline.

3. Under aseptic precautions, make an artificial fontanelle (2×2 cm) on the left half of the skull (see note 5).

4. At the posterior end of the incision over the parietal bone, perform twist-drill craniostomy. Perforate the duramater gently with a small hypodermic needle and insert a Codman intraparenchymal intracranial pressure (ICP) monitor into the brain.

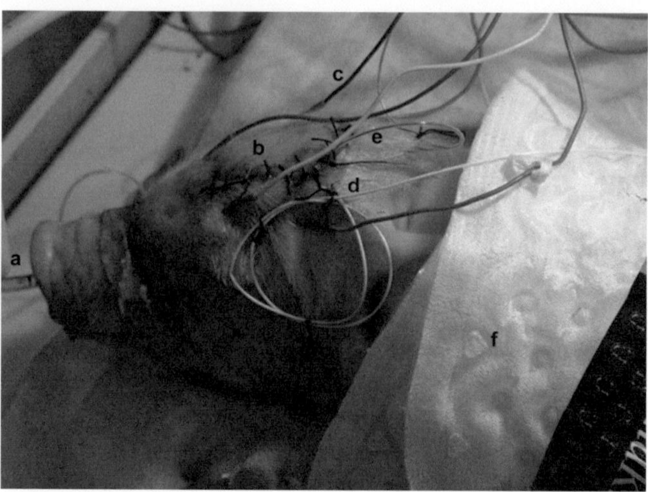

Fig. 2 Head preparation (**a**) Endotracheal tube; (**b**) Anterior fontanelle; (**c**) aEEG leads (two channel aEEG is monitored in this pig); (**d**) Intracranial pressure probe; (**e**) Brain temperature probe

Follow the manufacturer's instructions for zeroing and connecting to the express box and monitor. Secure the lead to the skin with a single 3/0 silk suture (Fig. 2).

5. Calibrate the Lectromed cerebral function monitor (see note 6).

6. Insert needle EEG electrodes with 3 cm inter-electrode length and the ground electrode in the frontal region and connect to the cerebral function monitor (Fig. 2).

3.1.5 Baseline Period

1. Place the pig in prone position with support rolls under the axilla and the groin. Obtain baseline blood gas.

2. Use a preterm infant nappy to cover the perianal region.

3. Swap isoflurane and nitrous oxide with 0.7 % halothane, 29 % oxygen, and 70.3 % nitrogen. Achieve stability for 30 min. Obtain blood gas, blood glucose and lactate.

4. Inflate the cuff of the ETT and ensure there is no leak around the ETT (see note 7).

3.1.6 Hypoxic Ischemic Insult

1. Administer hypoxic ischemic insult (HI) for 45 min monitoring the heart rate, blood pressure and amplitude integrated EEG (aEEG) activity by reducing the inhaled oxygen to 6 % [10, 11, 14, 15].

2. Alter the concentration of inspired gas, by reducing the inhaled oxygen to 400 ml/min and nitrogen to 5,550 ml with halothane of 0.7 %. Titrate oxygen and occasionally halothane concentration to maintain the aEEG <7 μV (Fig. 3) with variable

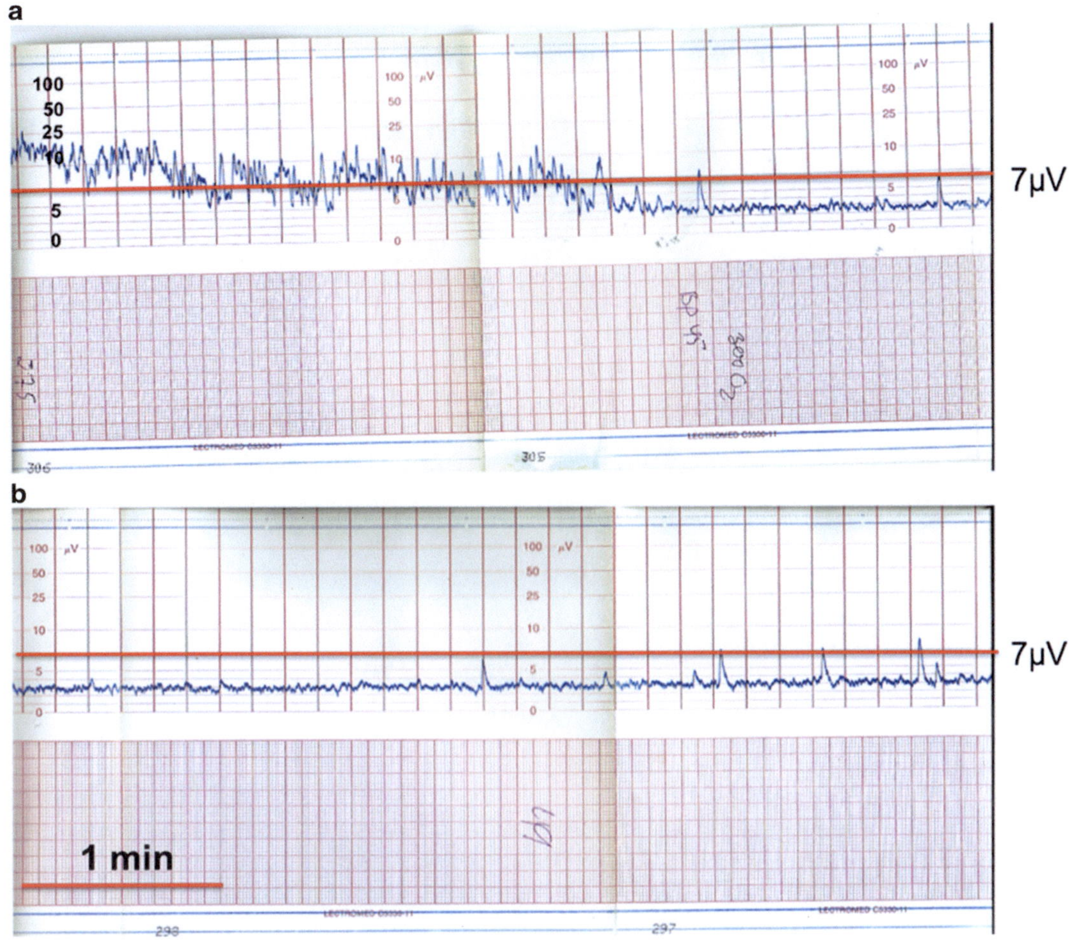

Fig. 3 aEEG during HI. The aEEG measured at 1 mm/s reduced from above 10 μV to <7 μV during hypoxic-ischemic insult (**a**). aEEG <7 μV during hypoxic-ischemic insult (**b**). The peaks in the aEEG correspond to the gasps

duration of hypotension (mean arterial blood pressure (MABP) <40 mmHg) and bradycardia (Fig. 4). If the MABP reduced below 20 mmHg or the heart rate reduced below 100/min, the FiO_2 was slightly increased. If pig sustained cardiac arrest, increase FiO_2 to 16 % and perform cardiac compressions (see note 8).

3. Monitor the blood gas, blood sugar and lactate every 15 min during the insult.

4. After 45 min of hypoxic-ischemic insult, increase the oxygen concentration to 21 %.

5. Swap the inhaled anesthesia regimen with intravenous anesthesia regimen (propofol bolus of 4 mg/kg followed by maintenance dose of 4–12 mg/kg/h and fentanyl dose of 1–3 μg/kg/h) (see note 9).

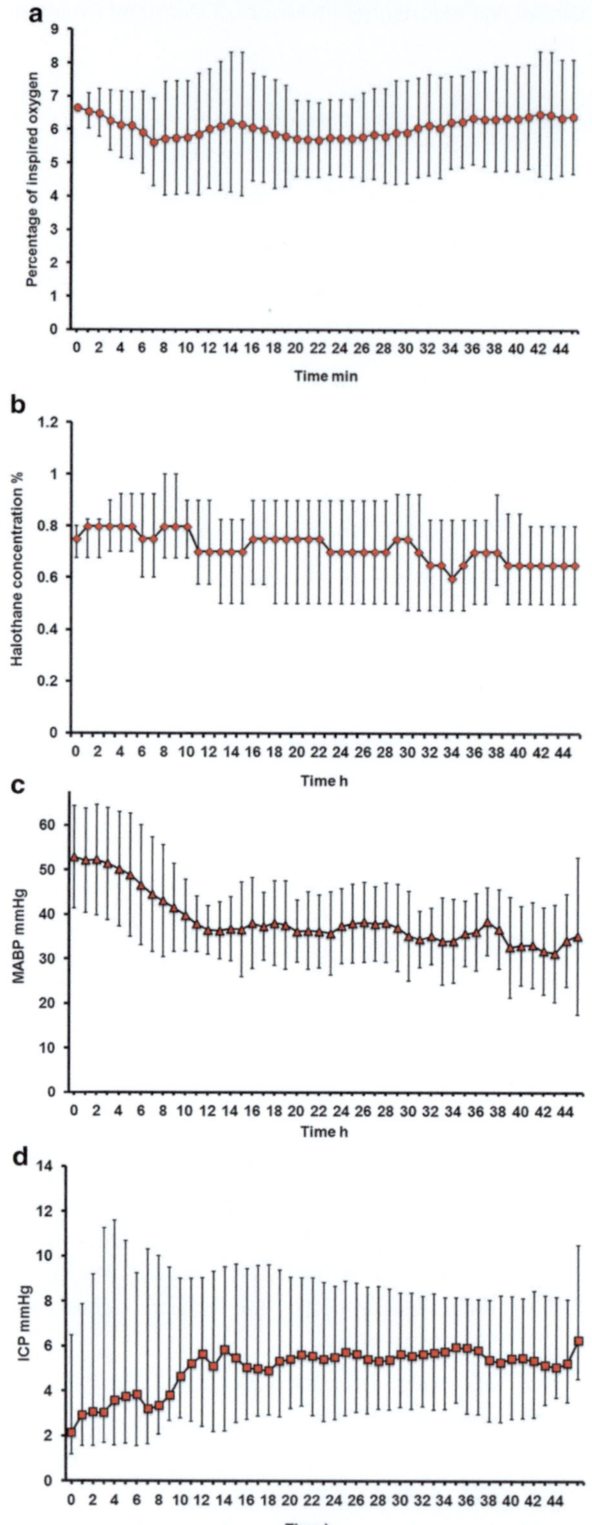

Fig. 4 Distribution of fraction of inspired oxygen (**a**), halothane concentration (**b**), mean arterial blood pressure (**c**), and intracranial pressure (**d**) during the hypoxic-ischemic insult in 18 pigs. Data represents mean (SD) for FiO_2 %, median (IQR) for halothane concentration %, mean (SD) for MABP mmHg and median (IQR) for intracranial pressure mmHg during hypoxic-ischemic insult. *Closed circles*: FiO_2; *diamonds*: halothane concentration; *triangle*: MABP; *square*: ICP

3.1.7 Recovery and Intensive Care Management

1. Ventilate the pig to maintain normocapnia and adjust inhaled oxygen concentration to maintain peripheral transcutaneous oxygen saturation between 95 and 98 %.

2. Maintain MABP >40 mmHg. Pig may require fluid bolus of 10 ml/kg of 0.9 % saline followed by inotropic support with dopamine infusion of 5–20 μg/kg/min, noradrenaline (20 ng–1 μg/kg/min) and hydrocortisone (2.5 mg/kg 6 hourly). The inotropic support is increased if the MABP is <40 consistently for 30 min despite reducing *i.v* anesthesia.

3. Wean inotropic support by 25 % when the MABP >40 mmHg for more than 30 min. Wean norepinephrine first followed by dopamine. After a stable 30 min, wean inotropes further by 25 % [16].

4. Treat electrical or clinical seizures lasting >5 min with phenobarbital (20 mg/kg×2 doses) followed by clonazepam (100 μg/kg) and lidocaine (2 mg/kg over 10 min), followed by a continuous infusion of 6 mg/kg/h (6 h), 4 mg/kg/h (12 h) and 2 mg/kg/h (12 h).

5. Failure of seizures to respond to treatment in 30 min entailed moving to the next step of the treatment protocol.

6. Hypoglycemia (blood sugar <54 mg/dl or <3 mmol/L) was treated with a 2.5 ml/kg bolus of 10 % dextrose, followed by increasing the glucose delivery up to 12 mg/kg/min.

7. Treat persistent hyperglycemia (blood glucose >180 mg/dl) by reducing the glucose delivery and monitor the blood glucose hourly.

8. Perform tracheal suction every 12 hourly.

9. Sodium bicarbonate is not used. Rarely an infusion of 1–2 mmol/kg/day of sodium bicarbonate is given if the serum bicarbonate was persistently low affecting the pH.

10. After weaning the inotropic support and the intravenous anesthesia, pigs will be weaned from the ventilator when breathing was adequate around 12 h after insult.

11. Oropharyngeal continuous positive airway pressure delivered by the ventilator (PEEP 6 cm H_2O) may have to be used after extubation.

12. We use intramuscular buprenorphine (3 μg/kg 8 hourly) for analgesia after extubation. Give the first dose prior to extubation.

13. Prior to extubation remove the intracranial probe, brain temperature sensor, and the aEEG electrodes. Perform subsequent aEEG recordings under analgesia keeping the pig comforted by wrapping the pig with appropriate sheets and held by an assistant at 24, 36, 48, 60, and 72 h after insult.

14. Secure UAC and UVC with elastoplast to the side of the body.

15. Repeat amoxicillin twice daily and gentamicin once daily.

16. Perform blood gas and blood glucose based on the clinical need.

17. From day 2, orally feed the pigs with pig formula every 3–4 hourly. Weigh the pigs daily.

18. Perform clinical neurological assessment every 24 h.

3.1.8 Autopsy

1. After 72 h survival, pigs were reintubated and under deep isoflurane anesthesia, underwent terminal perfusion fixation of the brain and full autopsy.

2. Make a midline longitudinal cervical incision and place a self-retaining retractor to expose the trachea and sternocleidomastoid muscles bilaterally. Identify the common carotid artery on either side, and dissect it free from the internal jugular vein and vagus nerve within the carotid sheath (see Fig. 5 in the chapter *A newborn piglet survival model of post-hemorrhagic ventricular dilatation (PHVD)*).

3. Pass a ligature around the carotid artery as a sling, to isolate it. Insert a large bore cannula into the carotid artery and confirm brisk arterial backflow. Secure the cannula to the artery distally and proximally with ligatures and ligate the distal carotid artery.

4. Once the cannula is secured in both carotid arteries, divide the internal jugular veins on either side. Connect the cannulae to 20 ml syringes filled with normal saline and flush slowly, manually, repeated until the venous effluent becomes clear.

5. Connect the cannulae to 20 ml syringes containing 4 % neutral phosphate buffered formaldehyde and flush several times until the resistance to flow increases significantly. This ensures satisfactory fixation of the brain.

6. Once complete, the cranium is opened and the brain can be removed for further analysis and storage (Fig. 5).

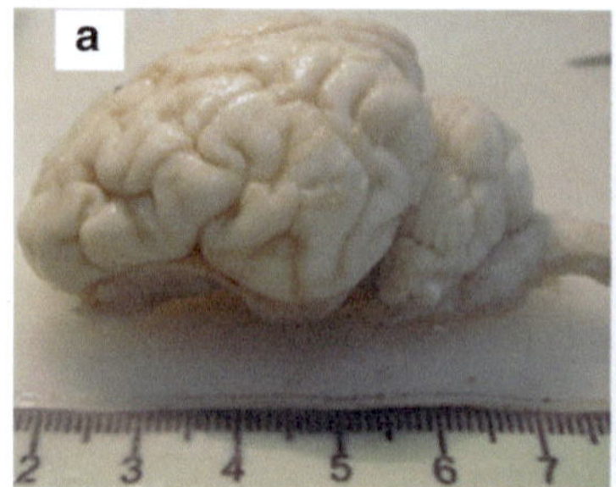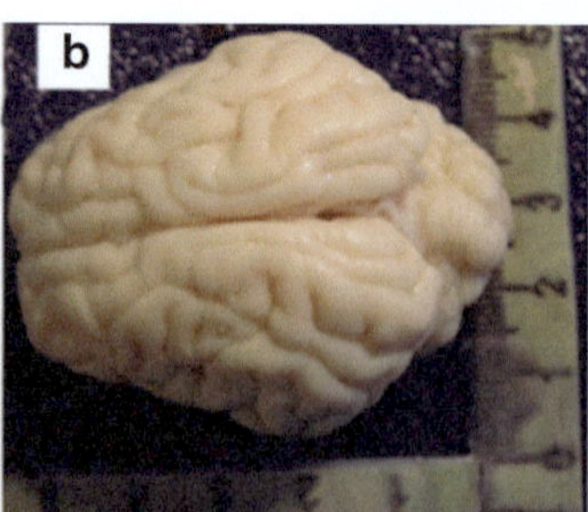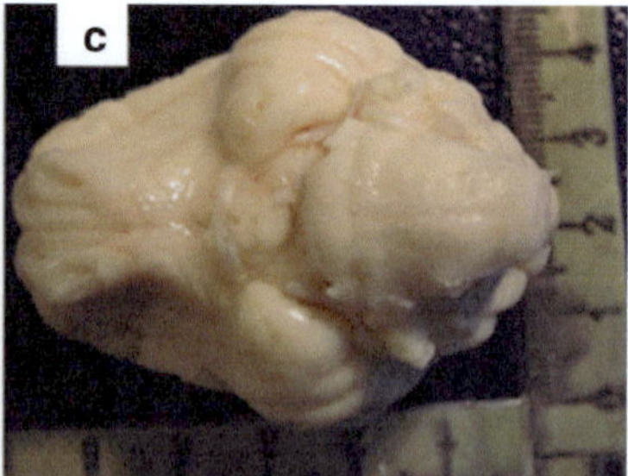

Fig. 5 Pig brain removed from the cranium after perfusion fixation (**a**): Side of the brain, (**b**) Brain from top; (**c**) Base of the brain

7. This is followed by removal of other organs. Make a vertical midline incision from the xiphisternum to the groin. Open the thorax by making a small incision in the anterior midline insertion of diaphragm to the chest wall. Cut the thoracic cage in the midline along the right sternal border in a cephalad direction. Open the rib cage and separate the lungs. Remove the heart after severing the pulmonary vessels and great arteries. Separate the liver from the diaphragm and the portal vessels. Dissect the abdominal skin from the peritoneum. Open the peritoneal cavity inspecting for hemorrhage. Dissect the spleen from its attachment to the mesentery and blood vessels. Dissect the pancreas hidden in between the intestines. Trace the UAC and UVC to the tip noting the presence of thrombus. Dissect the peritoneum from the posterior abdominal wall and remove the kidneys and adrenal gland. Fix the organs by immersion fixation in the 4 % neutral phosphate buffered formaldehyde.

3.1.9 Tissue Processing and Staining

1. After 48 h of further fixation in formaldehyde, 0.5 cm sections were made from the brain. The cerebellum and brain stem were separated from the cerebrum. The right uninstrumented cerebral hemisphere was separated from the left by cutting the corpus callosum. A transverse incision was made at the level of the temporal lobe. A total of 10 half cm sections were made from both halves of the right cerebrum. The sections were placed in the cassette in a way that the brain could be examined serially from frontal to occipital lobe (Fig. 6).

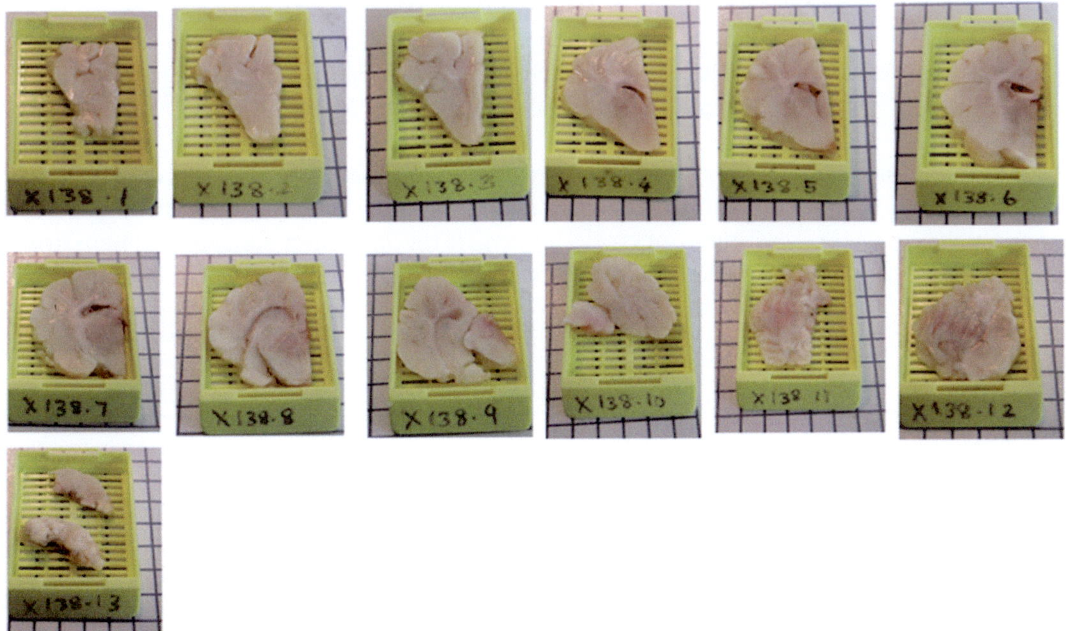

Fig. 6 Pig brain (right hemisphere) cut into 0.5 cm sections from frontal lobe to the brain stem

Table 2
Grading of neuropathology based on the proportion of area affected and the morphologic changes noted

Grade	Area affected (%)	Morphologic changes
1	≤10	Individual necrotic neurons, small patchy, complete or incomplete infarcts
2	20–30	Partly confluent complete or incomplete infarcts
3	40–60	Large confluent, complete infarcts
4	>75	In cortex: total disintegration In thalamus and basal ganglia: large complete infarcts In hippocampus: neuronal necrosis

Incomplete infarct describes a localized area where necrotic neurons are observed but other cell types including glia and vessels are preserved

2. Five tissue sections (0.5 cm) from the atrioventricular valve to the apex of heart; three 0.5 cm sections and two 0.5 cm sections from the right and left lungs; four 0.5 cm sections from four lobes of the liver; three 0.5 cm sections from the head, body and tail of spleen; two 0.5 cm sections from pancreas; two halves of the kidneys by a midline incision were made.

3. Tissue processing was carried out in an automated tissue processor (Tissue Tek VIP, Miles Scientific, Sakura, Torrance, CA, USA), which involved dehydration using ethanol followed by clearing with xylene. Sections made from processed tissue were embedded with paraffin. The embedded blocks were cut into 5-µm sections and mounted on to glass slides before staining with hematoxylin and eosin with automated stainer.

3.1.10 Neuropathology

1. The neuropathology was scored based on the proportion of area affected as shown in Table 2 (Fig. 7).

3.1.11 Organ Pathology

1. Organ pathology includes infarction in the heart; necrosis, tubular and ductal dilatation, interstitial hemorrhage in kidneys; necrotic lesions in the intestine; subcapsular hematoma, necrosis, thrombus, and focal infarcts in the liver; pneumonia, hemorrhagic necrosis, and hemorrhage in lungs and adrenal glands (Fig. 8) [11, 12].

4   Notes

1. Ensure the pig is not deeply anesthetized to avoid hypotension induced by anesthesia. Pigs will usually initially stand upright, followed by lying on their side. In approximately, 3–5 min, the pig will be calm having regular respirations and will not respond to pain. If the pig is deeply anesthetized, peripheral cannulation of the ear vein followed by infusion of 10 ml/kg 0.9 % saline bolus will be essential soon after intubation. On the other hand,

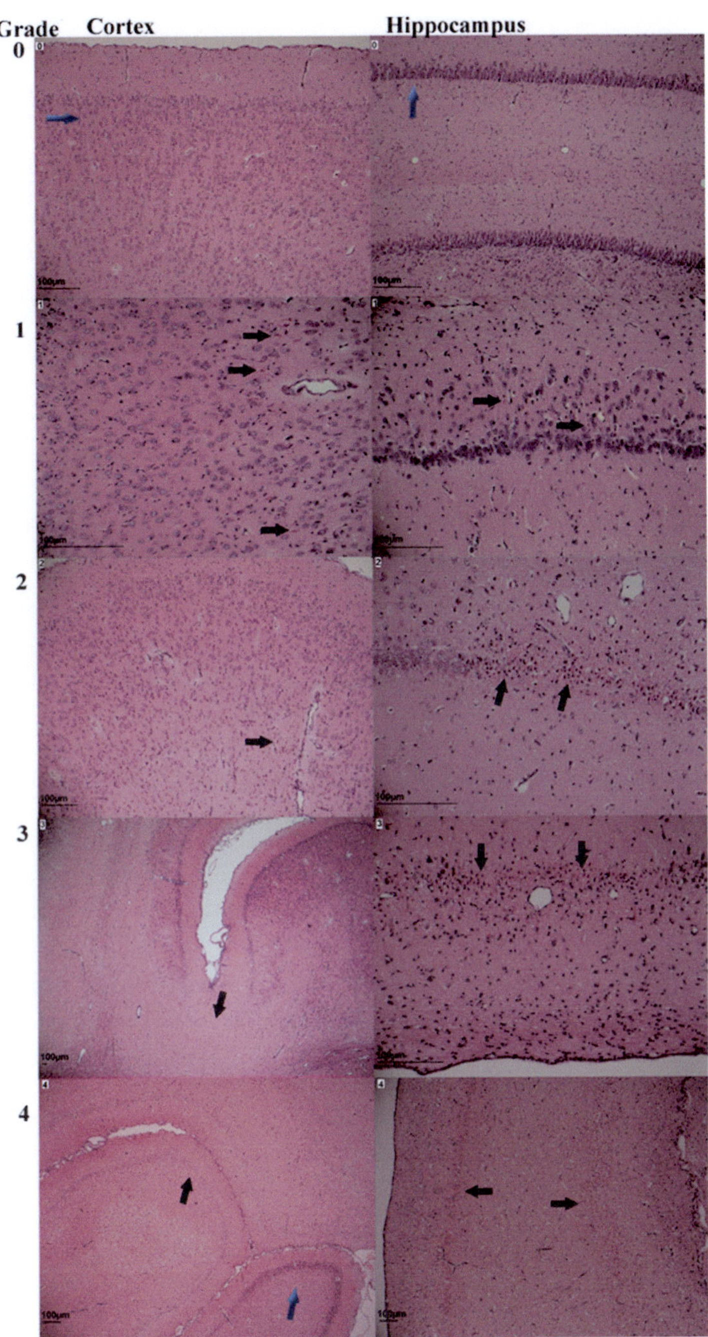

Fig. 7 Micrographs of cortex and hippocampus showing injury grades from 0 to 4. *Black arrows* indicate degenerating neurons and neuronal loss. *Blue arrows* indicate normal neurons in the cortex and hippocampus. Grade 0 shows normal cortex and hippocampus. Grade 1 shows scattered degenerating neurons in otherwise normal cortex and hippocampus. Grade 2 represents more degenerating neurons than Grade 1, some of which are in small groups. *Grade 3* in cortex shows neuronal degeneration and neuronal loss at the base of the sulcus (ulegyria). Grade 3 in hippocampus shows degeneration of most of the neurons in the section. Grade 4 in cortex shows extensive area of neuronal degeneration and neuronal loss with a small area of normal cortex. Grade 4 in hippocampus shows degeneration of all neurons. Reproduced from [Chakkarapani E, Dingley J, Liu X, Hoque N, Aquilina K, Porter H, Thoresen M (2010) Xenon enhances hypothermic neuroprotection in asphyxiated newborn pigs, Annals of neurology **68**, 330–341] with permission from [John Wiley and Sons]

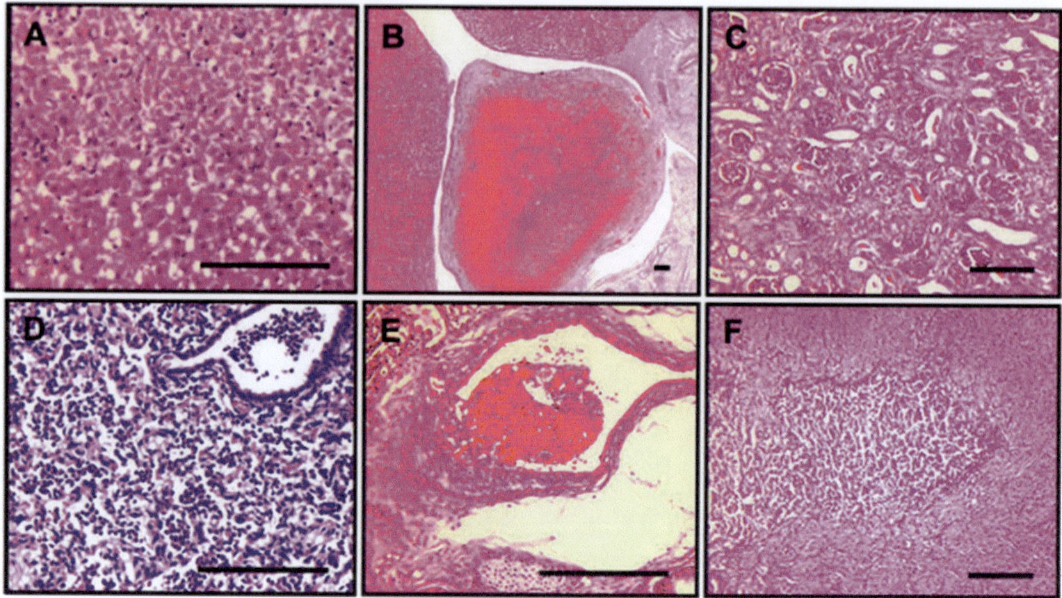

Fig. 8 Organ histology. Liver tissue with necrotic cells (**a** magnification ×400); thrombus in hepatic vein (**b**, ×50); Necrosis in kidney (**c**, ×200); Lung tissue with pneumonia (**d**, ×400) and thrombus in large pulmonary artery (**e**, ×400) and myocardial necrosis (**f**, ×200). Scale bar = 200 μm. Reproduced from [Karlsson M TJ, Satas S, Hobbs H, Chakkarapani E, Stone S, Porter H, Thoresen M (2008) Delayed hypothermia as selective head cooling or whole body cooling does not protect brain or body in newborn piglets, Pediatric research **64**, 74–80]

if the pig is light and waking up, use bag-valve-mask with gas flow from the anesthetic machine of 2 % isoflurane, 70 % nitrous oxide, and 30 % oxygen to re-anesthetize the pig.

2. While intubating the trachea, the pig should be held in supine position and with an elastic band the upper jaw should be stabilized to the incubator. With the miller laryngoscope blade, lift the lower jaw along with the tongue, and the blade tip should be beyond the epiglottis until the vocal cords are seen. The ETT should be inserted with the convexity facing up. After inserting the ETT few centimeters into the glottis, the ETT will have resistance, when the ETT should be rotated 90–180° until the resistance disappears. When the operator feels the change in resistance, the ETT tip should be advanced to 11 cm (Fig. 1).

3. The activities of connecting the end-tidal CO_2 monitor, saturation probe, connecting the ventilator to the ETT and securing the ETT with tape should occur simultaneously as the pig can destabilize quickly during this period. The ETT should be held at 11 cm firmly by the operator until it is secured with the tape to prevent dislodging of the ETT. The pig needs to be ventilated with inhalation anesthesia of 1–2 % isoflurane, 70–75 % nitrous oxide and 23–29 % oxygen to maintain transcutaneous peripheral oxygen saturation of 95–98 %. We replaced the

inspiratory circuit of the neonatal ventilator with inspired gas flow of (4–5 l of nitrous oxide, 1–2 l of oxygen and 1–2 % of isoflurane) from the anesthetic machine and the expiratory circuit is connected to the ventilator. This will cause the ventilator to alarm, which we glued to prevent it to be reset every 2 min. We correlated end-tidal CO_2 at the commencement of the experiment with the arterial PCO_2 in order to decide the range of end-tidal CO_2 that can be maintained during the recovery phase.

4. The pigs do not have ductus venosus at birth. The fetal ductus venosus changes to hepatic vein just before birth. Enthusiastic advancement of UVC should be avoided to prevent liver injuries. Aspirating free flowing blood from the catheters clinically checks line positions. We secure the catheters with 3-0 silk to the umbilical cord and then bring the catheters from the sides by securing the catheter to the sides of the body with elastoplast without causing any kinks.

5. On the left half of the skull, make a 5 cm linear parasagittal incision. Extend the incision medially from the posterior end for 5 cm. Retract the scalp skin medially with scalp tissue retractor followed by bunt dissection. Once the skull was visible, make a hole anterolateral to the left sagittal–coronal suture junction. Using bone or Kerrison rongeurs, make a 2×2 cm artificial fontanelle by stripping the skull bone from the burrhole leaving the duramater intact. Suture the scalp, so that an acoustic window is available. The anterior fontanelle makes the model similar to newborn babies and avoids brain injury induced by the raised intracranial pressure secondary to hypoxic-ischemic insult. A thin temperature sensor with multiple temperature sensors at the interval of 5 mm can be inserted into the brain to measure brain temperature (Fig. 2).

6. Connect the colored leads provided with the machine to the cable connector on one end and to the sockets on the front of the CFM on the other end. The connections are color coded: connect yellow-to-yellow, red-to red, and green-to-green. On the front panel of the CFM above the three sockets is a switch. With the power switched on turn the switch to the side marked 100 µVs. The needle that displays the amplitude should now be on the 100 µV mark on the scale printed on the printer paper. Use the needle-adjusting dial to adjust as needed. Turn the switch to the 25 kΩ mark, the needle that displays the impedance should rise a distance of five large squares on the printer paper. Adjust the needle-adjusting dial if necessary. If the digital CFM can have a custom made speed of 1 mm/s then it can used instead.

7. Listen to the airway for leak. We use artificial neonatal test lung to ventilate through the endotracheal tube, just for one breath cycle. The same volume of gas pushed through the ETT should be returned to the artificial lung by the recoil of the

chest. If there is leak, alveolar hypoxia will be difficult to achieve during hypoxic-ischemic insult.

8. The duration aEEG <7 µV should be above 30 min to induce brain injury. We do not alter the ventilation or use neuromuscular blocking agents during the hypoxic-ischemic insult. The pig may gasp or have seizures. Alter halothane concentration to keep the pig anesthetized. Pig becomes hypotensive due to the effect of halothane and hypoxia. The response to bradycardia had to be fast, within about 20 s, and the step increase in FiO_2 depended on the severity of bradycardia, usually 0.5–1.5 %. Maintain the level of increased FiO_2 until heart rate and mean arterial blood pressure starts to rise, which varies between 15 s and several minutes. The increased FiO_2 sometimes will transiently increase the aEEG amplitude to above 7 µV. If the pig's spontaneous circulation returned quickly without resorting to adrenaline, the insult was continued to 45 min. If the pig required adrenaline (0.1 ml of 1:10,000) to achieve return of spontaneous circulation, the insult was discontinued. Measure the total period of low amplitude aEEG (LAEEG) (<7.0 µV) by summating all individual LAEEG periods, and the summated duration of aEEG with amplitude ≤ 3.5µv and the longest single period of LA EEG. These are markers of insult severity.

9. After administering hypoxic-ischemic insult, connect the inspiratory circuit of the neonatal ventilator to the inspiratory outlet of the ventilator and remove the anesthetic circuit from the inspiratory circuit.

References

1. Levene ML, Kornberg J, Williams TH (1985) The incidence and severity of post-asphyxial encephalopathy in full-term infants. Early Hum Dev 11:21–26

2. Lawn JE, Bahl R, Bergstrom S, Bhutta ZA, Darmstadt GL, Ellis M, English M, Kurinczuk JJ, Lee AC, Merialdi M, Mohamed M, Osrin D, Pattinson R, Paul V, Ramji S, Saugstad OD, Sibley L, Singhal N, Wall SN, Woods D, Wyatt J, Chan KY, Rudan I (2011) Setting research priorities to reduce almost one million deaths from birth asphyxia by 2015. PLoS Med 8:e1000389

3. Edwards AD, Brocklehurst P, Gunn AJ, Halliday H, Juszczak E, Levene M, Strohm B, Thoresen M, Whitelaw A, Azzopardi D (2010) Neurological outcomes at 18 months of age after moderate hypothermia for perinatal hypoxic ischaemic encephalopathy: synthesis and meta-analysis of trial data. BMJ 340:c363, Clinical research ed

4. American College of Obstetricians and Gynecologists (Ed.) (1992) Guideline for perinatal care, 3rd ed. American Academy of Pediatrics ed., Elk Grove Village, IL

5. Low JA (1997) Intrapartum fetal asphyxia: definition, diagnosis, and classification. Am J Obstet Gynecol 176:957–959

6. Rice JE 3rd, Vannucci RC, Brierley JB (1981) The influence of immaturity on hypoxic-ischemic brain damage in the rat. Ann Neurol 9:131–141

7. Yue X, Mehmet H, Penrice J, Cooper C, Cady E, Wyatt JS, Reynolds EO, Edwards AD, Squier MV (1997) Apoptosis and necrosis in the newborn piglet brain following transient cerebral hypoxia-ischaemia. Neuropathol Appl Neurobiol 23:16–25

8. Williams CE, Gunn A, Gluckman PD (1991) Time course of intracellular edema and epileptiform activity following prenatal cerebral ischemia in sheep. Stroke 22:516–521

9. Dobbing J, Sands J (1979) Comparative aspects of the brain growth spurt. Early Hum Dev 3:79–83

10. Thoresen M, Haaland K, Loberg EM, Whitelaw A, Apricena F, Hanko E, Steen PA (1996) A piglet survival model of posthypoxic encephalopathy. Pediatr Res 40:738–748

11. Haaland K, Loberg EM, Steen PA, Satas S, Thoresen M (1997) The effect of mild posthypoxic hypothermia on organ pathology in a piglet survival model of global hypoxia. Prenat Neonatal Med 2:329–337

12. Karlsson TJ, Satas S, Hobbs H, Chakkarapani E, Stone S, Porter H, Thoresen M (2008) Delayed hypothermia as selective head cooling or whole body cooling does not protect brain or body in newborn piglets. Pediatr Res 64: 74–80

13. Tuchsherer M, Puppe B, Tuchsherer A, Tiemann U (2000) Early identification of Neonates at risk: traits of Newborn piglets with respect to survival. Theriogenology 54:371–388

14. Tooley JR, Satas S, Porter H, Silver IA, Thoresen M (2003) Head cooling with mild systemic hypothermia in anesthetized piglets is neuroprotective. Ann Neurol 53:65–72

15. Chakkarapani E, Dingley J, Liu X, Hoque N, Aquilina K, Porter H, Thoresen M (2010) Xenon enhances hypothermic neuroprotection in asphyxiated newborn pigs. Ann Neurol 68:330–341

16. Chakkarapani E, Thoresen M, Liu X, Walloe L, Dingley J (2012) Xenon offers stable haemodynamics independent of induced hypothermia after hypoxia-ischaemia in newborn pigs. Intensive Care Med 38:316–323

Chapter 13

Animal Models of Fetal Alcohol Spectrum Disorder

Wendy Comeau*, Tamara Bodnar*, Kristina Uban, Vivian Lam, Katarzyna Stepien, and Joanne Weinberg

Abstract

This chapter reviews the development and use of animal models in research on Fetal Alcohol Spectrum Disorder, highlighting methodological issues. The development of suitable animal models in this field has been crucial in investigating the wide range of effects of prenatal alcohol exposure under tightly controlled conditions, as well as factors that can influence the severity of effects observed, and the mechanisms mediating those effects. Animal models in this field have been critical in demonstrating that alcohol is a powerful teratogen with serious neurodevelopmental consequences, a fact that was met with skepticism by many in the scientific community when Fetal Alcohol Syndrome was first described. The review begins with a discussion of the various animal models in the field, highlighting the benefits and limitations of each model. Although it is not inclusive, this chapter aims to provide an overview of the challenges that FASD researchers have had in developing animal models that have construct, face and etiological validity. We then discuss the factors that must be addressed in the development of an animal model, including dose and timing of alcohol exposure, blood alcohol levels (BALs), mode of alcohol administration, nutritional considerations, maternal–pup interactions, housing considerations, and strain differences. The chapter then moves on to discuss in some detail the model of chronic, moderate prenatal alcohol exposure utilized in our research with Sprague–Dawley rats. A detailed discussion of our breeding, feeding, culling, and weaning procedures is provided, including the specifics of our liquid diets and feeding schedule, with particular consideration of the issue of pair-feeding. Finally, detailed notes are provided to explain the rationale underlying methodology and procedures described in our protocol.

Key words Fetal alcohol spectrum disorder (FASD), Prenatal alcohol exposure, Blood alcohol levels, Nutrition, Breeding, Pair-feeding, Cross-fostering, Maternal behavior

---

1 Introduction

Notable adverse effects of maternal alcohol use were first described by Lemoine and colleagues [1]. The term "Fetal Alcohol Syndrome" (FAS) was coined by Jones and Smith in 1973 [2] to describe the constellation of facial features, growth, and central nervous system effects that occur at the most severe end of the spectrum of alcohol's

---

*Equal first authors

Jerome Y. Yager (ed.), *Animal Models of Neurodevelopmental Disorders*, Neuromethods, vol. 104, DOI 10.1007/978-1-4939-2709-8_13, © Springer Science+Business Media New York 2015

adverse effects. More recently, the umbrella term "Fetal Alcohol Spectrum Disorder" (FASD) was adopted to describe the wide range of outcomes observed following prenatal exposure to alcohol.

In humans, there are obvious limitations in our ability to study the potential mechanism(s) mediating the adverse effects of prenatal alcohol exposure, as well as the individual and environmental factors that may influence outcome. The development of suitable animal models has been crucial in investigating the wide range of potential effects of prenatal alcohol exposure under tightly controlled conditions, as well as the factors that can influence the severity of effects observed, and the mechanisms mediating those effects. Moreover, animal models have been critical in demonstrating that alcohol is a powerful teratogen with serious neurodevelopmental consequences, a fact that was met with skepticism by many in the scientific community when FAS was first described.

As with other disorders, the investigation of FASD has utilized a range of animal models including, but not limited to, *Caenorhabditis elegans* (*C. elegans*) [3], *Drosophila* [4], *Danio rerio* (zebra fish) [5], dogs [6], sheep [7], rodents such as the mouse [8], rat [9, 10], and guinea pig [11, 12] and nonhuman primates [13–15], each of which can model particular aspects of the disorder. For instance, the simpler nervous systems of non-mammalian models have made them especially useful in elucidating the potential scope of alcohol-related outcomes on neural networks, nerve growth factors, signaling pathways, and genetic variables during development [16]. Their utility also comes, in part, from the ability to expose offspring to alcohol directly (e.g., zebra fish [17], *C. elegans* [3], and chick models [18]), which allows direct investigation of alcohol teratogenesis and mechanisms of action, without interference by maternal or placental factors. These models have also been useful for studies on genetics [19, 20], and have provided insight into the role of cell signaling pathways in alcohol's adverse effects [21, 22].

As important as non-mammalian models have been in FASD research, mammalian models are important for understanding alcohol's actions under conditions similar to those in the human situation. It is likely that interactions between direct and indirect (maternally mediated) effects of alcohol are responsible for its adverse effects in mammals [23]. Moreover, as alcohol is known to act on or modulate many different target molecules, multiple mechanisms, activated at different stages of development or at different dose thresholds of exposure, probably contribute to the diverse phenotypes seen in children with FASD [24]. For instance, alcohol may simultaneously and/or consecutively produce widespread cell death, disrupt axon migration, and alter neurochemical support necessary for 'typical' development. As well, alcohol could disrupt development through endocrine or neuroendocrine imbalance and altered maternal-fetal hormonal interactions (*see* Zhang [25] and Hellemans [26] for review). In this chapter, we focus on mammalian models of FASD with an emphasis on and examples from our own research with our well-established rat model.

2   Considerations

In developing mammalian models of FASD, a number of factors need to be addressed and/or considered. These include dose and timing of alcohol exposure, blood alcohol levels (BALs), mode of alcohol administration, nutritional considerations, maternal–pup interactions, housing considerations, and strain differences. As the neurodevelopmental ontogeny and physiology of different species vary considerably, these considerations will need to be adapted to suit the model organism of choice.

2.1   Timing and Amount of Alcohol Exposure

2.1.1   Timing

Much of the variability in the effects of alcohol on the developing embryo-fetus may be attributed to the timing of exposure. This has been underscored in the early research of Webster et al., [27], as well as Sulik and colleagues [28, 29], who demonstrated that development of craniofacial anomalies, a signature feature of FAS in humans, can be reproduced in mice if alcohol exposure is timed to occur exactly when the face is forming. These studies provided important evidence to confirm the hypothesis that alcohol itself is a teratogen capable of causing the characteristic features of FAS [30].

In humans, the development of FAS and other alcohol-related effects depends on alcohol exposure in utero. However, in many mammals, including the mouse and rat, offspring are born less developed than the human fetus, at a point in development, and particularly brain development, that is roughly equivalent to the end of the second trimester in humans [31]. The third trimester in humans is a critical developmental phase that includes the brain growth spurt [31] and the development of immunocompetency [32]. In rats this would encompass the postnatal period from birth to approximately postnatal day 15 [31]. Thus, if one's goal is to expose the fetus to alcohol during the brain growth spurt, a third trimester equivalent model requires alcohol exposure during the early postnatal period.

A number of methods have been employed to expose offspring to alcohol during postnatal life. One method involves lactational exposure, where pups nurse normally and thus consume milk from alcohol-consuming dams. Such models are beneficial in that they require only modest levels of disturbance to the dam and pups, thereby limiting additional stress. However, this model has a number of limitations. For example, blood alcohol levels in suckling pups have been shown to be considerably lower than the levels achieved in the dam as well as the levels achieved with exposure in utero [33]. In addition, maternal alcohol consumption might suppress lactation [reviewed in [34]] or alter maternal behavior [35]. An alternative method of postnatal alcohol exposure is through direct administration to the pup. This can be accomplished via a gastrostomy feeding tube (i.e., the "pup-in-cup" artificial rearing model [36]) or via intragastric intubation [37]. The latter methods also have their limitations. Artificial rearing involves separation of the pup from the dam for a prolonged time during a critical period of development (typically, postnatal days 4–9 or longer). Absence of the normal mother–infant interaction, including the normal licking/grooming by the mother, can alter development and may introduce an added confound of early life stress for the pups. Moreover, during separation pups must be stimulated to defecate and urinate, usually by anogenital stroking with a soft brush several times a day. While this keeps pups healthy, it in no way replaces the normal mother–infant interaction. For comparison, a normal suckling control group must be included in any study using the artificial rearing method. Intragastric intubation involves separation and handling of the pups for brief periods, multiple times daily. This allows pups to remain with the mother and receive normal maternal care, thus overcoming one serious confound of the artificial rearing model. As well, it allows one to control pup nutrition such that alcohol-exposed pups maintain body weights comparable to those of control pups, which is an issue in many models of prenatal alcohol exposure (*see* below). However, intubation introduces the element of daily handling, which in itself can alter many aspects of pup development [38]. Inclusion of a group that is handled alone and/or intubated with control diet is necessary with this model.

Several additional issues are also noteworthy. First, postnatal alcohol exposure in rats and mice, meant to simulate third trimester exposure in humans, does not involve the placenta. While controlling for placental effects on the fetus is a strength of these models, it is likely that effects on the developing fetus will be different from those in the typical mammalian model. For example, although alcohol crosses the placenta [39], placental aldehyde dehydrogenase activity allows metabolism of low levels of alcohol [40] (reviewed in [41]). While the impact of placental alcohol metabolism may be minor, these metabolic processes nevertheless change blood alcohol levels as well as produce acetaldehyde, a probable carcinogen that may in itself have harmful effects on the developing fetus [42]. Second, because the mother and fetus are an interrelated unit, alcohol-induced changes in the mother have implications for fetal development. Of particular relevance for our research, alcohol-induced maternal-fetal endocrine imbalances could possibly contribute to the etiology of FASD [43, 44]. Because alcohol readily crosses the placenta it can directly affect developing fetal tissues, including those involved in endocrine function. In addition, alcohol-induced changes in endocrine function can disrupt the hormonal interactions between the maternal and fetal systems, thus affecting the development of fetal metabolic, physiological, and endocrine functions.

In summary, artificial rearing models have been instrumental in providing crucial insight into alcohol's direct effects during the sensitive brain growth spurt period, while controlling for placental and/or maternal effects on the offspring. However, a limitation of these models is the absence of normal maternal–fetal or mother–pup interactions, which are critical aspects of mammalian development.

Another consideration related to timing of exposure in rodents is the accurate determination of age at the time of conception/birth. This is especially problematic in FASD research. For example, it is known that maternal alcohol use delays parturition. Indeed, research indicates that alcohol exposure may delay parturition by an average of 6–24 h in the rat, depending on the level of alcohol intake [45, 46]. While seemingly slight, this delay is not insignificant given the rat's 22–23 day gestation period. Moreover, it is likely that the differential 'experience' of these pups (time in utero, age post-conception at birth) has the potential to influence development. Furthermore, gestation length among litters can vary by several hours even for pups in the control condition. To address this issue, some researchers have incorporated a restricted mating schedule in which dam and sire pairing is limited to a 2–3 h window.

2.1.2   Levels of Alcohol Exposure: Blood Alcohol Levels (BALs)

In humans, an accurate determination of the amount of alcohol to which the fetus has been exposed at any given time is rarely, if ever, attainable. Moreover, variations in genetics, age, weight, food consumption, metabolism, concomitant drug use, and prior alcohol use all influence maternal blood alcohol levels and thus the level of alcohol to which the fetus will be exposed. Determining comparable levels of alcohol use across species poses similar problems including, but not limited to, variations in alcohol metabolism. Thus, even if administration of alcohol is based on maternal body weight (g/kg body weight), different blood alcohol levels will likely occur across species [47].

Together, the pattern of alcohol use and the timing of exposure may be the best determinants of the potentially detrimental effects on the fetus, with higher BALs resulting in potentially greater effects. Publications by Chernoff [48], and Randall and Taylor [49], demonstrate a positive correlation between BAL and developmental malformations, a finding elucidated further by the seminal research of West and colleagues [50–54]. Thus, the most consistent measure of alcohol consumption that can be extrapolated from animal research to humans has been the use of BALs [55].

Rodents rarely self-administer high enough alcohol concentrations to become intoxicated (with a few exceptions including animals bred to be "alcohol preferring"). Putting alcohol in the drinking water is thus an ineffective method in self-administration models. Typically, self-administration models have incorporated alcohol into a liquid diet or provided a two-bottle choice (alcohol and water) offered for a specified restricted period or have used a "drinking in the dark" paradigm. Despite implementation of such strategies, it may still be difficult to achieve high enough BALs using self-administration models. Rodents drink in spurts and tend to curtail their drinking as their alcohol intake approaches the maximum rate at which they are able to metabolize alcohol [56, 57]. It is therefore important to consider whether dams will voluntarily consume a diet with a high enough alcohol content to elevate BALs sufficiently to produce FASD-like effects similar to those found in humans. Strategies implemented to increase alcohol consumption include sweetening alcohol-containing liquids and limiting access. However, sucrose, similar to alcohol, is known to act on the brain's reward systems, which introduces potential confounds, thus making it difficult to differentiate between the effects of alcohol and the effects of sucrose exposure. As well, there is the potential for sucrose to produce unique physiological alterations in exposed offspring [58, 59].

Over the years, researchers have bred various strains of rats and mice to be alcohol-preferring or alcohol–nonpreferring. Recently, a mouse strain selectively bred for high alcohol consumption was shown to consume relatively high concentrations of alcohol (approximately 20 %) which is much higher than what can generally be achieved in rodents [57]. Similarly, there are a number of rat lines selectively bred for high vs. low alcohol consumption, including alcohol preferring (P) and non-preferring (NP) [60], Sardinian alcohol-preferring (sP) and non-preferring (sNP) [61], and Alko, Alcohol (AA) and Alko, Non-Alcohol (ANA) [62]. P and NP rats have been used in some studies of third trimester equivalent alcohol exposure to examine the effects of high levels of alcohol exposure on offspring behavioral development [63].

While many alcohol-exposure models aim at achieving high BALs, work by Daniel Savage and others would suggest that even low levels of in utero alcohol exposure have the potential to produce enduring changes in brain structure and function [64, 65]. For example, offspring from dams consuming a 3 or 5 % liquid alcohol diet throughout gestation, with peak BALs as low as 30 mg/dL and not higher than 83 mg/dL, showed deficits in a learning and memory task in adulthood [66]. Thus, in designing alcohol-exposure models, it is important to recognize species-associated limitations in alcohol consumption as well as to consider carefully the desired level of alcohol exposure and the corresponding BALs. The model chosen should be designed to fit the question being asked.

Of note, the process of obtaining BALs in and of itself can prove to be problematic, as taking a blood sample can be stressful and thus is a possible confounding factor in the investigation of alcohol's teratogenic effects. To avoid stress, one might breed a cohort of animals, in parallel with the study subjects, designated solely for measuring BALs. Another issue is the timing of blood collection. In acute models, blood alcohol concentrations peak soon after alcohol administration, at which time accurate and reproducible BALs are obtained. In the case of chronic exposure models, such as exposure throughout gestation, it is important to take blood samples at an appropriate time point based on drinking patterns [67]. However, due to the inherent variability in the amount and rate at which alcohol is consumed amongst dams, the timing of blood sampling needed to achieve an accurate representation of average or peak BALs is not always obvious. From our experience, sampling approximately 3 h after the presentation of the alcohol diet produces relatively consistent BALs, with maximal peaks occurring around 6 h. We have also measured relatively high BALs by sampling at lights-on, after animals have had ad libitum access to liquid alcohol diets for 12 h.

2.2 Modes of Administration

Miniature swine are one of the only mammals that will consume alcohol voluntarily at relatively high levels, even with available food and water [68, 69], although recently the vervet monkey has been shown to have a voluntary and naturalistic pattern of alcohol consumption, resulting in moderate, sustained blood alcohol levels [70]. In rodent models, various modes of administration have been utilized in FASD research, ranging from gavage or intubation, to intraperitoneal injections, alcohol vapor exposure [71, 72], and self-administration methods, including intravenous infusions, alcohol in the drinking water, and liquid diets (*see* [69, 73] and [74] for a review).

Alcohol administration using gavage/intubation or intraperitoneal injections allow for excellent control over timing and dose, and gavage/intubation is generally the method of choice for studies that demand consistent high BALs among subjects. It is important to note, however, that both ip injection and gavage/intubation may create some level of stress for the animals. Furthermore, although animals appear to show some habituation to the mild restraint necessary for intubation, they never fully habituate, and the intubation procedure itself may thus be somewhat stressful. The stress associated with repeated restraint has been linked to a high level of fetal loss [75], and may potentially exacerbate alcohol-associated deficits, illustrating clear limitations of models involving regular restraint. In addition, added stress to either the dam or offspring may make it difficult to parse out the effects of prenatal alcohol exposure from those of prenatal/early postnatal stress. Exposure using vapor chambers similarly provides excellent control over timing and dose, with minimal disturbance of the dam and/or offspring, and thus can be an excellent method for chronic administration during pregnancy [72]. A practical limitation of the vapor chamber method is that if one does large breedings, multiple chambers will be needed. Self-administration is a common and often preferred method for chronic exposure, particularly during pregnancy. Self-administration methods are minimally invasive, thus reducing additional stress and handling of the dams. Moreover, self-administration more closely reflects the human situation and thus has strong face validity. Nonetheless, each of these methods has limitations. For instance, self-administration via intravenous infusions has proven to be problematic, as rats appear to have a lower threshold for the aversive effects of alcohol relative to primates, and generally fail to self-administer in quantities high enough to produce even mild intoxication. The sucrose fading method developed by Samson and colleagues [76, 77], as well as the schedule-induced polydipsia (SIP) procedure increase the amount of alcohol intake beyond that which an animal would otherwise consume over a given period of time [78, 79]. However, increased consumption of alcohol is concomitant with decreases in food consumption, in part owing to the increased calories derived from alcohol itself [80]. As such, the SIP procedure is well suited to studies that investigate the effects of binge-like alcohol exposure but may be less attractive for chronic use or during pregnancy. In the liquid diet method developed by Sherwin and colleagues [81] and refined by Lieber, DiCarli, and colleagues [82], animals consume a liquid diet containing 30–40 % alcohol-derived calories in lieu of standard rat chow, with high protein variations most commonly used for pregnant and lactating animals [80, 83]. This paradigm provides for a relatively consistent and chronic alcohol intake throughout gestation while maintaining adequate nutritional status.

Each mode of administration has its benefits and drawbacks and as such, for the most part, the research question determines the method incorporated into each study. Further, the focus of the investigation (e.g., on a particular phase of development or on the effects of prolonged exposure) determines the usefulness of one model over another. For instance, investigations into the teratogenic effects of alcohol during specific phases of development and vulnerable periods require acute modes of administration with specific levels of alcohol needed to induce deficits. Self-administration methods are less advantageous for these types of studies as consistently high BALs are typically difficult to achieve and are restricted by the level of alcohol the dam will voluntarily consume. Increasing the percentage of alcohol in the diet to achieve higher BALs reduces consummatory behavior, and thus nutritional intake in both rats and mice, although the level varies across species and strains [84]. Nonetheless, the benefits of the minimally invasive liquid diet procedure make it an attractive model when the research question requires chronic, moderate exposure, such as during pregnancy.

2.3 Nutritional Issues

Alcohol effects on food intake, nutrient absorption and utilization, and blood flow in the placenta can produce nutritional deficiencies among other effects in humans as well as animal models of FASD. For example, chronic alcohol consumption increases intake of "empty" calories (calories not linked to specific nutrients) while decreasing food intake and interfering with the availability and utilization of essential minerals and vitamins including vitamin A, zinc, folate, thiamine, choline, and B6 (for a more complete review *see* [80]). These nutrition-related issues produce confounds in all prenatal alcohol-related models, and are a significant factor in the human situation as well. Although it is well recognized that an adequate control group that addresses these issues in their entirety does not exist, when possible it is prudent to include a pair-fed group that provides for some control related to the effects of dietary restriction induced by alcohol consumption.

2.4 Pair-feeding

Most researchers in FASD studies use pair-fed (PF) or yoked controls to control for the reduced food intake that typically occurs with alcohol consumption. However, the issue of pair-feeding is a complex one [85]. Although pair-feeding is the standard procedure utilized to separate nutritional effects of alcohol from its direct effects, it is at best an imperfect control and at worst, an experimental treatment in itself [80, 86, 87]. First, pair-feeding can only control for the reduced intake of the alcohol-consuming animals, but can never control for the effects of alcohol on absorption and utilization of nutrients. Indeed, alcohol intake always produces secondary nutritional effects that cannot be controlled and that are part of its effects on the body. Secondly, pair-feeding may be stressful to the dam if the reduced ration leaves her hungry, and thus may affect behavioral and physiological responses [87–89]. In contrast to alcohol-consuming dams, who consume their diet ad libitum, PF dams receive a reduced ration of food, less than they would consume if given the same diet without alcohol and therefore typically consume their entire daily ration within a few hours of the food presentation. Despite the fact that experimental diets are formulated to provide optimal nutrition during pregnancy, underfeeding constitutes a mild prenatal stressor, which can itself influence offspring behavioral and physiological responses [46, 90, 91]. Thus, depending on the outcome measure, unique effects of PAE may be observed and PF and ad libitum-fed control animals will be similar. On the other hand, if effects on outcome are due primarily to effects of reduced food intake, PAE and PF offspring may show similar changes in behavior or physiology, and both will differ from control animals. Alternatively, there may be unique effects of pair-feeding in itself, and PF and PAE offspring may differ both from each other and from controls [87, 92–94]. Furthermore, even if behavioral or hormonal responses are similar in PAE and PF offspring, central mechanisms underlying alterations in PAE and PF animals may differ, rather than occurring along a continuum of effects on the same pathway. Of note, pair-feeding may not always have adverse effects. If the alcohol-consuming dam to which a PF dam is yoked consumes a large amount of diet on a particular day, or is a "good" eater in general, PF dams may not be hungry and pair-feeding effects may be significantly attenuated.

In terms of methodology, optimally PF females are weight-matched to alcohol-consuming females. When modeling FASD, a PF animal is also matched by gestation day to a PAE dam and provided with the same daily ration of diet as that consumed by her PAE partner [80, 81]. One caveat to this method, however, is that occasionally a PAE dam will have days in which food consumption is uncharacteristically low. Infrequently, the drop in food intake by a dam may be due to a failure to sustain the pregnancy (pups will be resorbed). Irrespective of cause, the subsequent decrease in the PF dam's ration may result in additional stress. Therefore, on occasion when a particular PAE dam demonstrates reduced intake for several consecutive days it may be necessary to re-pair the PF female with another PAE dam (resulting in two PF dams yoked to a single PAE dam). Alternatively, PF dams may be nutritionally matched based on an average daily consumption of the PAE dams as a group, with daily averages typically calculated in advance over several breedings. This is, in fact, a practice in a number of laboratories and is accepted by the field. However, it should be noted that the latter method might artificially reduce the variability across PF litters whereas variability is inherent in the PAE group.

2.5 Alterations in Maternal–Pup Interactions Following Prenatal Alcohol Exposure

2.5.1 Maternal and Pup Behavior

Potential alcohol-induced alterations in the pup, which may affect its ability to elicit appropriate maternal behavior, should be considered in models of FASD. The relationship between the dam and pup is bidirectional with ultrasonic vocalizations eliciting maternal behavior from the dam and overall maternal behavior impacting the pup's response [95]. Maintenance of this behavioral synchrony between dam and pup is critical for normal development [95]. As a result, it has been necessary not only to assess maternal behavior but to monitor concomitantly dam–pup interactions in studies of prenatal alcohol exposure.

Ness and Franchina [96] have shown that pup ultrasonic vocalizations guide pup retrieval under conditions where pups become separated from the nest, initiate attentive maternal behaviors such as licking/grooming and arched back nursing, and may alter prolactin release in the dam [97, 98]. Therefore, altered vocalization could impact elicitation of maternal retrieval, potentially altering the offspring's developmental trajectory [96]. Some studies have found that pups prenatally exposed to alcohol show selective changes in isolation-induced ultrasonic vocalizations that may result in impaired maternal/infant communication following prenatal alcohol exposure [99, 100]. In addition, pup-based impairments in communication are suggested by the finding that pups prenatally exposed to alcohol demonstrate a decreased ability to elicit retrieval behavior from both control and alcohol-exposed dams [100]. In contrast to Ness and Franchina [96], who found that PF pups vocalize more than both alcohol-exposed and control pups, Marino and colleagues [100] found increased vocalizations by alcohol-exposed pups [96, 100, 101]. Importantly, neither group found preferential pup (either pair-fed or alcohol-exposed) retrieval related to a heightened level of ultrasonic vocalizations [96, 100]. However, it was shown that alcohol-exposed pups that were cross-fostered to surrogate, untreated dams, were retrieved at a slower rate [100], suggesting a possible deficit in the PAE pup's ability to elicit appropriate maternal behavior. Thus, while evidence varies to some extent, overall, the data suggest that prenatal exposure to alcohol may alter the ability of pups to elicit appropriate maternal behavior, although whether this is due to impaired pup vocalizations is not certain.

Whether alcohol consumption during pregnancy has long-term effects on maternal behavior postnatally is also unresolved. Ness and Franchina [96] found that alcohol-exposed dams retrieved fewer pups overall (regardless of prenatal treatment) and showed increased latency of response in all maternal competence measures, compared to both PF and ad libitum-fed control dams. On the other hand, Marino and colleagues [100] found that maternal care of pups by their own dams did not differ between prenatal treatment groups.

2.5.2 Cross-Fostering

In addition to the need to control for alcohol-related undernutrition of the pregnant dam, it is also important to consider the effects of the early postnatal environment on the offspring of dams exposed to alcohol during gestation. Alcohol-exposed pups have been shown to display alterations in feeding including increased latency in the initiation of suckling and decreased time spent suckling. These effects may be mediated by alcohol-induced changes in the pup or changes in maternal behavior as a result of alcohol administration [102, 103]. To reduce possible confounding effects of gestational alcohol consumption on postnatal maternal behavior as well as alterations in milk production that could influence the developmental outcome of the alcohol-exposed offspring, some studies have incorporated cross-fostering or surrogate-fostering procedures that entailed fostering pups from alcohol-exposed dams to unexposed control dams [104]. However, studies aimed specifically at dissociating the teratogenic effects of alcohol from those of postnatal factors showed that the teratogenic effects of alcohol on viability, growth and behavior were not attenuated with cross-fostering [105]. In other words, alcohol-exposed offspring raised by unexposed surrogate control mothers still showed decreased viability and body weights as well as alterations in behavior that were not rescued through cross-fostering [105]. In addition, by showing that control offspring raised by alcohol-exposed dams did not display developmental or behavioral deficits, it was concluded that maternal behavior was not significantly altered with alcohol exposure [105]. More recently, Marino and colleagues [100] confirmed these earlier conclusions, reporting that alcohol exposure does not produce measurable effects on attentive maternal behaviors. These results suggest that any observed effects in offspring prenatally exposed to alcohol may be a direct result of the alcohol exposure, putting into question the utility of the cross fostering paradigm [100]. Interestingly, under some circumstances, fostering may, in itself, serve as a type of treatment. In one study by Giberson and Weinberg [106], for example, it was shown that cross-fostering at birth has differential effects on splenic lymphocyte populations in offspring from prenatal alcohol, pair-fed and control conditions. Moreover, the effects of fostering varied with age and had differential long-term effects on male and female offspring. Thus, rather than serving as a control for indirect maternal effects of alcohol on offspring, fostering was a treatment that actually confounded the effects of prenatal alcohol exposure.

Because of the somewhat inconsistent findings on the effects of maternal alcohol exposure on maternal behavior as well as on the exposed pup's ability to elicit a maternal response, currently the majority of studies in the prenatal alcohol field do not involve cross-fostering. Although cross-fostering may offer the ability to firmly dissociate the effects of prenatal alcohol exposure from those of early life maternal factors, studies have shown prenatal alcohol effects on the offspring to be very stable and to date, many of the alterations seen in alcohol-exposed pups have reliably been linked directly to the alcohol exposure [96, 100, 105]. In addition, cross-fostering poses its own logistical challenges, notably the need to include an extra experimental group of pregnant dams that will act as surrogate mothers to the alcohol-exposed pups. As a result, maternal rearing of alcohol-exposed pups is currently an accepted methodology in the field, with the caveat that one must be aware of possible confounding effects of maternal behavior and other factors in the early environment.

2.6 Housing Considerations

Housing of experimental animals varies across laboratories and experiments. The size and type of the cage (e.g., single vs. multiple levels, height), amount and type of bedding (e.g., limited bedding is used in a naturalistic maternal abuse paradigm [107]), and the availability of enrichment tubes may impact the animals' response. Animals prenatally exposed to alcohol may also respond differently from controls to these variables; therefore, these environmental factors should be considered and kept constant across breedings. Another important issue is the number of animals housed in a cage. Single housing is often crucial when there is a need to monitor food and water consumption of individual animals, increasing accuracy over alternative estimations inherent with pair- or group-housing paradigms. Nonetheless, in studies that utilize social animals, as in the case of most rodent models including the rat, pair-housing is generally the preferred option as it reduces possible confounding factors such as stress related to social isolation. This is an especially pertinent issue when developing FASD models, as there is the potential for animals prenatally exposed to alcohol to show differential responses to environmental factors, such as stress [108–111]. Consequently, although the need to monitor and record food and alcohol consumption in individual dams necessitates a single-housing paradigm during gestation and lactation, we, as well as others, typically opt for pair-housing of the offspring.

2.7 Strain Differences

It has been established that the extent of alcohol-associated deficits is not only related to the pattern and timing of exposure but also to interactions between maternal factors, fetal genotype, and the environment (*see* [112] for review). Evidence as to the importance of genetics in the susceptibility to FASD stems from human twin studies where the concordance of FASD diagnosis was found to be 100 % in monozygotic twins, as compared to 64 % in dizygotic twins [113]. Variations in sensitivity to alcohol's teratogenic effects have long been noted in preclinical research. Notably, inbred mouse strains known as long-sleep and short-sleep mice that have been selectively bred for differential susceptibilities to alcohol's sedative properties [114], show very different profiles of teratogenic effects following in utero alcohol exposure [115, 116]. Long-sleep mice have been shown to be susceptible to deficits in body weight as well as skeletal malformations, whereas short-sleep mice are resistant to these changes despite achieving equivalent BALs [115–117]. Downing and colleagues [118] examined five inbred mouse strains for the teratogenic effects of in utero alcohol exposure following intubation with 5.8 g/kg of alcohol on gestational day 9, a sensitive period during which organogenesis is taking place. This study showed that inbred mouse strains show significant differences in susceptibility. B6 mice are the most sensitive to alcohol exposure, consistently showing weight deficits as well as organ and brain malformations, whereas the 129S6 mice are more resistant to alcohol's teratogenic effects [118]. Downing and colleagues [118] hypothesize that some of the genes and pathways involved in mediating these sleep-related effects may also underlie some of alcohol's teratogenic effects.

The underlying cause(s) or mechanism(s) mediating these strain-specific differences in susceptibilities and patterns of teratogenesis remain unclear. Although differences are likely mediated, at least in part, by fetal genetics, differences could also be mediated, in part or in combination, by maternal genetics and intrauterine/extrauterine environments. Based on the findings of Downing et al. [118] using embryonic day 8.25 embryos in culture and a 6-h binge-like alcohol exposure paradigm, Chen and colleagues [119] aimed to address this question using an embryo culture model involving three different mouse strains. Despite the limitations of extrapolating findings in culture to the in vivo situation, they had the advantage of being able to eliminate such confounding factors as intrauterine environment, maternal factors including genetics, physiology, and metabolism, as well as allowing for tight control over dose and timing of alcohol administration. Consistent with the findings of Downing et al. [118], they found more severe alcohol-induced teratogenic effects in B6 embryos relative to 129S6 embryos [119]. B6 embryos displayed alcohol-associated growth deficits, malformations in kidney, brain ventricles, and vertebrae, and apoptosis in the brain, neural tube, craniofacial region, and heart. Comparatively, 129S6 mice were highly resistant to the teratogenic effects of alcohol [119]. By comparing the effects of alcohol exposure on various strains using a culture model, in which maternal and environmental confounds are minimized or eliminated altogether, the researchers conclude that the differences in alcohol-induced teratogenic effects that exist between strains are highly correlated to differences in fetal genotype [119]. Using both culture models, where one can parse out direct effects of alcohol and examine more directly the influence of fetal genotype, and mammalian models, where maternal and intrauterine effects are likely of critical importance, one can gain critical insight into multiple aspects of alcohol teratogenesis and possible mechanisms underlying these effects.

3   A Model of Moderate, Chronic Prenatal Alcohol Exposure in Sprague Dawley Rats

The current model utilized in our laboratory is a model of moderate prenatal alcohol exposure throughout most of the gestational period (G1–G21). The time of conception (G1) is determined by vaginal smears, which confirm the presence of sperm on the morning following a successful mating. This model has been developed and refined over the past 35 years and is meant to simulate in utero exposure during the first two trimesters in humans. Alcohol is administered utilizing a liquid diet paradigm. The model has provided consistent results showing growth retardation, developmental delay, endocrine dysregulation, immune system alterations, learning, memory and behavioral abnormalities, increased vulnerability to addictive- and depressive-/anxiety-like behaviors, and deficits in higher cognitive functions [25, 26, 44, 45, 120–127]. Our working hypothesis is that many of these alcohol-related abnormalities, can, at least in part, be attributed to fetal programming, the concept that early life experiences can imprint behavioral and physiological systems, alter the developmental trajectory, and have long-term effects on vulnerability/resilience to illnesses or disorders later in life. The hypothalamic-pituitary-adrenal (HPA) or stress system is known to be particularly sensitive to programming by early life events, and it has been suggested that reprogramming of HPA activity by prenatal or early postnatal experiences or exposures may be one of the mechanisms linking early life events with long-term health consequences.

3.1 Breeding

The success in maintaining consistency across studies over the years is no doubt in part related to the fact that we have opted to maintain complete control over the experimental animals, including timed breeding. Charles River-derived virgin male and female Sprague Dawley rats are purchased from our vendor as young adults (males weighing an average of 350 g and females an average of 250 g). Animals are pair-housed in standard Plexiglas cages upon arrival and allowed to adapt to the new environment for a minimum of 1 week. Daily brief (5–10 min per rat) handling occurs during this time, as it allows the animals to habituate to the regular handling that occurs during breeding and throughout gestation. Animals are maintained on a regular 12 h light/dark cycle and provided ad libitum access to food (rat chow) and water.

Following the adaptation period, males are single housed and a single female is introduced into each cage (*see* **Note 1**). An even number of females are mated at any given time to eliminate the need for single housing of the remaining females that may induce undesirable stress prior to mating. Each morning thereafter, vaginal lavage is performed to check for the presence of sperm. A small smooth glass dropper (5 ml) with approximately 1 ml of sterile saline solution is inserted into the vagina (just far enough to expel the liquid). Once the fluid has been expelled into the vaginal opening, it is withdrawn and pipetted onto a glass slide, which is then placed under a light microscope. Although other methods have been used, this method is especially advantageous as it allows for the concurrent monitoring of successful mating and tracking of the estrous cycle. If pregnancy has not occurred, knowing the stage of the estrous cycle can inform when the optimal day will be for a subsequent pairing (e.g., optimally females should be in proestrus or estrus).

Once the presence of sperm has been confirmed, the female is single housed and assigned to one of three groups in a quasi-random manner: PAE (prenatal alcohol exposure), with ad libitum access to a liquid ethanol diet (36 % ethanol-derived calories); PF (pair-fed), yoked to a PAE dam (*see* **Note 2**) and fed a liquid control diet (maltose-dextrin isocalorically substituted for ethanol), with diet volume matched (g per kg body weight per day of gestation); and Control, with ad libitum access to the liquid control diet or a pelleted form of the liquid control diet (*see* **Note 3**). All dams are given ad libitum access to water. We have found that doing this helps prevent dehydration, particularly in the event that a female consumes only a small volume of diet in a particular day. We have never found that providing water in addition to the liquid diet results in overhydration or any adverse effects on the kidneys or on metabolism.

3.2 Feeding

3.2.1 Liquid Diets

We utilize liquid diets specially formulated by Dyets Inc. (Pennsylvania, USA), the Weinberg/Keiver High Protein Liquid Diet (experimental and control) (*see* **Note 3**), designed to provide optimal nutrition to the pregnant females. Ethanol is added to the experimental diet, and maltose-dextrin is isocalorically substituted for ethanol in the control diet, which is fed to the pair-fed group (*see* below). The ad libitum-fed control group receives ad libitum access to a pelleted version of the liquid control diet. This ensures that the control animals obtain a diet with the same nutritional composition as the experimental animals.

3.2.2 Feeding Protocol

Diet is prepared on a daily basis to maintain freshness (*see* **Note 3**). Diet is fed using a Dyets Inc. designed upright graduated cylinder that curves into a feeding trough at the lower end. The trough allows for easy access to the liquid diet but can fill with bedding if not attached at an appropriate height. Bottles with sipper tubes that have small ball-bearings to eliminate leakage are another alternative. These bottles eliminate the worry that bedding might block access to the liquid diet, but have their own drawback in that the diet is quite viscous and may clog the sipper tube. From our experience, although both systems are effective, both require close monitoring to ensure consistent access to diet.

Feeding tubes are removed just prior to lights off, and volume recorded for later calculation. Fresh diet is provided at this time as well. Pregnant dams are weighed weekly (gestation days 1, 7, 14, and 21) during weekly cage changes to reduce the amount of disturbance (*see* **Note 4**). This weight is then used to calculate the diet volumes provided to the PF females. This feeding schedule (feeding just prior to lights off) is designed to minimize shifts in circadian rhythms (particularly, for our studies, the HPA rhythm) that occurs in restricted or meal feeding paradigms, where circadian rhythms re-entrain to the feeding time rather than the light cycle, as feeding time becomes a more dominant cue than light when animals receive a reduced ration of food [88]. The feeding schedule is repeated daily, throughout gestation, until gestation day 22, at which time experimental diet is replaced with ad libitum access to laboratory rat chow (19 % protein) and water.

3.3 Culling and Weaning

3.3.1 Culling

The day of birth can be designated as postnatal day (PN) 1 or PN 0. Each practice has a valid rationale. For instance, PN 0 implies that the pup has just been born; PN 1 would thus be 24 h later. Alternatively, as dams often give birth during the evening hours, PN 1 is assigned as a method to correct for the fact that the pup may have been born anytime between the last check at 1700–1800 h and the time of discovery, ~0900 the following day. In either case, the actual age of the pup may vary by about 12 h. It is important to note the particular practice of each laboratory when citing or replicating protocols. Once birthing is complete (as determined by the presence of clean pups that have milk bands visible on the abdomen) the dams are weighed and pups are sexed. Overall litter weight and number of pups are recorded to determine average pup weight. One might also weigh females and males separately to obtain average pup weight by sex. The number of still-born pups, if any, as well as sickly or stunted pups is also recorded at this time. Litters are culled to an N of 10–12, with an attempt to have equal numbers of males and females per litter (*see* **Note 5**). Dams and pups (separate mean weights for culled males and females) are weighed again on Days 8, 15, and 22 of lactation. All dams are maintained on a 19 % protein lab chow and water, ad libitum, throughout the lactation period. Of note, in litters of less than 8 pups, pups tend to be heavier, and in litters of more than 12 pups, pups tend to be smaller than in litters of 10–12. Thus, culling of pups is critical so that there are equal numbers of pups in all litters and nutritional status of the pups is roughly similar across litters.

3.3.2 Weaning

On PN 22 (**Note 6**) litters are weighed, weaned, and housed with same sex litter mates for a minimum of 1 week before being further divided into groups of 2–3 per cage (necessary to eliminate overcrowding and to separate males and females prior to puberty). At the time that animals enter into the experiment, each animal is re-paired with a non-littermate, same-sex partner from the same treatment group (e.g., PAE paired with PAE). This is a standard procedure when the litter, rather than the animal, is the N used within each subsequent treatment group (**Note 7**). Thus, each cage, rather than animal, is randomly assigned to an experimental condition for subsequent testing.

---

4   Notes

1. If males and females weigh less than expected upon arrival to our facility, breeding is delayed until females weigh a minimum of 250 g. We have found that this increases the likelihood of impregnation upon pairing with the male. Although this may mean females are not entered into breeding all at the same time, we have not found this to be problematic. Indeed, this is actually an advantage as the breeding needs to be spread out over a number of days to ensure equal numbers of animals in each group (pair-feeding requires that animals are matched to an alcohol-consuming dam on G1), and to avoid having to test all offspring in an experiment on exactly the same day.

2. As PF dams are yoked to PAE dams beginning on G1 and feeding begins in the late afternoon of G1, pairing of a PF dam requires that the PAE dam be at least at G2 such that calculations

relating to food consumption can be obtained from the PAE partner and the equivalent volume per g body weight can be calculated for the PF animal. To calculate, the volume of diet consumed by a PAE dam on each day is divided by the dam's weight, providing an ml/g ratio, which is then used to calculate the volume of food provided to each PF dam: vol for PAE/wt of PAE × wt of PF. It is important to note that the type of bottle may impact the amount of available diet. For example, if a standard bottle with a sipper tube is utilized, and bottle sits on the cage top at an angle, the Pair-fed dam will likely not have access to the last 5–10 ml of diet, whereas if a graduated cylinder is used, the Pair-fed dam will likely be able to consume everything. In the case of the standard bottle, we typically add 5–10 ml extra to the bottle (amount not accessible calculated previously) so that the dam can receive her full ration.

3. The liquid diets, including those obtained from Dyets Inc., are typically reconstituted from powder. In order to make 1 L of experimental diet, 820 ml of cold water is blended for 30 s with 0.50 g of xanthan gum (to ensure that the powder fully dissolves). Next, 156.1 g of powdered experimental diet and 67 ml of 95 % ethanol are added and the mixture is blended for an additional 30 s. When reconstituted, this diet contains 1.0 kcal/ml of which 16.4 % are derived from fat, 23 % are derived from carbohydrate, 35.5 % are derived from ethanol, and 25.1 % are derived from protein. The control (PF) diet is also reconstituted. Xanthan gum (0.5 g) is added to 810 ml of water (note volume difference compared to the experimental diet) and blended for 30 s. Next, 245.8 g of powdered control diet is added and blended for an additional 30 s. When reconstituted, this diet contains 1.0 Kcal/ml of which 16.4 % are derived from fat, 58.6 % are derived from carbohydrate and 25 % are derived from protein.

    ***To reduce waste and associated cost, left-over diet may be used the following day but any additional diet not used by the end of the second day is discarded.

    The C dams have ad libitum access to liquid control diet or consume a pelleted form of the liquid control diet in order to provide nutritional consistency across prenatal treatment groups. However, another option for the C dams is to provide them with standard laboratory chow in order to replicate the nutrient intake of the typical laboratory rat, as well as to save on costs. The latter is an acceptable procedure in the field, provided the focus of the study is not on nutritional issues per se.

4. To reduce stress to the dam prior to birthing, cage changing is not performed on G21 but rather nesting material such as sterile paper towel is provided.

5. To equalize litters as much as possible and thereby control for potential confounds, cross-fostering may be done at the time of culling to foster in additional pups. This can only be achieved successfully, however, if fostering occurs between dams from the same prenatal treatment group (e.g., PF pup to PF dam) born on the same day (or within a 24 h period). Litters can be culled to as few as 8 or as many as 12 pups/litter. Typically, for adequate and relatively equal nutrition we do not go below 8 or above 12 pups/litter. If, during the lactation period, pups within a litter die and the n drops below eight, we eliminate that litter from our study. Owing to evidence that indicates that maternal behavior differs in response to male and female pups, it should also be noted that the ratio of males and females should be approximately the same for all litters [128, 129]. Accordingly, even in cases where the research focus is on one sex, it is important to raise a mixed litter.

6. Weaning varies slightly among laboratories but is typically performed on either PN 21 or 22, although in some laboratories it may occur as late as PN 25. Although these differences are often attributable to slight variations in protocol, they may also reflect differences in the determination of day of conception/birth discussed earlier.

7. As pups within a given litter are not truly independent subjects, it is customary to consider the litter as the 'N' for each sex. Therefore, each subsequent treatment or condition within an experiment can potentially contain one male and one female from each litter: i.e., for an n of ten per prenatal group, ten litters must be included in each prenatal treatment group, with one male and one female from each litter available for assignment to each experimental condition. Treating pups within a litter as independent subjects can bias the results and is not typically an acceptable practice. In our laboratory, pups from each breeding are shared among studies occurring simultaneously so that all offspring can be utilized. Spare offspring not assigned into an experiment are typically used for pilot studies.

Acknowledgments

The research in Dr. Weinberg's laboratory is supported by grants from the National Institutes of Health/National Institute on Alcohol Abuse and Alcoholism, NeuroDevNet (Canadian Network of Centres of Excellence), and Canadian Foundation on Fetal Alcohol Research. Student authors were supported by NSERC (TB, VL), IMPART (CIHR STIHR administered by BC Centre of Excellence for Women's Health) (KU), and CFRI/NeuroDevNet Graduate Studentship (KAS).

References

1. Lemoine P et al (1968) Children of alcoholic parents: abnormalities observed in 127 cases. Ouest Med 8:476–482

2. Jones K, Smith D (1973) Recognition of the fetal alcohol syndrome in early infancy. Lancet 2:999–1001

3. Davis J, Li Y, Rankin C (2008) Effects of developmental exposure to ethanol on Caenorhabditis elegans. Alcohol Clin Exp Res 32(5):853–867

4. McClure KD, French RL, Heberlein U (2011) A Drosophila model for fetal alcohol syndrome disorders: role for the insulin pathway. Dis Model Mech 4(3):335–346

5. Bilotta J et al (2002) Effects of embryonic exposure to ethanol on zebrafish visual function. Neurotoxicol Teratol 24(6):759–766

6. Ellis FW, Pick JR (1980) An animal model of the fetal alcohol syndrome in beagles. Alcohol Clin Exp Res 4(2):123–134

7. Lafond JS et al (1985) Effects of maternal alcohol intoxication on fetal circulation and myocardial function: an experimental study in the ovine fetus. J Pediatr 107(6):947–950

8. Lochry EA et al (1982) Effects of acute alcohol exposure during selected days of gestation in C3H mice. Neurobehav Toxicol Teratol 4(1):15–19

9. West JR et al (1984) Prenatal and early postnatal exposure to ethanol permanently alters the rat hippocampus. Ciba Found Symp 105:8–25

10. Weinberg J, Nelson LR, Taylor AN (1986) Hormonal effects of fetal alcohol exposure. In: West J (ed) Alcohol and brain development. Oxford University Press, New York, NY, pp 310–342

11. Dobson CC et al (2012) Sensitivity of modified Biel-maze task, compared with Y-maze task, to measure spatial learning and memory deficits of ethanol teratogenicity in the guinea pig. Behav Brain Res 233(1):162–168

12. Abdollah S, Brien JF (1995) Effect of chronic maternal ethanol administration on glutamate and N-methyl-D-aspartate binding sites in the hippocampus of the near-term fetal guinea pig. Alcohol 12(4):377–382

13. Altshuler HL, Shippenberg TS (1981) A sub-human primate model for fetal alcohol syndrome research. Neurobehav Toxicol Teratol 3(2):121–126

14. Clarren SK, Bowden DM (1982) Fetal alcohol syndrome: a new primate model for binge drinking and its relevance to human ethanol teratogenesis. J Pediatr 101(5):819–824

15. Fisher SE et al (1983) Selective fetal malnutrition: the effect of in vivo ethanol exposure upon in vitro placental uptake of amino acids in the non-human primate. Pediatr Res 17(9):704–707

16. Heaton MB et al (1992) Ethanol exposure affects trophic factor activity and responsiveness in chick embryo. Alcohol 9(2):161–166

17. Lockwood B et al (2004) Acute effects of alcohol on larval zebrafish: a genetic system for large-scale screening. Pharmacol Biochem Behav 77(3):647–654

18. Means LW, Burnette MA, Pennington SN (1988) The effect of embryonic ethanol exposure on detour learning in the chick. Alcohol 5(4):305–308

19. Gerlai R (2003) Zebra fish: an uncharted behavior genetic model. Behav Genet 33(5):461–468

20. Su B et al (2001) Genetic influences on craniofacial outcome in an avian model of prenatal alcohol exposure. Alcohol Clin Exp Res 25(1):60–69

21. Ahlgren SC, Bronner-Fraser M (1999) Inhibition of sonic hedgehog signaling in vivo results in craniofacial neural crest cell death. Curr Biol 9(22):1304–1314

22. Morgan PG, Sedensky MM (1995) Mutations affecting sensitivity to ethanol in the nematode, Caenorhabditis elegans. Alcohol Clin Exp Res 19(6):1423–1429

23. Randall CL, Ekblad U, Anton RF (1990) Perspectives on the pathophysiology of fetal alcohol syndrome. Alcohol Clin Exp Res 14(6):807–812

24. Goodlett CR, Horn KH, Zhou FC (2005) Alcohol teratogenesis: mechanisms of damage and strategies for intervention. Exp Biol Med (Maywood) 230(6):394–406

25. Zhang X, Sliwowska JH, Weinberg J (2005) Prenatal alcohol exposure and fetal programming: effects on neuroendocrine and immune function. Exp Biol Med (Maywood) 230(6):376–388

26. Hellemans KG et al (2010) Prenatal alcohol exposure: fetal programming and later life vulnerability to stress, depression and anxiety disorders. Neurosci Biobehav Rev 34(6):791–807

27. Webster WS et al (1980) Teratogenesis after acute alcohol exposure in inbred and outbred mice. Neurobehav Toxicol 2(3):227–234

28. Sulik KK, Johnston MC (1983) Sequence of developmental alterations following acute ethanol exposure in mice - craniofacial

features of the fetal alcohol syndrome. Am J Anat 166(3):257–269

29. Sulik KK, Schoenwolf GC (1985) Highlights of craniofacial morphogenesis in mammalian embryos, as revealed by scanning electron-microscopy. Scan Electron Microsc (Pt 4): 1735–1752

30. Sinden JD, Le Magnen J (1982) Parameters of low-dose ethanol intravenous self-administration in the rat. Pharmacol Biochem Behav 16(1):181–183

31. Dobbing J, Sands J (1979) Comparative aspects of the brain growth spurt. Early Hum Dev 3(1):79–83

32. Kimura S et al (1985) Immunoregulation in the rat: ontogeny of B cell responses to types 1, 2, and T-dependent antigens. J Immunol 134(5):2839–2846

33. Gottesfeld Z, LeGrue SJ (1990) Lactational alcohol exposure elicits long-term immune deficits and increased noradrenergic synaptic transmission in lymphoid organs. Life Sci 47(5):457–465

34. Mennella J (2001) Alcohol's effect on lactation. Alcohol Res Health 25(3):230–234

35. Pepino MY et al (2002) Disruption of maternal behavior by alcohol intoxication in the lactating rat: a behavioral and metabolic analysis. Alcohol Clin Exp Res 26(8):1205–1214

36. West JR (1993) Use of pup in a cup model to study brain development. J Nutr 123(2 Suppl):382–385

37. Goodlett CR, Johnson TB (1997) Neonatal binge ethanol exposure using intubation: timing and dose effects on place learning. Neurotoxicol Teratol 19(6):435–446

38. Levine S (1967) Maternal and environmental influences on the adrenocortical response to stress in weanling rats. Science 156(772): 258–260

39. Idanpaan-Heikkila J et al (1972) Elimination and metabolic effects of ethanol in mother, fetus, and newborn infant. Am J Obstet Gynecol 112(3):387–393

40. Meier-Tackmann D et al (1985) Human placental aldehyde dehydrogenase. Subcellular distribution and properties. Enzyme 33(3): 153–161

41. Burd L et al (2007) Ethanol and the placenta: a review. J Matern Fetal Neonatal Med 20(5):361–375

42. Karl PI et al (1988) Acetaldehyde production and transfer by the perfused human placental cotyledon. Science 242(4876):273–275

43. Anderson RA Jr (1981) Endocrine balance as a factor in the etiology of the fetal alcohol syndrome. Neurobehav Toxicol Teratol 3(2):89–104

44. Sliwowska J, Zhang X, Weinberg J (2006) Prenatal ethanol exposure and fetal programming: implications for endocrine and immune development and long-term health. In: Miller MW (ed) Brain development: normal processes and the effects of alcohol and nicotine. Oxford University Press, Oxford [u.a.]

45. Gallo PV, Weinberg J (1986) Organ growth and cellular development in ethanol-exposed rats. Alcohol 3(4):261–267

46. Weinberg J (1989) Prenatal ethanol exposure alters adrenocortical development of offspring. Alcohol Clin Exp Res 13(1):73–83

47. Driscoll CD, Streissguth AP, Riley EP (1990) Prenatal alcohol exposure: comparability of effects in humans and animal models. Neurotoxicol Teratol 12(3):231–237

48. Chernoff GF (1977) The fetal alcohol syndrome in mice: an animal model. Teratology 15(3):223–229

49. Randall CL, Taylor WJ (1979) Prenatal ethanol exposure in mice: teratogenic effects. Teratology 19(3):305–311

50. West JR, Kelly SJ, Pierce DR (1987) Severity of alcohol-induced deficits in rats during the third trimester equivalent is determined by the pattern of exposure. Alcohol Alcohol Suppl 1:461–465

51. West JR et al (1989) Manipulating peak blood alcohol concentrations in neonatal rats: review of an animal model for alcohol-related developmental effects. Neurotoxicology 10(3):347–365

52. Pierce DR, West JR (1986) Blood alcohol concentration: a critical factor for producing fetal alcohol effects. Alcohol 3(4):269–272

53. Pierce DR, West JR (1986) Alcohol-induced microencephaly during the third trimester equivalent: relationship to dose and blood alcohol concentration. Alcohol 3(3): 185–191

54. Bonthius DJ, West JR (1988) Blood alcohol concentration and microencephaly: a dose-response study in the neonatal rat. Teratology 37(3):223–231

55. Maier SE, West JR (2001) Drinking patterns and alcohol-related birth defects. Alcohol Res Health 25(3):168–174

56. Dole VP, Ho A, Gentry RT (1985) Toward an analogue of alcoholism in mice: criteria for recognition of pharmacologically motivated drinking. Proc Natl Acad Sci U S A 82(10): 3469–3471

57. Crabbe JC et al (2009) A line of mice selected for high blood ethanol concentrations shows drinking in the dark to intoxication. Biol Psychiatry 65(8):662–670

58. Carelli RM (2002) The nucleus accumbens and reward: neurophysiological investigations in behaving animals. Behav Cogn Neurosci Rev 1(4):281–296

59. Frazier CR et al (2008) Sucrose exposure in early life alters adult motivation and weight gain. PLoS One 3(9):e3221

60. Roman E et al (2012) Behavioral profiling of multiple pairs of rats selectively bred for high and low alcohol intake using the MCSF test. Addict Biol 17:33

61. Colombo G et al (2006) Phenotypic characterization of genetically selected Sardinian alcohol-preferring (sP) and -non-preferring (sNP) rats. Addict Biol 11(3–4):324–338

62. Sinclair JD, Le AD, Kiianmaa K (1989) The AA and ANA rat lines, selected for differences in voluntary alcohol consumption. Experientia 45(9):798–805

63. Riley EP et al (1993) Alterations in activity following alcohol administration during the third trimester equivalent in P and NP rats. Alcohol Clin Exp Res 17(6):1240–1246

64. Hamilton DA et al (2010) Prenatal exposure to moderate levels of ethanol alters social behavior in adult rats: relationship to structural plasticity and immediate early gene expression in frontal cortex. Behav Brain Res 207(2):290–304

65. Hamilton DA et al (2010) Patterns of social-experience-related c-fos and Arc expression in the frontal cortices of rats exposed to saccharin or moderate levels of ethanol during prenatal brain development. Behav Brain Res 214(1):66–74

66. Savage DD et al (2002) Dose-dependent effects of prenatal ethanol exposure on synaptic plasticity and learning in mature offspring. Alcohol Clin Exp Res 26(11):1752–1758

67. Wiener SG et al (1981) Interaction of ethanol and nutrition during gestation: influence on maternal and offspring development in the rat. J Pharmacol Exp Ther 216(3):572–579

68. Dexter JD et al (1976) Sinclair(S-1) miniature swine as a model for the study of human alcoholism. Ann N Y Acad Sci 273:188–193

69. Riley EP, Meyer LS (1984) Considerations for the design, implementation, and interpretation of animal models of fetal alcohol effects. Neurobehav Toxicol Teratol 6(2):97–101

70. Burke MW et al (2009) Neuronal reduction in frontal cortex of primates after prenatal alcohol exposure. Neuroreport 20(1):13–17

71. Lee S et al (1990) Effect of prenatal exposure to ethanol on the activity of the hypothalamic-pituitary-adrenal axis of the offspring: importance of the time of exposure to ethanol and possible modulating mechanisms. Mol Cell Neurosci 1(2):168–177

72. Kang SS et al (2004) Development of individual alcohol inhalation chambers for mice: validation in a model of prenatal alcohol. Alcohol Clin Exp Res 28(10):1549–1556

73. Lieber CS, DeCarli LM, Sorrell MF (1989) Experimental methods of ethanol administration. Hepatology 10(4):501–510

74. Ponnappa BC, Rubin E (2000) Modeling alcohol's effects on organs in animal models. Alcohol Res Health 24(2):93–104

75. Sugino N et al (1994) Effects of restraint stress on luteal function in rats during mid-pregnancy. J Reprod Fertil 101(1):23–26

76. Grant KA, Samson HH (1985) Oral self administration of ethanol in free feeding rats. Alcohol 2(2):317–321

77. Grant KA, Samson HH (1985) Induction and maintenance of ethanol self-administration without food deprivation in the rat. Psychopharmacology (Berl) 86(4):475–479

78. Samson HH, Falk JL (1974) Schedule-induced ethanol polydipsia: enhancement by saccharin. Pharmacol Biochem Behav 2(6):835–838

79. Meisch RA, Thompson T (1972) Ethanol intake during schedule-induced polydipsia. Physiol Behav 8(3):471–475

80. Weinberg J (1984) Nutritional issues in perinatal alcohol exposure. Neurobehav Toxicol Teratol 6(4):261–269

81. Sherwin BT, Jacobson S, Zagorski D (1981) A rat model of the fetal alcohol syndrome – preliminary histological findings. Curr Alcohol 8:495–510

82. Lieber CS, DeCarli LM (1989) Liquid diet technique of ethanol administration: 1989 update. Alcohol Alcohol 24(3):197–211

83. Lieber CS, DeCarli LM (1982) The feeding of alcohol in liquid diets: two decades of applications and 1982 update. Alcohol Clin Exp Res 6(4):523–531

84. George FR (1987) Genetic and environmental factors in ethanol self-administration. Pharmacol Biochem Behav 27(2):379–384

85. Weinberg J (1985) Effects of ethanol and maternal nutritional status on fetal development. Alcohol Clin Exp Res 9(1):49–55

86. Glavas MM et al (2007) Effects of prenatal ethanol exposure on basal limbic-hypothalamic-pituitary-adrenal regulation: role of corticosterone. Alcohol Clin Exp Res 31(9):1598–1610

87. Hofmann C et al (1999) Glucocorticoid fast feedback is not altered in rats prenatally exposed to ethanol. Alcohol Clin Exp Res 23(5):891–900

88. Gallo PV, Weinberg J (1981) Corticosterone rhythmicity in the rat: interactive effects of dietary restriction and schedule of feeding. J Nutr 111(2):208–218

89. Weinberg J, Gallo PV (1982) Prenatal ethanol exposure: pituitary-adrenal activity in pregnant dams and offspring. Neurobehav Toxicol Teratol 4(5):515–520

90. Kwong WY et al (2000) Maternal undernutrition during the preimplantation period of rat development causes blastocyst abnormalities and programming of postnatal hypertension. Development 127(19):4195–4202

91. Levine S, Wiener S (1976) A critical analysis of data on malnutrition and behavioral deficits. Adv Pediatr 22:113–136

92. Redila VA et al (2006) Hippocampal cell proliferation is reduced following prenatal ethanol exposure but can be rescued with voluntary exercise. Hippocampus 16(3):305–311

93. Taylor AN et al (1988) Maternal alcohol consumption and stress responsiveness in offspring. Adv Exp Med Biol 245:311–317

94. Yirmiya R et al (1998) Effects of prenatal alcohol and pair feeding on lipopolysaccharide-induced secretion of TNF-alpha and corticosterone. Alcohol 15(4):327–335

95. Smotherman WP et al (1978) Orientation to rat pup cues - effects of maternal experiential history. Anim Behav 26(Feb):265–273

96. Ness JW, Franchina JJ (1990) Effects of prenatal alcohol exposure on rat pups ability to elicit retrieval behavior from dams. Dev Psychobiol 23(1):85–99

97. Terkel J, Damassa DA, Sawyer CH (1979) Ultrasonic cries from infant rats stimulate prolactin-release in lactating mothers. Horm Behav 12(1):95–102

98. Sewell GD (1970) Ultrasonic communication in rodents. Nature 227(5256):410

99. Barron S, Gilbertson R (2005) Neonatal ethanol exposure but not neonatal cocaine selectively reduces specific isolation-induced vocalization waveforms in rats. Behav Genet 35:93–102

100. Marino MD et al (2002) Ultrasonic vocalizations and maternal-infant interactions in a rat model of fetal alcohol syndrome. Dev Psychobiol 41(4):341–351

101. Fernandez K et al (1983) Effects of prenatal alcohol on homing behavior, maternal responding and open-field activity in rats. Neurobehav Toxicol Teratol 5(3):351–356

102. Barron S, Kelly SJ, Riley EP (1991) Neonatal alcohol exposure alters suckling behavior in neonatal rat pups. Pharmacol Biochem Behav 39(2):423–427

103. Rockwood GA, Riley EP (1990) Nipple attachment behavior in rat pups exposed to alcohol in utero. Neurotoxicol Teratol 12(4):383–389

104. Abel EL, Dintcheff BA (1978) Effects of prenatal alcohol exposure on growth and development in rats. J Pharmacol Exp Ther 207(3):916–921

105. Osborne GL, Caul WF, Fernandez K (1980) Behavioral effects of prenatal ethanol exposure and differential early experience in rats. Pharmacol Biochem Behav 12(3):393–401

106. Giberson PK, Weinberg J (1993) Fetal alcohol syndrome and functioning of the immune system. Alcohol Health Res World 16:29–38

107. Molet J et al (2014) Naturalistic rodent models of chronic early-life stress. Dev Psychobiol 56:1675

108. Schneider ML et al (2002) The impact of prenatal stress, fetal alcohol exposure, or both on development: perspectives from a primate model. Psychoneuroendocrinology 27(1–2):285–298

109. Schneider ML, Moore CF, Kraemer GW (2004) Moderate level alcohol during pregnancy, prenatal stress, or both and limbic-hypothalamic-pituitary-adrenocortical axis response to stress in rhesus monkeys. Child Dev 75(1):96–109

110. Riley EP (1990) The long-term behavioral effects of prenatal alcohol exposure in rats. Alcohol Clin Exp Res 14(5):670–673

111. Weinberg J et al (2008) Prenatal alcohol exposure: foetal programming, the hypothalamic-pituitary-adrenal axis and sex differences in outcome. J Neuroendocrinol 20(4):470–488

112. Abel EL, Hannigan JH (1995) Maternal risk factors in fetal alcohol syndrome: provocative and permissive influences. Neurotoxicol Teratol 17(4):445–462

113. Streissguth AP, Dehaene P (1993) Fetal alcohol syndrome in twins of alcoholic mothers - concordance of diagnosis and Iq. Am J Med Genet 47(6):857–861

114. McClearn GE, Kakihana R (1981) Selective breeding for ethanol sensitivity: short-sleep and long-sleep mice. In: McClearn DE, Deitrich RA, Erwin VG (eds) Development of animal models as pharmacogenetic tools, vol 6, NIAAA research monograph. U.S. Government Printing Office, Washington, DC

115. Gilliam DM et al (1988) Ethanol teratogenesis in mice selected for differences in alcohol sensitivity. Alcohol 5(6):513–519

116. Gilliam DM et al (1989) Ethanol teratogenesis in selectivity bred long-sleep and short-sleep mice: a comparison to inbred C57BL/6J mice. Alcohol Clin Exp Res 13(5):667–672

117. Gilliam DM, Kotch LE (1996) Dose-related growth deficits in LS but not SS mice prenatally exposed to alcohol. Alcohol 13(1): 47–51

118. Downing C et al (2009) Ethanol teratogenesis in five inbred strains of mice. Alcohol Clin Exp Res 33(7):1238–1245

119. Chen Y et al (2011) Strain differences in developmental vulnerability to alcohol exposure via embryo culture in mice. Alcohol Clin Exp Res 35(7):1293–1304

120. Comeau WL, Winstanley CA, Weinberg J (2014) Prenatal alcohol exposure and adolescent stress - unmasking persistent attentional deficits in rats. Eur J Neurosci 40:3078

121. Gabriel KI et al (2006) Prenatal ethanol exposure alters sensitivity to the effects of corticotropin-releasing factor (CRF) on behavior in the elevated plus-maze. Psychoneuroendocrinology 31(9):1046–1056

122. Hellemans KGC et al (2008) Prenatal alcohol exposure increases vulnerability to stress and anxiety-like disorders in adulthood. Ann N Y Acad Sci 1144(1):154–175

123. Keiver K, Weinberg J (2004) Effect of duration of maternal alcohol consumption on calcium metabolism and bone in the fetal rat. Alcohol Clin Exp Res 28(3):456–467

124. Kim CK et al (1999) Effects of prenatal exposure to alcohol on the release of adenocorticotropic hormone, corticosterone, and proinflammatory cytokines. Alcohol Clin Exp Res 23(1):52–59

125. O'Neil R, Lan N, Innis S, Devlin A, Ellis L, Chan B, Weinberg J (2007) Metabolic effects of prenatal ethanol exposure and epigenetic reprogramming of the HPA axis. Alcohol Clin Exp Res 31:100A

126. Raineki C et al (2014) Neurocircuitry underlying stress and emotional regulation in animals prenatally exposed to alcohol and subjected to chronic mild stress in adulthood. Front Endocrinol (Lausanne) 5:5

127. Uban KA et al (2014) Amphetamine sensitization and cross-sensitization with acute restraint stress: impact of prenatal alcohol exposure in male and female rats. Psychopharmacology. doi:10.1007/s00213-014-3804-y

128. Moore C (1992) The role of maternal stimulation in the development of sexual behavior and its neural basis. Ann N Y Acad Sci 662: 160–177

129. Moore CL, Chadwick-Dias AM (1986) Behavioral responses of infant rats to maternal licking: variations with age and sex. Dev Psychobiol 19(5):427–438

Chapter 14

Modeling Intellectual Disability in Drosophila

Alaura Androschuk and Francois V. Bolduc

Abstract

Intellectual disability (ID) is a common neurodevelopmental disorder affecting 3 % of the population. At this time, no pharmacological treatment has been identified for the treatment of ID patients. An increasing number of genes are identified in patients with ID but a significant gap persists between gene discovery and treatment. Lack of treatment can be explained by our poor understanding of the molecular mechanisms linking cognition and genes. We present here a model of learning and memory in Drosophila that has now been used by several groups to gain genetic understanding in memory defects related to ID genes. We review the pertinent background about the assay and its relation to ID. We also review the assay in detail and provide some advice about troubleshooting tips. Finally, we discuss how flies and other animal models could be used synergistically to develop new treatment for ID patients.

Key words Intellectual disability, Genetics, Drosophila, Learning and memory, Classical conditioning, Learning disability

1 Background and Historical Overview

The last decade has witnessed an explosion in the number of novel genetic defects identified in patients with intellectual disability (ID), previously known as mental retardation (MR). This has been, in part, due to the development of guidelines suggesting that genetic investigations be part of the investigation of patients affected with ID [1], as well as increased use of novel genetic tests, such as array-based comparative genomic hybridization and more recently whole-exome sequencing, in the clinical setting. Such techniques have provided an unbiased, more in-depth genome-wide analysis of patients with ID [2]. This greater genetic understanding has made the need for treatment even more striking to families and physicians. The important gap in the knowledge of how ID genes lead to cognitive defects has made it difficult to develop a pharmacological treatment [3–7]. Nonetheless, experimental work in animal models of gene-specific ID has shown promising

Jerome Y. Yager (ed.), *Animal Models of Neurodevelopmental Disorders*, Neuromethods, vol. 104, DOI 10.1007/978-1-4939-2709-8_14, © Springer Science+Business Media New York 2015

avenues of treatment that have already translated to clinical trials [8, 9]. Here we focus on how the fruit fly model of memory has helped us to understand how to treat ID.

1.1 Intellectual Disability

ID is clinically defined as impairment in intelligence (as measured by intellectual quotient (IQ)), communication, self-care, and ability to function independently. ID affects 3 % of the population [10–14] or about 560,000 Canadian children under 14 years of age [15], and costs $1.3 billion each year for schooling and specialized medical services [16]. Although many etiologies have been linked to ID, severe ID is most often associated with genetic defects [17]. ID causes significant impairments in quality of life because in addition to the decreased IQ, patients also often present with comorbid conditions such as epilepsy, autism, obsessive-compulsive disorder, sleep disturbances, and attention-deficit and hyperactivity syndrome [17].

1.2 Memory Is Commonly Affected in ID Patients

One approach to understanding complex neurological disorders such as ID is to identify specific symptoms and signs shared by patients and then divide the disorder based on the presence of such symptoms and signs. This approach has been labeled endophenotyping. One of the common characteristics of many ID syndromes has been the presence of learning and memory defects. Learning is generally defined as the acquisition of new knowledge (either facts or skills). On the other hand, memory refers to the long-term storage of that information in ways that lead to persistent modification [6]. Interestingly, defects in various steps of memory ranging from memory formation to consolidation have been established in several forms of ID [18–20]. For instance, in Fragile X syndrome, the most common single-gene type of ID in boys, patients present stereotyped visuospatial memory defects [21, 22]. Interestingly, memory defects were also identified in the Fragile X knockout mouse when it was tested in a Morris water maze [23–25]. Memory defects in various paradigms were also observed in fruit flies, Drosophila, in which the homologue of the Fragile X gene, known as dfmr1, was deleted [26, 27]. Specific memory defects are also observed in patients with Down [28, 29] and Williams syndromes [30, 31]. These results taken altogether suggest that, for an increasing number of ID genes, their function is conserved throughout evolution and that the study of simpler behaviors in model systems can serve to test genetic and pharmacological rescue approach.

1.3 Experimental Understanding of Memory

Most of the experimental models and behavioral tests used today in laboratories have been derived from principles identified by experimental psychologists. Some of the early insights on the process of memory formation came from experiments conducted by Herman Ebbinghaus in 1885 [32]. Ebbinghaus studied memory formation by generating nonsense words (to remove the effect of previous familiarity) and testing his recall of those words at various time points. He showed that memory recall was initially close to 100 % but that long-term memory performance stabilized at around 30 % recall. Surprisingly similar "forgetting curves" have since been identified in several other animal models, including Drosophila [33]. Memories were therefore labile. This supported the work of Theodule Ribot, who had shown that memories are initially labile before stabilizing [34]. Finally, Georg Muller and Alfons Pilzecker showed that memory became more stable through the process of consolidation [35].

The next step in understanding the biological basis of memory came via a combination of human cases and work in animal models. Ivan Pavlov discovered that animals could be conditioned to respond to a stimulus in a novel way [36]. Indeed, he trained dogs by presenting them with food and the sound of a bell simultaneously. Prior to training the food would trigger salivation but the bell would not trigger any response. After conditioning, the animal started responding to the bell by salivating even in the absence of food. He termed this: the principle of stimulus substitution. This form of training has been since known as classical conditioning. John Watson brought interest into Pavlov work in America and translated Pavlov's findings from dogs to human in his famous study of learning in children. Indeed, Watson had showed that classical conditioning could also occur in humans, and constituted a basic building block of human epistemology. A young child (Little Albert) was exposed to a series of toy animals and did not present any specific response. Among them was a toy rat that was then later presented with a strong noise. The child later exhibited fear when presented with the same toy rat. In addition, he reinforced the need for repetition in training for long-term memory [38]. He also developed the concept of comparative psychology. Edward Thorndike studied animals and found that they displayed features of learning as indicated by the progressive decrease in response time [37]. He found that animals were capable of more than automatisms, a prevailing concept since Descartes. He also localized the process to the synapse. Donald Hebb expanded on the role of synaptic modification as an important site for learning and memory. He formalized a theoretical model in which changes in synaptic strength were related to repeated simultaneous firing of neuron pairs [39]. He also developed the idea that neuronal networks could become linked together if they were activated together.

Finally, the case of patient H.M. who had each hippocampus successively removed as part of the surgical management of his temporal lobe epilepsy, resulting in profound anterograde memory loss helped localized memory formation. This unfortunate patient revealed the importance of the hippocampus as the initial bed for memory and led to a vast interest into the in vitro study of synaptic plasticity in the hippocampus (long-term potentiation, long-term depression) [40]. A student of Hebb, Brenda Millner studied patient H.M. extensively and showed that several forms (episodic and procedural) of memory could be observed in human and that each type was localized in different neuroanatomical structures [41, 42].

1.4 Drosophila Is a Well-Established Animal Model to Study the Genetics of Memory

The fruit fly Drosophila has contributed significantly to the understanding of the genetic basis of memory. Seymour Benzer, at the California Institute of Technology, pioneered the use of Drosophila for research of the genetic basis of memory in the 60s [43, 44]. He developed an assay whereby flies learned to associate an odor with a foot shock. The assay was based on flies' normal attraction to light (phototaxis) to drive them in a tube where an odor was presented in association with a foot shock provided by a grid inside the tube [45]. The flies were then brought back into a resting tube and then allowed to climb (with the help of a light source again) in another tube where they were exposed to a second odor, this time without the application of a foot shock. Flies were then brought back to the resting tube before being allowed to climb in the tube containing the shocked odor or a tube containing the non-shocked odor. Flies then learned to avoid the odor that was associated with a foot shock. To discover genes responsible for this behavior, Benzer used a forward genetic approach where mutants were produced and then screened for their learning performance [45]. They used a chemical named ethane methyl sulfonate (EMS) to induce mutations in the DNA of flies. They initially identified two very important mutants: *dunce* [45] and *rutabaga* [46]. *Dunce* was later found to be a mutant for the gene encoding phosphodiesterase, the enzyme that degrades cAMP [47]. On the other hand, *rutabaga* is a mutant for the enzyme that synthesizes cAMP, adenylyl cyclase [46, 48]. These results supported earlier findings by Eric Kandel that cAMP was involved in the sensitization of the gill withdrawal reflex in the sea slug Aplasia [49].

Tim Tully and William Quinn then modified this assay to perform classical conditioning (or Pavlovian) using an olfactory cue [50]. They redesigned the training apparatus so that flies would be placed in a training chamber and then presented each odor successively without having to change chambers. Only one odor (to be remembered) was paired with a foot shock. Several genes have now been identified for each step of memory formation from learning to long-term memory (LTM) [7, 51–53].

Tully showed that, as in the humans, repeated training was required for long-term memory formation in Drosophila [33]. He showed that flies exposed to 10 training sessions separated by rest intervals would have memory lasting for over 1 week. Together with Jerry Yin, he showed that long-term memory was dependent on the synthesis of new protein and transcription factors such as the cAMP-responsive-element-binding protein (CREB) [54].

LTM lasts for at least 1 week but is blocked in wild-type (WT) flies by the protein synthesis inhibitor cycloheximide [33]. Alternatively, flies trained for the same amount of time without rest intervals form a memory that lasts at most 4 days. Importantly, at 1 day both forms of memory are present, allowing one to dissect protein synthesis-dependent versus -independent memory in the memory mutant.

Several other Drosophila research groups have also greatly contributed to our understanding of memory by using Drosophila memory assays based on other behaviors such as courtship behavior, visual conditioning, or appetitive conditioning. In courtship memory testing, we assess if a male remembers not to court a female that has been previously courted. In visual learning a fly has to remember where to go to avoid a heat beam from a laser. Finally, in appetitive learning, conditioning is based on positive reward from food as opposed to punitive conditioning from a shock. Similarly, several of the genes identified as required for memory in Drosophila have also been involved in human neurological disorders (Table 1). This reinforced the notion that an overlap exists between the molecular processes involved in memory and ID.

Table 1
Genes involved in memory formation in Drosophila and human mental retardation

Memory phase	Human disorder	Fly gene	Reference
Learning	Neurofibromatosis	*neurofibromin*	[73]
	Pseudohypoparathiroidism	*Gs Alpha*	[74]
	Congenital muscular dystrophy	*Integrin*	[75, 76]
	Hydrocephalus	*neuroglian*	[77]
	Down syndrome	*minibrain*	[78, 79]
	Epilepsy X-linked with variable learning disability	*synapsin*	[80, 81]
Short-term memory	Schizophrenia	*dunce*	[45, 47]
	Down syndrome	*leonardo*	[82, 83]
	Down syndrome	*nebula*	[84]
	CHARGE syndrome	*kismet*	[85]
Long-term memory	Autosomal recessive mental retardation	*tequila*	[59]
	Fragile X syndrome	*dfmr1*	[26, 27]
	Angelman syndrome	*dube3a*	[58]
	Periventricular nodular heterotopia	*cheerio*	[57]
	Coffin–Lowry syndrome	*rsk2*	[86]
	Alzheimer disease	*crammer*	[87]
	Alzheimer disease	*notch*	[88]

The genes are grouped based on which phase of memory formation is affected in Drosophila learning and memory

1.5 Memory Genes Identified in Drosophila Overlap with Genes Found in Human Patients with Cognitive Dysfunction

Several genes identified in memory assays in Drosophila are also responsible for ID in humans. One of the first examples recognized was linked to CREB: indeed, the CREB binding-protein (CBP) has been linked to an ID syndrome known as Rubinstein–Taybi syndrome (RTS) [55]. Enhancement of the cAMP-CREB pathway using phosphodiesterase inhibitors (PDE) have now been shown to reverse the memory defects of mice with CBP mutations, and are being developed as drug treatments for patients with RTS and other ID syndromes [56]. In addition, to studying genes already identified in human (termed reverse genetics), the Drosophila model allows investigators to mutagenize wild-type flies and then screen for learning or memory defects (termed forward genetics). The Tully laboratory conducted a large forward genetic screen for long-term memory specific mutants. Several of the mutants identified were linked to RNA binding proteins and translation control [57]. Interestingly, one of the mutants identified (*joy*) coded for the gene *cheerio* (homologue of filamin A). Filamin A mutations are found in patients with Periventricular Nodular Heterotopia which is associated with epilepsy and cognitive defects. As mentioned above, alternatively, Drosophila carrying mutation in ID gene homologue can be tested for learning and memory defects. So far, examples include Angelman syndrome [58], Fragile X syndrome [26, 27], and X-linked mental retardation linked to Neurotrypsin [59], among others (Table 1). The Drosophila's genome contains most (87 %) of the genes involved in neurological function [60], proving that Drosophila is an ideal model for the study of neurological disorders.

1.6 Drosophila as an Emerging Model for ID Research

Drosophila are increasingly accepted as an ideal animal model to understand the genetic basis of ID. (1) First, considering the complexity of cognition and the impressive number of genes now linked to it, a genetically tractable animal is essential. Several mutant lines already exist in Drosophila. (2) Second, RNA interference transgenic lines have been developed for most genes and are readily available through stock centers. These flies allow for the reproduction of the effect of loss-of-function of the gene of interest. (3) Thirdly, Drosophila has a fast life cycle (2 weeks) which allow us to perform genetic interaction experiments rapidly and to screen for multiple candidate genes relatively fast. (4) In addition, genetic tools (such as the GAL4-UAS system) [61] available in Drosophila allow one to spatially or temporally (or both) restrict the expression of transgenes. These transgenes can lead to overexpression (by using wild-type construct) or loss-of-function (by using RNA interference construct). This can help to understand the effect of gene dosage, a common theme in copy number variation detected by array-based comparative genomic hybridization (CGH). In addition, these tools can be used to dissect the temporal requirement of a gene in brain function. A gene may be functioning during development and/or at postnatal stages [62]. The GAL4 driver can be controlled by temperature acutely to distinguish the effect of a gene during development or in adults. In other words, this helps to dissect between brain development and postnatal neuronal dysfunction. This is extremely important for two reasons: (1) it provides a strong foundation for the probability of a treatment given to children after diagnosis to be successful and (2) it may change the perception that all manifestations of genetically linked ID are irreversible. Finally, Drosophila has become well accepted as a valid model for neurological disorders. Drosophila has been used to study a variety of neurological disorders such as Alzheimer's [63], Parkinson's [64], Huntington's [65], and more recently Rett syndrome [66]. Moreover, in the last 5 years, Drosophila has emerged as a model for the study of ID genes (reviewed in ref. [7]). As mentioned above, several experimental paradigms have been developed to study the genetic basis of memory, including visual and courtship memory, but we will focus our chapter here on the methodology used in the olfactory-based classical conditioning.

1.7 Establishing a Genetic Network

Dysfunction of a single ID gene has repercussion on several other genes belonging to the same pathway or to parallel pathways. This has been well-established in cancer but is only now emerging in neurological disorders. Flies can be used to test in vivo the interaction between genes. Homozygous mutant flies for a given ID gene can be mated to flies homozygous mutant for a candidate interacting gene (Fig. 1). The progeny will contain one copy of

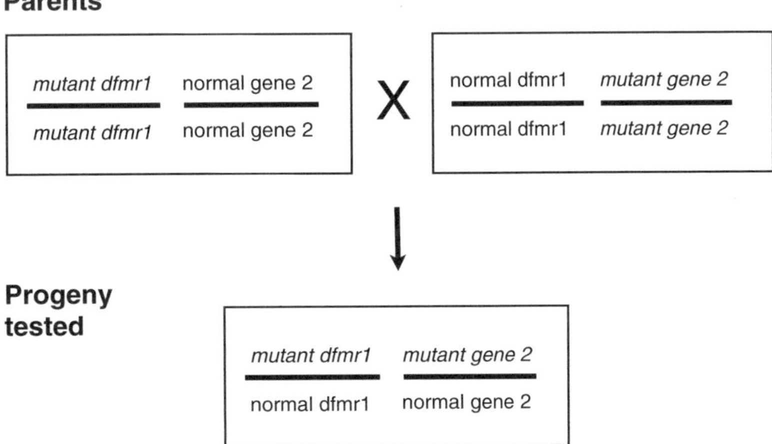

Fig. 1 Scheme of genetic crosses used for interaction studies. Representation of the genetic cross required to obtain flies heterozygous for *dfmr1* and the gene of interest. Negative genetic controls will include the heterozygous mutants for *dfmr1* alone; heterozygous mutants for the gene of interest (*gene 2*) alone; and wild-type flies

each mutation. The performance of those flies can be compared to flies heterozygous mutant for the ID gene or the candidate interactor alone. This constitutes a sensitive screen that can be used to rapidly identify enhancers of a gene of interest.

1.8 Neuroanatomical Divergence

It is important to remember that despite the impressive genetic conservation, the neuroanatomy of the fly brain and peripheral nervous system is very different from that of mammals. After all, the flies and human diverged a few million years ago. Nonetheless, the notion of neuronal networks in memory processing seems to be conserved in Drosophila. The antenna, antenna brain lobe (where olfactory information is first received) and the mushroom bodies (the secondary structure for information processing) have been shown to be essential for early olfactory memory processing. They could be compared functionally to the hippocampus for instance but share no structural homology. The site of long-term storage of memory, which corresponds to the cerebral cortex in human, is still unclear in flies. A potential candidate includes a deep central structure in the fly brain, the central complex, but this is still a matter of investigation [67]. Nonetheless, the in vivo olfactory memory paradigm in flies offers an intact brain circuitry with multiple stages of information processing (as in human) that can be dissected genetically. All in all, flies can be used to test genetic and neuronal network hypotheses that can be later translated to mammalian animal models before being potentially applied to the human brain.

2   Equipment, Materials, and Setup

2.1 Fly Stocks

Drosophila melanogaster, also known commonly as fruit flies, are the most commonly used species in the laboratory and are raised usually in vials or bottles containing a cornmeal and yeast based food. In general, the flies are raised at 25 °C on fly food media that has been standardized for olfactory memory experiments. The *Drosophila* mutants can be obtained from several stock centers (listed on www.flybase.org) as well as from the individual researchers who published original research on the mutants. Several wild-type strains have been established and include the canton-S and Berlin strains. Flies can be shipped in the mail or using courier and will usually survive if properly packaged. Some countries will require an import permit. It is therefore important to verify prior to ordering flies with the shipping and receiving department and the stock center itself. When shipping flies, it is best to allow 2–3 days for the flies to lay eggs prior to mailing them. When receiving flies from another lab, flies should be isolated for two generations to make sure that the vials did not contain mites or other parasites. During that time, the flies are transferred to a new food vial every 3–4 days.

At 25 °C flies will take 2 weeks to mature to the adult stage. Adult flies will mate; the female will lay eggs that will transform into larva that will then pupate to produce an adult fly. In our memory assay, only adult flies are used. This short generation time allows testing for genetic interactions rapidly. At 25 °C, flies will live for about 4 weeks. It is important to grow the flies in a room without odors and ensure a constant temperature is maintained. The flies also need to be raised with a 12 h light–dark cycle.

2.2 Type of Mutants Used in Behavior Experiments

Several types of mutagens have been used throughout the years. X-ray induced mutations and chemical (EMS) mutations were traditionally used and create deletion or smaller point mutations, respectively. These mutations are harder to localize in the genome and require sequencing or genetic complementation studies, which can be time-consuming. For more recent studies this has prompted the use of a novel mutagenesis tool, transposable elements (transposons). Transposons are pieces of DNA flanked by repetitive sequences that are usually fixed within the genome in the absence of an enzyme required for their mobilization, known as transposase. The transposons start mobilizing within the genome when in the presence of the transposase which leads to de novo genetic mutation. With transposons, the mutations can be secondary to the insertion within a gene or in its neighboring regulating region. Mutations can also be caused by imprecise excision of the transposon deleting surrounding gene DNA. The location of the transposon can be mapped using directed PCR followed by sequencing of the genomic regions flanking the transposon. This sequence is then blasted against the Drosophila genome. Some transposons will tend to localize in certain regions of the genome (hotspot), so a variety of different transposons have been used.

2.3 Mutant Used for Memory Experiments

Memory experiments targeting the understanding of ID can be conducted with several types of flies. Mutant fly lines where a complete deficiency (null) or loss-of-function for the gene of interest is present can be used to establish a role for the gene in memory. The level of protein expression, as measured on Western Blot, will determine the nature of the mutants (hypomorph or null). Sometimes, a gene will be essential or play a key role in development, making it very difficult for the mutant to develop to the adult stage (which is required for memory testing in our assay). In those situations, we will use a "conditional expression approach" where the gene of interest will be expressed normally during the developmental stage and then restricted only passed the lethality period. This can be accomplished in a number of ways in Drosophila. First, we can take advantage of temperature-sensitive mutant alleles. In these cases, mutation will exist leading to misfolding or inactivation of the protein of interest at restrictive temperatures. In contrast, the protein will be fully functional at permissive temperatures. These temperatures will be defined for each mutant. These mutants will be very informative for the discovery of what stage of development a gene is required. This approach was previously used successfully by Dubnau to establish the role of *staufen* in memory independently of its role in development [57]. Sometimes, there is no temperature-sensitive mutant available. So one can then use the RNA interference tool to decrease the level of a protein of interest or of overexpression of the normal protein. RNAi lines can be obtained from the Bloomington (http://flystocks.bio.indiana.edu/) or the VDRC stock center (http://stockcenter.vdrc.at/control/main). In addition, one can also overexpress a dominant negative protein to cancel the effect of the normal native protein. Another option includes the overexpression of the wild-type protein to assess the role of gain-of-function in memory.

2.4 The GAL4-UAS System

The Galactose 4-Upstream Activating Sequence (or GAL4-UAS) system provides Drosophila with an advantage for studying the role of genes in brain function. The GAL4-UAS system is a "lock and key" system used in Drosophila, where two constructs (GAL4 and UAS) need to be present together in the fly to turn transgenic expression "on" (Fig. 2). These constructs do not naturally exist in Drosophila as they are derived from yeast. They can be used to express the wild-type transgenes or the RNAi transgenes. Flies exist with the GAL4 activator of transcription being expressed in a known region of the brain, while other flies have an upstream activator sequence (UAS) with a genetic construct (coding sequence of the gene or RNAi). When parents from each stock are mated together, the progeny contain both GAL4 and UAS, and expresses the genetic construct specifically in the region determined by the GAL4 chosen. Several GAL4 lines exist for each region of the brain involved in olfactory memory: the antennal lobe (AL, information encoding), the mushroom bodies (MB, processing of information into LTM) and the central complex (CC, downstream storage from the MB). Usually, at least two lines will be used in order to show that the overlapping brain region is involved in a given result. This will rule out the possibility of GAL4 strain-specific artifact. On the other hand, strains with partly overlapping, partly different patterns of expression can be used in order to dissect the spatial requirement for a gene. In addition, at least two transgenic UAS lines will be used to avoid the possibility that the behavior is due to the site of insertion of the transgene.

2.5 Genetic Background Effects

For all olfactory memory experiments, transgenic and mutant flies are crossed for six generations to the wild-type controls, w^{1118}(isoCJ1) in our case. This leads to sufficient recombination of

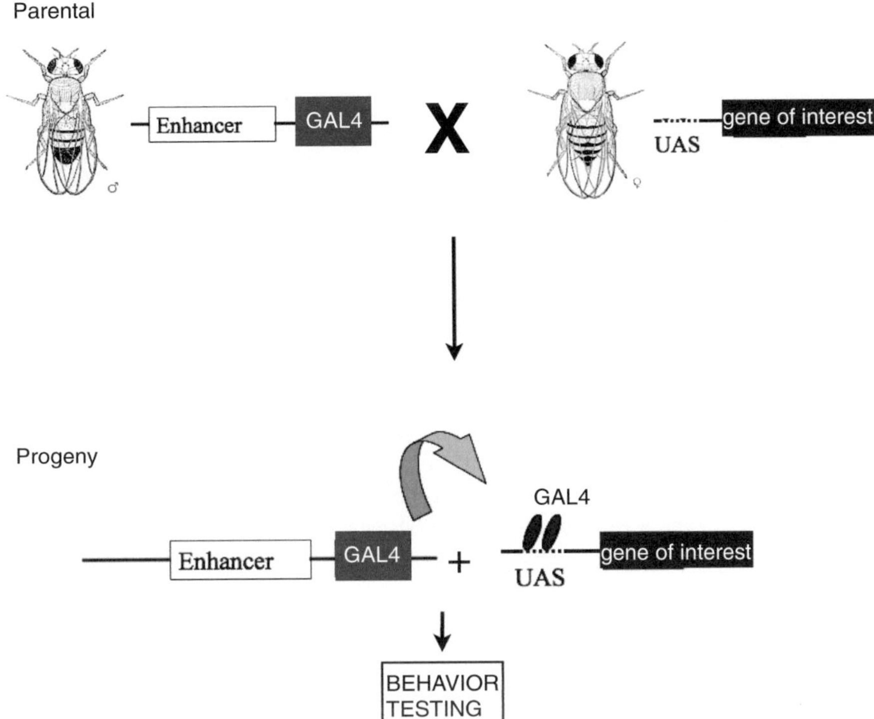

Fig. 2 The GAL4-UAS system can be used to study temporal and spatial roles of genes. Flies carrying a GAL4 transgene inserted in their genome will produce GAL4 in the location where the neighboring gene is usually expressed (hence the term enhancer trap). In addition, any given gene can be placed downtream of an Upstream Activating Sequence (UAS) and then injected to flies. Transgenic flies carrying either the GAL4 transgene or the UAS transgene will not express anything different than wild-type flies. But, when mating flies homozygous for either the GAL4 or the UAS construct together, their progeny will contain the GAL4 and the UAS transgenes and will express the transgene in the location dictated by the GAL4 line. Temporal control can be achieved (but not spatial) when the GAL4 line selected is for the heat shock protein 70. That GAL4 will be turned on with heat shock. The progeny will be tested for the effect of the transgene

the genome and allows us to eliminate genetic background variations that could differ from strain to strain and influence memory performance [26]. Variation in genetic background can greatly influence phenotypes as was shown for neurodevelopmental malformations [68]. At the behavior level, variations can exist despite the absence of visible malformation. The importance of genetic background effect has also been shown in mice where memory and social behaviors varied in function of the mice strain used [69]. This also reflects the interindividual variability one observes in the clinic between patients with the same syndrome.

2.6 Equipment

The equipment used for Drosophila olfactory memory training consists of a training apparatus made of cast acrylic (Fig. 3). The training chamber consists of a 10 mL Falcon plastic tube that is cut to the proper length and then fitted with a grid that delivers the

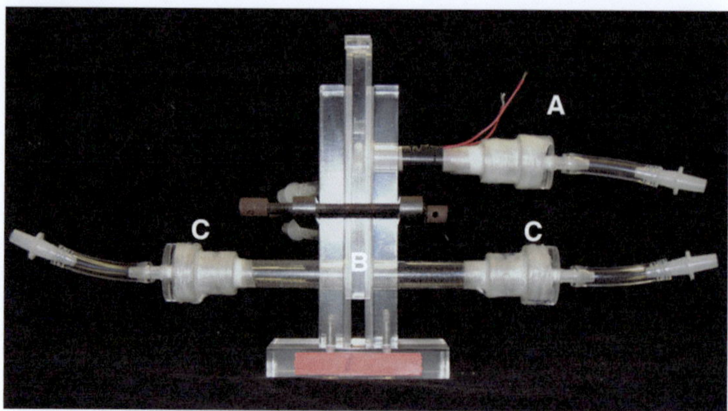

Fig. 3 Photograph of the apparatus used for olfactory classical conditioning. The apparatus is composed of a electrified training chamber (**a**) where flies are exposed successively to two odors, one of which is associated with an electrical shock, the other not. (**b**) Illustration of the choice point where once trained, the flies are brought. (**c**) Represents the tubes where each odor (shocked and not shocked) is provided

electrical shock to the flies feet. Two other tubes of the same length (but without grid) are used as testing chambers for each odor. These tubes need to be replaced regularly. The odors are provided using a device known as the bubbler, which consist of an aluminum block holding two stems and vials. The odors are dissolved in mineral oil and placed in glass vials that screw into the bubbler. A tube is then connected to the training chamber from each odor successively. The electrical shock is provided using an electrophysiology stimulator (Grass). Tim Tully and William Quinn established a response curve for shock and showed that 11 shocks at 60 V of direct current produced the maximal effect without causing significant damage to the flies [50].

2.7 Odors

The odors used traditionally for olfactory memory in Drosophila have included methylcyclohexanol (MCH), benzaldehyde, and 3-octanol (OCT). These odors are naturally aversive to Drosophila. In each laboratory, the concentrations to be used have been standardized so that flies would avoid them equally [50].

2.8 Controlled Environment Room

Tim Tully and William Quinn reported that variability in temperature and humidity could have significant effect on the assay [50]. The assay has therefore, since been performed in rooms with very strict environmental control. The temperature (25 °C) and humidity (70 %) are controlled constantly, preventing variation between experimental days and allowing optimal shock conductivity to the flies in the grid tube. The room is also completely sealed, which allows for complete darkness required during the experiments.

3 Procedures

3.1 Planning an Olfactory Conditioning Experiment

Several possibilities can be considered to gain knowledge about a gene involved in ID using Drosophila learning and memory. First, it is important to verify if the gene of interest has a homologue in Drosophila. It is also important to note if the important functional domains are conserved between humans and fly. Many times, the name of the gene in flies and human will be different. Second, it is important to know precisely the effect at the protein level of the genetic defect. For instance, if the human defect is related to a loss-of-function or gain-of-function of the protein encoded by the gene; or, if there is any accumulation of toxic or misfolded protein. This will become useful in deciding if the experiment will be conducted on null mutants or in transgenic flies where the gene is overexpressed. Sometimes, if a known point mutation has been linked to the disease, transgenic flies overexpressing the same point mutation could test if a mirror cognitive defect can be observed in flies and then exploited for further molecular, biochemical, or pharmaceutical studies. Third, any information known about the developmental role of the gene will also need to be taken into account in planning the experiments. If a gene is essential for development to adulthood, conditional approaches will be required (*see* Sect. 2).

Having established a role for the gene in the memory assay, a better understanding of the gene network of the candidate can be established. Indeed, using genetic crosses, mutant alleles can be combined together to see if one gene can rescue or enhance the mutation in another gene. Finally, pharmacological interventions can be performed to test the effect of specific inhibitors during development, in early adult flies or in aging flies. This can allow us to gain a better understanding of the pathway but also to test small molecules candidate for potential therapeutic intervention [26].

3.2 General Methods for the Behavior Experiments

The olfactory classical (Pavlovian) conditioning paradigm [43, 44, 70, 71] (Fig. 4) consists of training flies to associate a painful unconditioned stimulus (US), a foot shock, to a conditioned stimulus (CS+), an odor, that is presented simultaneously for 1 min. A second odor (CS–) is presented after a 45 s rest interval, for 1 min but without a foot shock. Flies are then transported to a choice point where each odor is presented on each side. Wild-type flies will remember the odor paired with the shock whereas learning mutants will not. All the experiments are performed blind to the genotype. In every experiment, wild-type flies will be used as control. In addition, for experiments involving the UAS-GAL4 system, flies containing only GAL4 or only UAS will be used as control for the flies containing both UAS and GAL4.

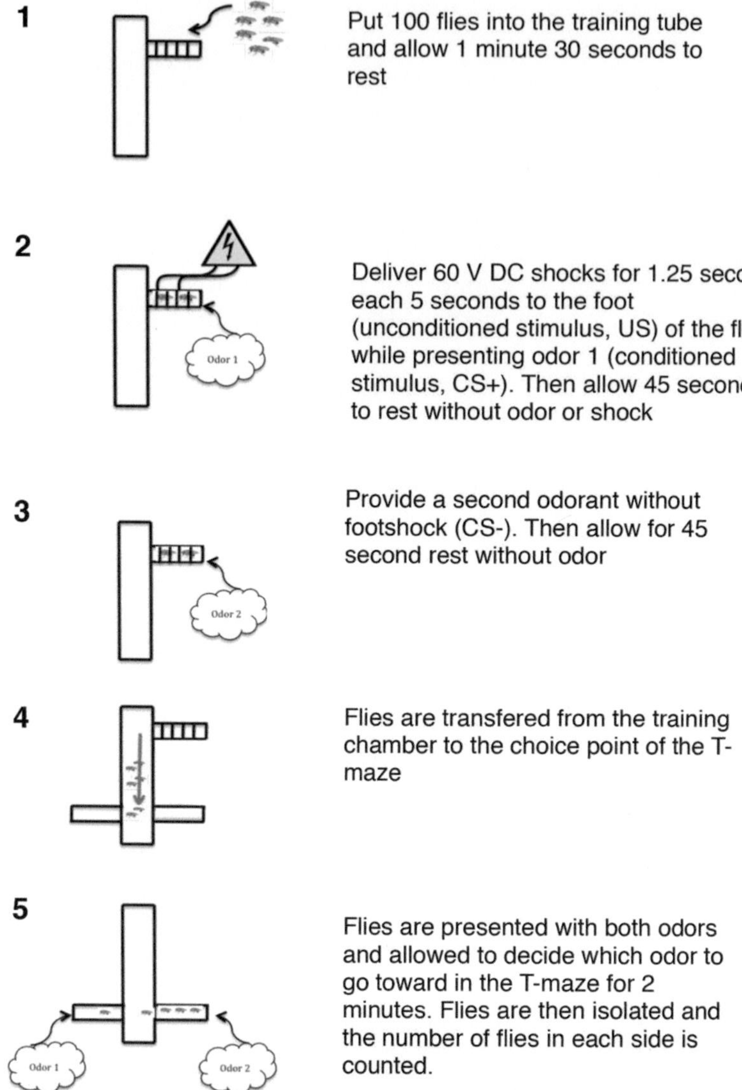

Fig. 4 Schematic representation of the steps in the learning paradigm. An experiment consist of steps 1–5, with odor 1 associated with shock. These steps are then repeated with shock associated with odor 2

3.3 Preparation of the Flies

Twenty-four hours before training, one needs to prepare ½ pint glass milk bottles containing food at the bottom for the flies. One round piece of Whatman filter paper is placed on the bottom of a glass bottle. With a metal spatula, the filter is tapped down lightly so that the paper adheres to the food. Then a ¼ piece of paper towel is rolled up and inserted in the food. This allows the flies to stay relatively dry while having access to the food. Approximately 400 flies of the same genotype are placed into each bottle.

The flies are put at 23 °C overnight. On the day of the experiment, about 1 h prior to the experiment, the flies are placed inside the environment chamber at 25 °C and 70 % humidity to acclimate.

3.4 Short-Term Memory Procedure

The behavioral experiments are performed at 25 °C and 70 % humidity in a controlled environmental chamber [50]. In brief, a single training session (learning) consists of 100 genetically identical flies of the same age (between 1 and 3 days) being placed inside a training chamber. The flies are presented for 1 min with the first odor, Octanol (CS+), paired with an electrical foot shock (US). Flies rest for 45 s before receiving a different odor, Methylcyclohexanol (CS−) for 1 min *without* a foot shock. Flies are then transported to the choice point where both odors are on either side. Flies distribute toward each odor depending on their conditioned response. Flies are then isolated and counted. Another 100 flies are then trained similarly but with odor 2 as CS+ and odor 1 as CS−. This then constitutes one experimental "N." After even a single trial (learning), about 95 % of flies will learn to avoid the CS+, but their memory decays rapidly. In general an $N = 4$ will be the minimal number obtained for a given experiment. In addition, two to three replicates will be obtained to verify the consistency of results.

3.5 Long-Term Memory Procedure

To form a longer-lasting memory, repeated training is required. Tim Tully showed that long-term memory (LTM) forms when flies are trained ten times with a 15 min rest between sessions and is called spaced training [33] (Fig. 5). In general LTM is tested 1 day after training. This form of memory depends on protein synthesis and transcription. On the other hand, the flies can be trained with ten sessions without rest intervals thus generating massed training. Massed training produces a transient form of memory present at 1 day but decays completely by 4 days. The latter form of memory does not depend on transcription or translation. In general, the experiments will include an $N = 8$ per genotype and should be replicated two to three times.

3.6 Sensory Control Experiments

In cases where the flies do not perform as well as control flies, it is important to perform control experiments where the olfactory capacity and shock reactivity of the flies are assessed. This allows us to know that the performance defect observed is due to learning and memory and not to a sensory dysfunction leading to inadequate input for normal memory to form. For shock reactivity, the flies will be brought to the choice point and allowed to go on either an electrified shocking tube or a non-electrified shocking tube. Flies are given 2 min to distribute and are then counted for their avoidance of the tube with shock being presented. Typically about 90 % of flies will avoid the tube where shocks are being provided. An $N = 6$ is obtained for each group. For olfactory controls,

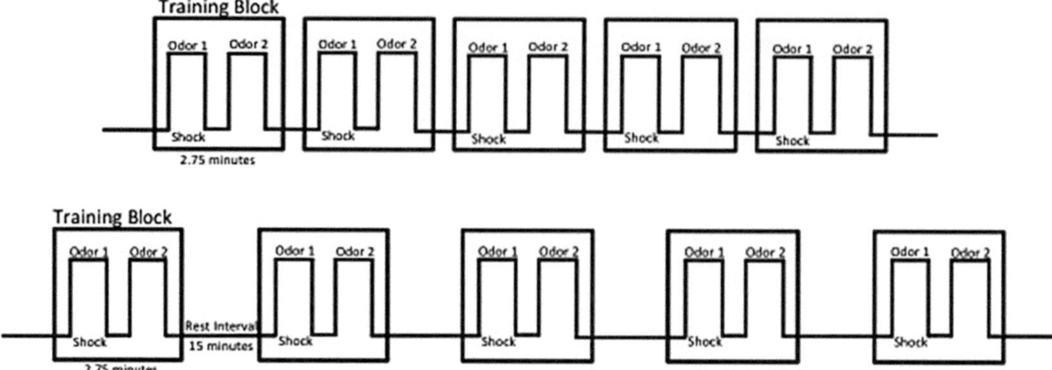

Fig. 5 Schematic representation of the training protocols used for long-term memory. Each training block (*in box*) consists of the simultaneous presentation of an odor with a foot shock for 1 minute followed by a 45 s rest interval. A second odor is then presented without foot shock. The first training displayed is called massed training and consists of 10 consecutive training sessions represented without rest interval between each training block. In the second experimental setting, called Spaced training, 10 training sessions are presented but spaced by 15 min rest interval between each training block. (Modified after Tully, Cell 1994;79:1:35)

flies are presented with one odor on one side and pure air on the other side of the choice point. As mentioned above, the odors used in the assay are naturally aversive to the flies. Therefore, usually 70 % of the flies will avoid the odor when given a choice with pure air. One odor is tested at a time. This is repeated for each odor used in the conditioning assay. A similar level of avoidance to each odor should be observed. An $N=6$ is obtained for each odor tested for each genotype.

3.7 Calculation of the Performance Index (PI) and Statistical Analyses

Each trial consists of training 100 flies to associate an odor with a foot shock. In one trial, the flies will be trained to the first odor. In a second trial a new group of 100 flies of the same genotype will be trained to the second odor. A composite of both trials is used to make one experimental unit ($N=1$). This allows one to balance the experiment for any odor or side bias that may be present in the flies. The results are expressed as a performance index (PI) (Fig. 6). The PI consists of an average of performance index for each odor (odor 1 and 2). An individual odor PI is calculated by subtracting the proportion of flies avoiding the non-shocked odor to the proportion of flies avoiding the shocked odor. Different statistical analyses are performed depending on the number of groups. For planned comparisons between two groups, a two-tailed student t-test with a confidence interval of 95 % is used. For unplanned multiple comparisons among more than two groups, a one-way analysis of variance (ANOVA) followed by a Tukey test are performed using JMP software (SAS).

Calculating the Performance Index (PI)

$$PI\ odor\ 1 = \frac{Number\ flies\ avoiding\ shocked\ odor\ 1}{Total\ number\ of\ flies} - \frac{Number\ of\ flies\ avoiding\ odor\ 2}{Total\ number\ of\ flies}$$

$$PI\ odor\ 2 = \frac{Number\ flies\ avoiding\ shocked\ odor\ 2}{Total\ number\ of\ flies} - \frac{Number\ of\ flies\ avoiding\ odor1}{Total\ number\ of\ flies}$$

$$PI = \frac{PI\ odor\ 1\ +\ PI\ odor\ 2}{2}$$

Fig. 6 Performance Index calculation. The performance index represents the average of two groups for their avoidance of a shock-paired odor (CS+) compared to the fraction of flies that avoid the control odor (CS−). Each group represents training to odor 1 (PI odor 1) and training to odor 2 (PI odor 2)

4 Notes

4.1 Experimental Variables

4.1.1 Fly Stocks

For each experiment, several genotypes will be tested. The experimenter is blind to the genotype of the flies. An important variable is the age of the flies. Flies should be 1–3 days old when the assay is performed. The cultures therefore need to be synchronized. All flies used for one experiment should be raised in the same room and at the same temperature. Moreover, culturing the flies in the same room will control for potential unplanned odor exposure during development of the flies.

4.1.2 Mutant Alleles

It is important to verify if the fly mutant alleles used for the study correspond to the ortholog human gene of interest. It is also important to verify the level of gene expression using a Western blot (protein level) or quantitative PCR (mRNA level) in the case where there is no antibody available for Western blotting. Various alleles may have different levels of disruption of the mutant protein from mild hypomorph to null and thus influence the behavior results. It is also important to verify if the flies are homozygous or heterozygous for the mutant gene. Many stocks are kept over balancer chromosomes (chromosome engineered to prevent recombination and containing specific phenotypic traits) as heterozygous.

4.1.3 Training Machine

The training/testing apparatus can significantly affect the performance of the flies. It is therefore important to verify the age and wear of the training and testing tubes. The tubes will progressively crack or be soiled by the flies with repeated usage. The manual machine or the bubbler may develop air leaks with time due to the numerous connections within the system. Air leaks can cause turbulence within the system. All genotypes should be tested with the same apparatus to prevent bias related to equipment.

4.1.4 Temperature and Humidity

The temperature (25 °C) and humidity (70 %) in the training chamber are crucial in the level of performance of the flies. The temperature will affect the level of activity of the flies and their metabolism. This is especially important when performing drug-feeding experiments. Drug consumption will be higher when flies are dehydrated and incubated at higher temperature. The level of humidity will have an impact on fly activity level but more importantly will affect the conductivity of the foot shock in the training chamber. Ensure that the flies are not too wet since they will then receive too much current and may die on the grid.

4.2 Typical/ Anticipated Results

Wild-type fly performance may vary depending on the genetic background selected. Most olfactory behavior laboratories will use the Canton-S flies in order to obtain the best results. After a single training cycle, a performance index (PI) of around 85–90 is usually obtained. Learning mutants will score below that. The PI for memory after massed training at 1 day will be around 20–25. On the other hand, memory performance at 1 day after spaced training will give a PI around 40–45. The memory after a single training session will rapidly decay after 1 day. The memory after massed training will last up to 4 days. In contrast, memory after spaced training will last for at least 1 week with a PI of 35. This is a significant duration corresponding to about one third of the fruit fly lifespan.

4.3 Troubleshooting

4.3.1 Fly Stocks

Drosophila can be infected with bacteria and viruses. A common virus is the JC virus and has been previously identified by Tully (unpublished data) as a significant reason for lower performance. A procedure involving bleach aiming at removing the chorion of the freshly laid Drosophila eggs has been established. Bacteria are treated by the use of antibiotics that are routinely put in the fly food.

Inconsistent performance can also occur when flies of different ages are used in an experiment. In some cases memory performance will deteriorate more rapidly in mutants than in controls as the flies age [72].

Some potential problems exist with the use of transposons for mutagenesis. First, as mentioned earlier, some genomic regions may be a "hot spot" for transposon insertion and therefore, the mutagenesis may not cover the entire genome (i.e., not be random). A variety of transposons have been designed to address that problem. Another risk is that a transposon may insert in multiple genomic sites. Finally, the transposon may have inserted and then jumped out of a gene leaving it with a deletion that will not show up with the PCR. These caveats can be ruled out by testing the failure of complementation between multiple alleles. In the case where the behavior defect is due to a second site mutation, the behavior defect will not be seen when the allele of interest is studied heterozygous with an allele affecting the location where the mutation should be.

For all experiments, it is important to make sure that the flies are dry, not too crowded, and well fed. It is also important to make sure that the flies from one genotype are not contaminated with another.

4.3.2 Procedures and Equipment

The odor presentation is dependent on vacuum pressure. It is therefore necessary to make sure that the pressure of air in the shocker tube and each collection tube is at 750 mm using a manometer. In the case of an air flow leak, one needs to check the manual machine and bubbler parts one by one. The holes through which the vacuum is applied in the training apparatus can also get plugged by fly debris. The shocker tube may become less efficient with time, and should be checked for good conductivity of current all the way from the stimulator to the grid. Alternatively, the setting of the stimulator box may have been changed. Odor contamination can occur due to misplacement of the tubes containing the diluted odors in the manual bubbler or contamination at the source. Use fresh odors for each experiment to obtain the most consistent results. Finally, it is important to always replace both odors at the same time when obtaining fresh stocks bottles of odors.

5 Conclusion

The unprecedented gain in the understanding of the genetic of intellectual disability (ID) has made obvious the need for animal studies aimed at uncovering the molecular mechanisms linking genetic and cognitive behavior. The learning and memory model in Drosophila is well positioned to help understand the genotype–phenotype correlations in ID. Several patients with ID have been shown to have significant memory defects, and several of the genes involved in ID have produced memory defects in animal models, including Drosophila. The high degree of genetic conservation between Drosophila and human, combined with the abundance of Drosophila mutant lines as well as the short life cycle has already provided the medical community with important insight into how gene mutation leads to complex neurobehavioral defects such as dementia, ID, and learning disability. More recently, pharmacological targets identified in Drosophila have been used for clinical trials in humans. Considering the recent development of novel tools (optogenetics) in mice for the study of conditional gene expression and the well established value of the mammalian models, future work combining the screening abilities of Drosophila to the large capabilities of other animal models will allow us to gain a better understanding of the biology of ID and accelerate the pace of development of novel therapeutic interventions for patients.

Acknowledgements

We would like to thank Dr. Tim Tully for his mentorship. We would also like to thank the Canadian Child Health Clinician Scientist Program (CCHCSP) for the Career Development Award, the Woman and Children Health Research Institute (WCHRI) and Canadian Institute of Health Research (CIHR) and Dart Neuroscience for their financial support to my laboratory. Finally, we would like to thank the members of the Tully and Bolduc lab, Dr. Andrew Simmonds and Sarah Hughes, Dr. Steven De Belle, Dr. Terry Klassen, Dr. Thiery Lacaze, Dr. Peter Nguyen, Dr. Jerry Yager and Dr. Susan Gilmour and Dr. Lori West as well as the members of the University of Alberta Fly Group and of the Neuroscience and Mental Health Institute at the University of Alberta for stimulating discussions.

References

1. Shevell M, Ashwal S, Donley D et al (2003) Practice parameter: evaluation of the child with global developmental delay: report of the Quality Standards Subcommittee of the American Academy of Neurology and The Practice Committee of the Child Neurology Society. Neurology 60:367–380

2. Jacob FD, Ramaswamy V, Andersen J, Bolduc FV (2009) Atypical Rett syndrome with selective FOXG1 deletion detected by comparative genomic hybridization: case report and review of literature. Eur J Hum Genet 17:1577

3. Flint J, Wilkie AO, Buckle VJ, Winter RM, Holland AJ, McDermid HE (1995) The detection of subtelomeric chromosomal rearrangements in idiopathic mental retardation. Nat Genet 9:132–140

4. Ravnan JB, Tepperberg JH, Papenhausen P et al (2006) Subtelomere FISH analysis of 11 688 cases: an evaluation of the frequency and pattern of subtelomere rearrangements in individuals with developmental disabilities. J Med Genet 43:478–489

5. Cook EH Jr, Scherer SW (2008) Copy-number variations associated with neuropsychiatric conditions. Nature 455:919–923

6. Bolduc FV, Tully T (2009) Molecular biology of learning. In: Shevell M (ed) Clinical and scientific aspects of neurodevelopmental disabilities. MacKeith Press, London

7. Bolduc FV, Tully T (2009) Fruit flies and intellectual disability. Fly (Austin) 3

8. Paribello C, Tao L, Folino A et al (2010) Open-label add-on treatment trial of minocycline in fragile X syndrome. BMC Neurol 10:91

9. Krueger DA, Care MM, Holland K et al (2010) Everolimus for subependymal giant-cell astrocytomas in tuberous sclerosis. N Engl J Med 363:1801–1811

10. Ropers HH, Hamel BC (2005) X-linked mental retardation. Nat Rev Genet 6:46–57

11. Hagberg B, Hagberg G, Lewerth A, Lindberg U (1981) Mild mental retardation in Swedish school children. II. Etiologic and pathogenetic aspects. Acta Paediatr Scand 70:445–452

12. McLaren J, Bryson SE (1987) Review of recent epidemiological studies of mental retardation: prevalence, associated disorders, and etiology. Am J Ment Retard 92:243–254

13. Yeargin-Allsopp M, Boyle C (2002) Overview: the epidemiology of neurodevelopmental disorders. Ment Retard Dev Disabil Res Rev 8:113–116

14. Roeleveld N, Zielhuis GA, Gabreels F (1997) The prevalence of mental retardation: a critical review of recent literature. Dev Med Child Neurol 39:125–132

15. Shea SE (2006) Mental retardation in children ages 6 to 16. Semin Pediatr Neurol 13:262–270

16. Honeycutt AA, Grosse S, Dunlap LJ, Chen H, Al Homsi G, Schendel D (2003) Economic cost of mental retardation, cerebral palsy, hearing loss, and vision impairment. Elsevier, London

17. Aicardi J (1998) The etiology of developmental delay. Semin Pediatr Neurol 5:15–20

18. Katz ER, Ellis NR (1991) Memory for spatial location in retarded and nonretarded persons. J Ment Defic Res 35(Pt 3):209–220

19. McCartney JR (1987) Mentally retarded and nonretarded subjects' long-term recognition memory. Am J Ment Retard 92:312–317

20. Winters JJ Jr, Semchuk MT (1986) Retrieval from long-term store as a function of mental age and intelligence. Am J Ment Defic 90:440–448

21. Cornish KM, Munir F, Cross G (1999) Spatial cognition in males with Fragile-X syndrome: evidence for a neuropsychological phenotype. Cortex 35:263–271

22. Kogan CS, Boutet I, Cornish K et al (2009) A comparative neuropsychological test battery differentiates cognitive signatures of Fragile X and Down syndrome. J Intellect Disabil Res 53:125–142

23. The Dutch-Belgian Fragile X Consortium (1994) Fmr1 knockout mice: a model to study fragile X mental retardation. Cell 78:23–33

24. Maes B, Fryns JP, Van Walleghem M, Van den Berghe H (1994) Cognitive functioning and information processing of adult mentally retarded men with fragile-X syndrome. Am J Med Genet 50:190–200

25. Kooy RF, D'Hooge R, Reyniers E et al (1996) Transgenic mouse model for the fragile X syndrome. Am J Med Genet 64:241–245

26. Bolduc FV, Bell K, Cox H, Broadie KS, Tully T (2008) Excess protein synthesis in Drosophila fragile X mutants impairs long-term memory. Nat Neurosci 11:1143–1145

27. McBride SM, Choi CH, Wang Y et al (2005) Pharmacological rescue of synaptic plasticity, courtship behavior, and mushroom body defects in a Drosophila model of fragile X syndrome. Neuron 45:753–764

28. Marcell MM, Harvey CF, Cothran LP (1988) An attempt to improve auditory short-term memory in Down's syndrome individuals through reducing distractions. Res Dev Disabil 9:405–417

29. Marcell MM, Weeks SL (1988) Short-term memory difficulties and Down's syndrome. J Ment Defic Res 32(Pt 2):153–162

30. Vicari S, Bellucci S, Carlesimo GA (2006) Evidence from two genetic syndromes for the independence of spatial and visual working memory. Dev Med Child Neurol 48:126–131

31. Vicari S, Carlesimo GA (2006) Short-term memory deficits are not uniform in Down and Williams syndromes. Neuropsychol Rev 16:87–94

32. Ebbinghaus H (1885) Memory: a contribution to experimental psychology. Teachers College, Columbia University, New York

33. Tully T, Preat T, Boynton SC, Del Vecchio M (1994) Genetic dissection of consolidated memory in Drosophila. Cell 79:35–47

34. Ribot T (1882) L'Heredite psychologique. Baillière, Paris

35. Mueller G, Pilzecker A (1900) Experimentelle Beitrage zur Lehre vom Gedachtniss. Z Psychol Ergänzungsband 1:1

36. Pavlov I (1927) Conditioned reflexes. International Publishers, New York, NY

37. Thorndike E (1901) The influence of improvement in one mental function upon the efficiency of other functions. Psychol Rev 8:247–261

38. Watson JB (1920) Conditioned emotional reactions. J Exp Psychol 3:1–14

39. Hebb DO (1949) The organization of behavior; a neuropsychological theory. Wiley, New York

40. Scoville WB, Milner B (1957) Loss of recent memory after bilateral hippocampal lesions. J Neurol Neurosurg Psychiatry 20:11–21

41. Milner B, Penfield W (1955) The effect of hippocampal lesions on recent memory. Trans Am Neurol Assoc 42–48

42. Milner B (1954) Intellectual function of the temporal lobes. Psychol Bull 51:42–62

43. Benzer S (1967) Behavioral mutants of Drosophila isolated by countercurrent distribution. Proc Natl Acad Sci U S A 58:1112–1119

44. Benzer S (1973) Genetic dissection of behavior. Sci Am 229:24–37

45. Dudai Y, Jan YN, Byers D, Quinn WG, Benzer S (1976) Dunce, a mutant of Drosophila deficient in learning. Proc Natl Acad Sci U S A 73:1684–1688

46. Livingstone MS, Sziber PP, Quinn WG (1984) Loss of calcium/calmodulin responsiveness in adenylate cyclase of rutabaga, a Drosophila learning mutant. Cell 37:205–215

47. Davis RL, Davidson N (1984) Isolation of the Drosophila melanogaster dunce chromosomal region and recombinational mapping of dunce sequences with restriction site polymorphisms as genetic markers. Mol Cell Biol 4:358–367

48. Levin LR, Han PL, Hwang PM, Feinstein PG, Davis RL, Reed RR (1992) The Drosophila learning and memory gene rutabaga encodes a Ca2+/Calmodulin-responsive adenylyl cyclase. Cell 68:479–489

49. Brunelli M, Castellucci V, Kandel ER (1976) Synaptic facilitation and behavioral sensitization in Aplysia: possible role of serotonin and cyclic AMP. Science 194:1178–1181

50. Tully T, Quinn WG (1985) Classical conditioning and retention in normal and mutant Drosophila melanogaster. J Comp Physiol A 157:263–277

51. Dubnau J, Chiang AS, Tully T (2003) Neural substrates of memory: from synapse to system. J Neurobiol 54:238–253

52. Pitman JL, Dasgupta S, Krashes MJ, Leung B, Perrat PN, Waddell S (2009) There are many ways to train a fly. Fly (Austin) 3

53. Margulies C, Tully T, Dubnau J (2005) Deconstructing memory in Drosophila. Curr Biol 15:R700–R713

54. Yin JC, Wallach JS, Del Vecchio M et al (1994) Induction of a dominant negative CREB transgene specifically blocks long-term memory in Drosophila. Cell 79:49–58

55. Bourtchouladze R, Lidge R, Catapano R et al (2003) A mouse model of Rubinstein-Taybi syndrome: defective long-term memory is ameliorated by inhibitors of phosphodiesterase 4. Proc Natl Acad Sci U S A 100:10518–10522

56. Tully T, Bourtchouladze R, Scott R, Tallman J (2003) Targeting the CREB pathway for memory enhancers. Nat Rev Drug Discov 2:267–277

57. Dubnau J, Chiang AS, Grady L et al (2003) The staufen/pumilio pathway is involved in Drosophila long-term memory. Curr Biol 13:286–296

58. Wu Y, Bolduc FV, Bell K et al (2008) A Drosophila model for Angelman syndrome. Proc Natl Acad Sci U S A 105:12399–12404

59. Didelot G, Molinari F, Tchenio P et al (2006) Tequila, a neurotrypsin ortholog, regulates long-term memory formation in Drosophila. Science 313:851–853

60. Inlow JK, Restifo LL (2004) Molecular and comparative genetics of mental retardation. Genetics 166:835–881

61. Brand AH, Perrimon N (1993) Targeted gene expression as a means of altering cell fates and generating dominant phenotypes. Development 118:401–415

62. McGuire SE, Le PT, Osborn AJ, Matsumoto K, Davis RL (2003) Spatiotemporal rescue of memory dysfunction in Drosophila. Science 302:1765–1768

63. Bonini NM (2000) Drosophila as a genetic tool to define vertebrate pathway players. Methods Mol Biol 136:7–14

64. Feany MB, Bender WW (2000) A Drosophila model of Parkinson's disease. Nature 404: 394–398

65. Li Z, Karlovich CA, Fish MP, Scott MP, Myers RM (1999) A putative Drosophila homolog of the Huntington's disease gene. Hum Mol Genet 8:1807–1815

66. Cukier HN, Perez AM, Collins AL, Zhou Z, Zoghbi HY, Botas J (2008) Genetic modifiers of MeCP2 function in Drosophila. PLoS Genet 4, e1000179

67. Xia S, Miyashita T, Fu TF et al (2005) NMDA receptors mediate olfactory learning and memory in Drosophila. Curr Biol 15:603–615

68. de Belle JS, Heisenberg M (1996) Expression of Drosophila mushroom body mutations in alternative genetic backgrounds: a case study of the mushroom body miniature gene (mbm). Proc Natl Acad Sci U S A 93:9875–9880

69. Pietropaolo S, Guilleminot A, Martin B, D'Amato FR, Crusio WE (2011) Genetic-background modulation of core and variable autistic-like symptoms in Fmr1 knock-out mice. PLoS One 6, e17073

70. Hotta Y, Benzer S (1973) Mapping of behavior in Drosophila mosaics. Symp Soc Dev Biol 31:129–167

71. Quinn WG, Harris WA, Benzer S (1974) Conditioned behavior in Drosophila melanogaster. Proc Natl Acad Sci U S A 71:708–712

72. Choi CH, McBride SM, Schoenfeld BP et al (2010) Age-dependent cognitive impairment in a Drosophila fragile X model and its pharmacological rescue. Biogerontology 11:347–362

73. Guo HF, Tong J, Hannan F, Luo L, Zhong Y (2000) A neurofibromatosis-1-regulated pathway is required for learning in Drosophila. Nature 403:895–898

74. Wolfgang WJ, Hoskote A, Roberts IJ, Jackson S, Forte M (2001) Genetic analysis of the

Drosophila Gs(alpha) gene. Genetics 158:1189–1201

75. Connolly JB, Tully T (1998) Integrins: a role for adhesion molecules in olfactory memory. Curr Biol 8:R386–R389

76. Grotewiel MS, Beck CD, Wu KH, Zhu XR, Davis RL (1998) Integrin-mediated short-term memory in Drosophila. Nature 391:455–460

77. Godenschwege TA, Kristiansen LV, Uthaman SB, Hortsch M, Murphey RK (2006) A conserved role for Drosophila Neuroglian and human L1-CAM in central-synapse formation. Curr Biol 16:12–23

78. Guimera J, Casas C, Pucharcos C et al (1996) A human homologue of Drosophila minibrain (MNB) is expressed in the neuronal regions affected in Down syndrome and maps to the critical region. Hum Mol Genet 5:1305–1310

79. Tejedor F, Zhu XR, Kaltenbach E et al (1995) Minibrain: a new protein kinase family involved in postembryonic neurogenesis in Drosophila. Neuron 14:287–301

80. Garcia CC, Blair HJ, Seager M et al (2004) Identification of a mutation in synapsin I, a synaptic vesicle protein, in a family with epilepsy. J Med Genet 41:183–186

81. Godenschwege TA, Reisch D, Diegelmann S et al (2004) Flies lacking all synapsins are unexpectedly healthy but are impaired in complex behaviour. Eur J Neurosci 20:611–622

82. Broadie K, Rushton E, Skoulakis EM, Davis RL (1997) Leonardo, a Drosophila 14-3-3 protein involved in learning, regulates presynaptic function. Neuron 19:391–402

83. Skoulakis EM, Davis RL (1996) Olfactory learning deficits in mutants for leonardo, a Drosophila gene encoding a 14-3-3 protein. Neuron 17:931–944

84. Chang KT, Shi YJ, Min KT (2003) The Drosophila homolog of Down's syndrome critical region 1 gene regulates learning: implications for mental retardation. Proc Natl Acad Sci U S A 100:15794–15799

85. Melicharek DJ, Ramirez LC, Singh S, Thompson R, Marenda DR (2010) Kismet/CHD7 regulates axon morphology, memory and locomotion in a Drosophila model of CHARGE syndrome. Hum Mol Genet 19:4253

86. Putz G, Bertolucci F, Raabe T, Zars T, Heisenberg M (2004) The S6KII (rsk) gene of Drosophila melanogaster differentially affects an operant and a classical learning task. J Neurosci 24:9745–9751

87. Comas D, Petit F, Preat T (2004) Drosophila long-term memory formation involves regulation of cathepsin activity. Nature 430:460–463

88. Ge X, Hannan F, Xie Z et al (2004) Notch signaling in Drosophila long-term memory formation. Proc Natl Acad Sci U S A 101:10172–10176

Chapter 15

Animal Models of Cerebral Dysgenesis: Excitotoxic Brain Injury

Luigi Titomanlio, Leslie Schwendimann, and Pierre Gressens

Abstract

Brain damage through excitotoxic mechanisms is a major cause of pediatric neurologic diseases. These include hypoxic-ischemic encephalopathy, periventricular white matter damage (PWMD), stroke, meningoencephalitis, traumatic brain injury, and neurodegenerative disorders. In the present chapter, we describe the procedure to generate a model of excitotoxic injury by intra-cerebrally injecting glutamate analogues. This model provides tools for investigating excitotoxic influences at various stages of neural development and for identifying protective substances against excitotoxicity.

Key words Excitotoxicity, Cerebral palsy, Ibotenate

1 Introduction

Brain injury in children is a major health problem worldwide: improvements in survival rates, particularly in preterm infants, have outpaced a concomitant decrease in long-term neurodevelopmental disability rates [1–4]. Cerebral injury in preterm and newborn babies is usually secondary to focal infarction, brain malformations, infections, or PWMD. Acquired injuries can also result from multiple causes, including trauma, infections, noninfectious disorders (epilepsy, hypoxia/ischemia, genetic/metabolic disorders), tumors, and vascular abnormalities. Despite the personal as well as the economic burdens of long-term neurological morbidity of brain injuries, there is to date no effective treatment [5, 6]. Excitotoxicity is a key factor involved in the pathogenesis of many pediatric brain diseases [7] and a main contributor to the evolution of PWMD [8–10].

Well-characterized animal models of excitotoxic damage are generated by the activation of the excitotoxic cascade via the NMDA and metabotropic glutamate receptors through ibotenate,

Jerome Y. Yager (ed.), *Animal Models of Neurodevelopmental Disorders*, Neuromethods, vol. 104, DOI 10.1007/978-1-4939-2709-8_15, © Springer Science+Business Media New York 2015

quinolinate [11], and S-bromowillardiine (acting on AMPA-kainate receptors) [12, 13]. The ibotenate-lesioned mouse is a classic excitotoxic model for the study of neurodevelopmental diseases in humans, and particularly of PWMD. Ibotenate activates N-methyl-D-aspartate (NMDA) and metabotropic glutamatergic receptors but not α-3-amino-hydroxy-5-methyl-4-isoxazole propionic acid (AMPA) and kainate receptors. When injected at the time of completion of neuronal migration (P0) ibotenate induces isolated depopulation of deep cortical layers V–VI, resulting in a microgyria-like lesion pattern [14]. Injected after completion of migration (P5–P10) ibotenate produces severe neuronal loss in layers II, III, IV, V, and VI. Periventricular white matter lesions are observed after ibotenate injection at P2–P10, with a peak of occurrence at P5 [15]. These lesions are characterized by the formation of focal necrosis within the white matter surrounding the lateral ventricles and/or the subsequent appearance of more widespread, diffuse lesions [16] involving the apoptotic death of late oligodendrocyte progenitors [17]. The main neuropathological finding is substantial hypomyelination [18] accompanied by neuronal loss and impaired neuronal guidance [10], supporting the view that some of the dysfunctions seen in PWMD infants reflect a reduction in "connectivity," needed for the integration of information arriving from different areas of the brain [19–21]. The ibotenate-induced lesion progressively increases in size during the first 24 h after the insult with a significant increase in the density of activated microglia at the level of the lesion 4 h following injection. The lesion then remains stable for 3–4 weeks, eventually being replaced by a glial scar [15, 22].

The ibotenate lesions may be induced in other rodents or even in rabbits to analyze different developmental effects. Hamsters, at birth are at a stage similar to E17 mice, resembling human fetal brain at 15 weeks of gestation. In these animals, ibotenate injection at P0 induces disorders of neuronal migration, caused by arrests of migrating neurons at different distances from the germinative zone within the radial migratory corridors [23]. The resulting cyto-architectonic patterns include periventricular nodular heterotopias and subcortical band heterotopias. In rats, ibotenate injection may be performed at P0 and at P4 [24–26]. Newborn rats reflect an intermediate stage of brain development between hamsters and mice, and the pattern of cortical malformations found in P0 lesioned rats reflects a combination of disturbed neuronal migration molecular ectopias accompanied by depopulation of deep cortical layers. If ibotenate is injected at P4, the lesion resembles human PWMD. Injection of ibotenate at adult stages of development in rodents causes focal cortical necrosis.

In rabbits, the period from P7 to P9 corresponds to the time of maximum vulnerability of the brain to excitotoxin-induced white matter damage [27], which neuropathologically simulates

PWMD. The advantage of using these animals is that their brain is larger and the neuroanatomic organization of the subcortical white matter more closely resembles that of the human.

To attenuate ibotènate-induced excitotoxic damage, an allosteric modulator of AMPA receptors may be employed, such as S18986 which provides a dose-dependent and long-lasting protection of developing white matter and cortical grey matter [13] or the noncompetitive NMDA antagonist MK-801 (1)-5-methyl-10,11-dihydro-5H-dibenzo[a,d]cyclohepten-5,10-imine hydrogen maleate [14, 15].

2 Materials

2.1 Animals

As all experimental protocols involving animals, this protocol should also be carried out in accordance with local ethical guidelines following approval by the institutional review committee. Wild-type mouse pups are obtained from pregnant mice. Animals are housed at 24 °C with a 12 h light–dark cycle and free access to food and water.

As detailed above, this procedure may be performed at different developmental ages, and in several animal species, including transgenic mice.

2.2 Products

The products used are listed below. Comparable products from other suppliers should also be effective.

- Isoflurane (for anesthesia).
- Ibotenate (5 μg/μl, Sigma, St.-Quentin Fallavier, France).
- 26-gauge beveled needles.
- 50 μl Hamilton syringe (Hamilton, Massay, France).
- Calibrated micro-dispenser.
- Phosphate buffered saline (PB; 0.12 M TPO4, pH 7.4).
- 4 % paraformaldehyde (PFA).
- 10 % sucrose.
- 7.5 % gelatine.
- Superfrost slides.
- Material for paraffin staining.
- Material for cresyl violet staining.

3 Methods

3.1 Excitotoxic Lesions

Excitotoxic brain lesions are induced by a unique intra-cerebral injection of 10 μg of the glutamate analogue ibotenate at P5 fluorine-anesthetized pups, as described previously [28, 29].

This dose consistently causes brain damage in P5 mice [22]. Anesthetized mouse pups are injected intra-cerebrally into the neopallial parenchyma on day P0 or P5 using a 25-gauge needle on a 50-µl Hamilton syringe mounted on a calibrated micro-dispenser. The needle is inserted 2 mm under the external surface of the scalp skin in the fronto-parietal area of the right hemisphere, 2 mm from the midline in the lateral-medial plane, and 2 mm (at P0), and 3 mm (at P5) from the junction between the sagittal and lambdoid sutures [30]. Two 1 µl boluses are injected with a 20-s interval. The needle should be left in place for an additional 30 s. After the injection, pups are allowed to recover from anesthesia and returned to their dams.

3.2 Tissue Processing

Injected mice are sacrificed by isoflurane inhalation followed by transcardiac perfusion of phosphate buffer saline prior to whole body fixation with PFA in PBS.

Brains are gently removed and post-fixed overnight in PFA in 0.12 M PB at 4 °C and then cryoprotected in sucrose in 0.12 M PB for 2 days.

Subsequently, brains are immersed in sucrose and gelatine in 0.12 M PB for 1 h at 37 °C and then embedded in a block of the same solution for 1 h at 4 °C.

Parasagittal sections, 10 µm thick are cut using a cryostat, mounted on Superfrost Plus slides and stored at –80 °C.

3.3 Lesion Size Determination

Brains dedicated to the determination of lesion size should be fixed in PFA for 5 days and then embedded in paraffin.

Lesion size is determined in pups sacrificed on P10, i.e., 5 days after treatment, by transcardiac perfusion with PFA. Brains should be gently removed.

After post-fixation in PFA for an additional 24 h at 4 °C, the brains are dehydrated in alcohol and embedded in paraffin using standard protocols.

Brains are serially sectioned and volumes may be measured using the Neurolucida software-controlled computer system (MicroBright-Field Europe, Magdeburg, Germany) [31]. Because of the good correlation between the volume of the lesion and its maximal sagittal fronto–occipital diameter, a reproducible and accurate index of the volume of the lesion can be calculated by multiplying the number of coronal sections where the lesion is present multiplied by the thickness of each section [15, 31]. Each brain is completely and serially sectioned from the frontal pole to the occipital lobes at 15 µm intervals in the coronal plane. One every third section is stained with cresyl violet and observed under microscopy. Cyst, cell death, cell rarefaction, gliosis, and disrupted cytoarchitecture are considered as criteria of lesion. Sections containing any of these abnormalities in the cortical plate or white matter are included in the calculation of lesion size of cortical plate and white matter, respectively.

3.4 Neuroimaging of the Ibotenate Model

In experimental PWMD following ibotenate injection, MRI diffusion-weighted imaging detects cortical and white matter lesions [27]. Coupled with ultra-small superparamagnetic particles of iron oxide (USPIO), a contrast agent taken up by macrophages [32, 33], MRI successfully visualizes neuroinflammation. Post-USPIO T2-weighted MRI demonstrate negative enhancement of excitotoxic lesions 72 h after intraperitoneal injection without exacerbating brain lesions on histology [34].

3.5 Behavioral Deficits

Ibotenate-induced excitotoxic lesions at P5 do not produce any significant alterations at the openfield evaluation in mice [35]. The number of rears and of contacts with objects (testing exploratory behavior and sensory-motor activity, respectively) as well as the total number of crossed squares (motor activity) inside (inversely related to anxious behavior) and outside crossings (related to motor activity) are similar to controls. On the other hand, ibotenate lesions disrupt odor preference conditioning in newborn mice [36] and cause significant deficit in spatial and temporal memory function as investigated by the Morris water maze (Bouslama M, personal communication) and by the novel object recognition (NOR) test when performed at age 3 and 6 weeks [35].

4 Notes

- Usually, about 3–5 % of pups die following injection (minutes to hours later). This could be due to the injection technique, but also to the extent of the excitotoxic injury. So it is recommended to verify the correct positioning of the needle during ibotenate injection by injecting some animals with toluidine blue. This is mandatory if the mortality rate is more than expected.

- After sucrose and gelatine embedding, brains can be flash frozen in isopentane at −70 °C and stored at −80 °C until further use.

- Tissues adjacent to the site of ibotenate injection may be collected for RNA extraction and quantification of gene expression in the injured area. As an example, gene involved in the inflammatory response (e.g., IL-1β, IL-1-ra, IL-6, IL-10, TNFα, TNFR1, TNFR2, and BDNF) may be studied using standard techniques [37].

- An association between pro-inflammatory cytokines (for example, IL-1β, IL-6, and TNFα), perinatal brain damage and adverse outcome has been demonstrated in children with PWMD in clinical trials [38]. The sensitizing effect of systemic inflammation operates through the induction of an imbalance between pro- and anti-inflammatory factors following the insult [37]. Pretreatment with proinflammatory Th1 cytokines significantly exacerbates the severity of ibotenate-induced brain lesions [39]. To produce an inflammatory insult, mice pups can be injected intraperitoneally (i.p.) with recombinant mouse interleukin-1 beta (IL-1β; R&D systems, Europe Ltd., Oxon, UK). A 5-μl volume of PBS containing IL-1β (2.5 μg/kg) is injected i.p. twice a day (between 8 and 10 a.m. and again between 6 and 8 p.m.) on days P1 to P4 and once a day (between 8 and 10 a.m.) on day P5. Pups are then injected 2 h after the last injection of IL-1β, by intracerebral (i.c.) ibotenate as detailed above.

- Data from animal models also support the link between maternal–fetal infection and PWMD [40, 41]. Administration of the endotoxin lipopolysaccharide (LPS), a potent inducer of innate immune response and inflammation [42–44], either intracerebrally during early neonatal period or peritoneally to a pregnant mother results in inflammation and hypomyelination [45, 46]. LPS may be injected intraperitoneally at days 19 and 20 of gestation [26] and then followed by ibotenate injection at P5. Animals in the LPS group show larger cortical and white matter lesions than the control group with greater microglial activation and astrogliosis and less white matter myelination.

- We also shown that exposure to chronic stress throughout gestation sensitize the offspring to neonatal excitotoxic brain lesions [47]. In this model, mothers receive a daily subcutaneous injection of 3 mg kg^{-1} d^{-1} corticosterone (Sigma-Aldrich, St. Quentin Fallavier, France) diluted in 100 μl of DMSO or DMSO alone between embryonic day 13 (E13) and E18. Ibotenate is then injected intracerebrally to P5 pups as described.

- Ibotenate injection can be used to test the effect of drugs or physiological substances (recombinant growth factors, blocking antibodies, etc.) in decreasing or increasing lesion size in other models [31, 48].

References

1. Allen MC (2008) Neurodevelopmental outcomes of preterm infants. Curr Opin Neurol 21(2):123–128

2. Robertson CM, Watt MJ, Yasui Y (2007) Changes in the prevalence of cerebral palsy for children born very prematurely within a population-based program over 30 years. JAMA 297(24):2733–2740

3. Vincer MJ, Allen AC, Joseph KS, Stinson DA, Scott H, Wood E (2006) Increasing prevalence of cerebral palsy among very preterm infants: a population-based study. Pediatrics 118(6):e1621–e1626

4. Wilson-Costello D, Friedman H, Minich N, Siner B, Taylor G, Schluchter M, Hack M (2007) Improved neurodevelopmental outcomes for extremely low birth weight infants in 2000-2002. Pediatrics 119(1):37–45

5. O'Shea M (2008) Cerebral palsy. Semin Perinatol 32(1):35–41

6. Johnson AR, DeMatt E, Salorio CF (2009) Predictors of outcome following acquired brain injury in children. Dev Disabil Res Rev 15(2):124–132

7. Moritani T, Smoker WR, Sato Y, Numaguchi Y, Westesson PL (2005) Diffusion-weighted

imaging of acute excitotoxic brain injury. AJNR Am J Neuroradiol 26(2):216–228

8. Deng W, Pleasure J, Pleasure D (2008) Progress in periventricular leukomalacia. Arch Neurol 65(10):1291–1295

9. Johnston MV (2005) Excitotoxicity in perinatal brain injury. Brain Pathol 15(3):234–240

10. Volpe JJ (2005) Encephalopathy of prematurity includes neuronal abnormalities. Pediatrics 116(1):221–225

11. Inglis WL, Semba K (1997) Discriminable excitotoxic effects of ibotenic acid, AMPA, NMDA and quinolinic acid in the rat laterodorsal tegmental nucleus. Brain Res 755(1):17–27

12. Sfaello I, Baud O, Arzimanoglou A, Gressens P (2005) Topiramate prevents excitotoxic damage in the newborn rodent brain. Neurobiol Dis 20(3):837–848

13. Destot-Wong KD, Liang K, Gupta SK, Favrais G, Schwendimann L, Pansiot J, Baud O, Spedding M, Lelievre V, Mani S, Gressens P (2009) The AMPA receptor positive allosteric modulator, S18986, is neuroprotective against neonatal excitotoxic and inflammatory brain damage through BDNF synthesis. Neuropharmacology 57(3):277–286

14. Marret S, Mukendi R, Gadisseux JF, Gressens P, Evrard P (1995) Effect of ibotenate on brain development: an excitotoxic mouse model of microgyria and posthypoxic-like lesions. J Neuropathol Exp Neurol 54(3):358–370

15. Tahraoui SL, Marret S, Bodenant C, Leroux P, Dommergues MA, Evrard P, Gressens P (2001) Central role of microglia in neonatal excitotoxic lesions of the murine periventricular white matter. Brain Pathol 11(1):56–71

16. Volpe JJ (2001) Neurobiology of periventricular leukomalacia in the premature infant. Pediatr Res 50(5):553–562

17. Craig A, Ling Luo N, Beardsley DJ, Wingate-Pearse N, Walker DW, Hohimer AR, Back SA (2003) Quantitative analysis of perinatal rodent oligodendrocyte lineage progression and its correlation with human. Exp Neurol 181(2): 231–240

18. Inder TE, Warfield SK, Wang H, Huppi PS, Volpe JJ (2005) Abnormal cerebral structure is present at term in premature infants. Pediatrics 115(2):286–294

19. Kesler SR, Vohr B, Schneider KC, Katz KH, Makuch RW, Reiss AL, Ment LR (2006) Increased temporal lobe gyrification in preterm children. Neuropsychologia 44(3):445–453

20. Leviton A, Gressens P (2007) Neuronal damage accompanies perinatal white-matter damage. Trends Neurosci 30(9):473–478

21. Marlow N, Wolke D, Bracewell MA, Samara M (2005) Neurologic and developmental disability at six years of age after extremely preterm birth. N Engl J Med 352(1):9–19

22. Husson I, Rangon CM, Lelievre V, Bemelmans AP, Sachs P, Mallet J, Kosofsky BE, Gressens P (2005) BDNF-induced white matter neuroprotection and stage-dependent neuronal survival following a neonatal excitotoxic challenge. Cereb Cortex 15(3):250–261

23. Marret S, Gressens P, Evrard P (1996) Arrest of neuronal migration by excitatory amino acids in hamster developing brain. Proc Natl Acad Sci U S A 93(26):15463–15468

24. Fontaine RH, Olivier P, Massonneau V, Leroux P, Degos V, Lebon S, El Ghouzzi V, Lelievre V, Gressens P, Baud O (2008) Vulnerability of white matter towards antenatal hypoxia is linked to a species-dependent regulation of glutamate receptor subunits. Proc Natl Acad Sci U S A 105(43):16779–16784

25. Redecker C, Hagemann G, Witte OW, Marret S, Evrard P, Gressens P (1998) Long-term evolution of excitotoxic cortical dysgenesis induced in the developing rat brain. Brain Res Dev Brain Res 109(1):109–113

26. Rousset CI, Kassem J, Olivier P, Chalon S, Gressens P, Saliba E (2008) Antenatal bacterial endotoxin sensitizes the immature rat brain to postnatal excitotoxic injury. J Neuropathol Exp Neurol 67(10):994–1000

27. Sfaello I, Daire JL, Husson I, Kosofsky B, Sebag G, Gressens P (2005) Patterns of excitotoxin-induced brain lesions in the newborn rabbit: a neuropathological and MRI correlation. Dev Neurosci 27(2–4):160–168

28. Gressens P, Marret S, Hill JM, Brenneman DE, Gozes I, Fridkin M, Evrard P (1997) Vasoactive intestinal peptide prevents excitotoxic cell death in the murine developing brain. J Clin Invest 100(2):390–397

29. Olney JW, Ikonomidou C, Mosinger JL, Frierdich G (1989) MK-801 prevents hypobaric-ischemic neuronal degeneration in infant rat brain. J Neurosci 9(5):1701–1704

30. Sokolowska P, Passemard S, Mok A, Schwendimann L, Gozes I, Gressens P (2010) Neuroprotective effects of NAP against excitotoxic brain damage in the newborn mice: implications for cerebral palsy. Neuroscience 173:156–168

31. Medja F, Lelievre V, Fontaine RH, Lebas F, Leroux P, Ouimet T, Saria A, Rougeot C, Dournaud P, Gressens P (2006) Thiorphan, a neutral endopeptidase inhibitor used for diarrhoea, is neuroprotective in newborn mice. Brain 129(Pt 12):3209–3223

32. Schroeter M, Saleh A, Wiedermann D, Hoehn M, Jander S (2004) Histochemical detection of ultrasmall superparamagnetic iron oxide (USPIO) contrast medium uptake in experimental brain ischemia. Magn Reson Med 52(2):403–406

33. Dousset V, Ballarino L, Delalande C, Coussemacq M, Canioni P, Petry KG, Caille JM (1999) Comparison of ultrasmall particles of iron oxide (USPIO)-enhanced T2-weighted, conventional T2-weighted, and gadolinium-enhanced T1-weighted MR images in rats with experimental autoimmune encephalomyelitis. AJNR Am J Neuroradiol 20(2):223–227

34. Alison M, Azoulay R, Chalard F, Gressens P, Sebag G (2010) In vivo assessment of experimental neonatal excitotoxic brain lesion with USPIO-enhanced MR imaging. Eur Radiol 20(9):2204–2212

35. Titomanlio L, Bouslama M, Le Verche V, Dalous J, Kaindl A, Tsenkina Y, Lacaud A, Peineau S, Elghouzzi V, Lelievre V, Gressens P (2011) Implanted neurosphere-derived precursors promote recovery after neonatal excitotoxic brain injury. Stem Cells Dev 20(5):865–879

36. Bouslama M, Renaud J, Olivier P, Fontaine RH, Matrot B, Gressens P, Gallego J (2007) Melatonin prevents learning disorders in brain-lesioned newborn mice. Neuroscience 150(3):712–719

37. Favrais G, Schwendimann L, Gressens P, Lelievre V (2007) Cyclooxygenase-2 mediates the sensitizing effects of systemic IL-1-beta on excitotoxic brain lesions in newborn mice. Neurobiol Dis 25(3):496–505

38. Nelson KB, Dambrosia JM, Grether JK, Phillips TM (1998) Neonatal cytokines and coagulation factors in children with cerebral palsy. Ann Neurol 44(4):665–675

39. Dommergues MA, Patkai J, Renauld JC, Evrard P, Gressens P (2000) Proinflammatory cytokines and interleukin-9 exacerbate excitotoxic lesions of the newborn murine neopallium. Ann Neurol 47(1):54–63

40. Yoon BH, Kim CJ, Romero R, Jun JK, Park KH, Choi ST, Chi JG (1997) Experimentally induced intrauterine infection causes fetal brain white matter lesions in rabbits. Am J Obstet Gynecol 177(4):797–802

41. Romero R, Avila C, Santhanam U, Sehgal PB (1990) Amniotic fluid interleukin 6 in preterm labor. Association with infection. J Clin Invest 85(5):1392–1400

42. Pang Y, Cai Z, Rhodes PG (2003) Disturbance of oligodendrocyte development, hypomyelination and white matter injury in the neonatal rat brain after intracerebral injection of lipopolysaccharide. Brain Res Dev Brain Res 140(2):205–214

43. Fan LW, Tien LT, Mitchell HJ, Rhodes PG, Cai Z (2008) Alpha-phenyl-n-tert-butyl-nitrone ameliorates hippocampal injury and improves learning and memory in juvenile rats following neonatal exposure to lipopolysaccharide. Eur J Neurosci 27(6):1475–1484

44. Lehnardt S, Lachance C, Patrizi S, Lefebvre S, Follett PL, Jensen FE, Rosenberg PA, Volpe JJ, Vartanian T (2002) The toll-like receptor TLR4 is necessary for lipopolysaccharide-induced oligodendrocyte injury in the CNS. J Neurosci 22(7):2478–2486

45. Cai Z, Pan ZL, Pang Y, Evans OB, Rhodes PG (2000) Cytokine induction in fetal rat brains and brain injury in neonatal rats after maternal lipopolysaccharide administration. Pediatr Res 47(1):64–72

46. Wang X, Rousset CI, Hagberg H, Mallard C (2006) Lipopolysaccharide-induced inflammation and perinatal brain injury. Semin Fetal Neonatal Med 11(5):343–353

47. Rangon CM, Fortes S, Lelievre V, Leroux P, Plaisant F, Joubert C, Lanfumey L, Cohen-Salmon C, Gressens P (2007) Chronic mild stress during gestation worsens neonatal brain lesions in mice. J Neurosci 27(28):7532–7540

48. Laudenbach V, Fontaine RH, Medja F, Carmeliet P, Hicklin DJ, Gallego J, Leroux P, Marret S, Gressens P (2007) Neonatal hypoxic preconditioning involves vascular endothelial growth factor. Neurobiol Dis 26(1):243–252

Chapter 16

The Effect of Age on Brain Plasticity in Animal Models of Developmental Disability

Bryan Kolb and Deborah Saucier

Abstract

Brain development is a complex interaction of environmental experiences and genetic influences. Experiences include both prenatal (gestational), perinatal, and later events including stress, gonadal hormones, drugs (prescription and others), sensory stimulation (e.g., tactile stimulation), and sensory deprivation. By manipulating these factors it is possible to get a better understanding of how brain and behavioral phenotypes emerge. Methods are outlined on how to study and assess these early experiences.

Key words Development, Brain plasticity, Gestational stress, Cortex, Gonadal hormones, Brain injury, Sensory experience, Postnatal stress, Tactile stimulation, Sensory deprivation, Handling, Complex housing

1   Introduction

There is a long history of theory and experimentation asking the question, "When is it best to have brain injury?" The subtext to this question is "When is the brain most plastic?" It was generally assumed up until the 1970s that the earlier brain injury occurred during development, the better the outcome would be. This idea dates back to the late nineteenth century when Broca and others noticed that children rarely had persistent aphasia after damage to the cortical language zones. Studies of the motor behavior of infant monkeys with motor cortex injuries in the 1930s and 1940s by Margaret Kennard [1] confirmed this general notion of "earlier is better." The behavioral impairments in the infant monkeys were milder than those in the adults, which led Kennard to hypothesize that there had been a change in cortical organization in the infants and these changes supported the behavioral recovery. Although they are intuitively appealing, Donald Hebb's studies of children with early brain injury in the 1940s led to a different conclusion [2]. Hebb noticed that children with frontal lobe injuries often had worse outcomes than adults with similar injuries and proposed

Jerome Y. Yager (ed.), *Animal Models of Neurodevelopmental Disorders*, Neuromethods, vol. 104, DOI 10.1007/978-1-4939-2709-8_16, © Springer Science+Business Media New York 2015

that the early injuries prevented a normal initial organization of the brain, thus making it difficult for the child to develop many behaviors, especially socioaffective behaviors. Beginning in the 1970s and continuing until today, extensive studies of both cats and rats with cortical injuries have shown that both views are partially correct (e.g., 3–6). It is the precise developmental age at injury that predicts the Kennard or Hebb outcome.

Over the past 20 years studies of early brain plasticity have expanded to look at the effects of a wide range of experiences on brain development and later plasticity. These experiences include prenatal and postnatal sensory experiences, psychoactive drugs, gonadal hormones, and play, among others.

Before we discuss the details of methodological approaches, we should briefly consider brain development. Mammalian brain development is characterized by a series of stages beginning with neural generation as summarized in Table 1. Cells are generated in the subventricular zone, they migrate to their presumptive location where they differentiate and form connections. Mammalian brains overproduce both neurons and connections and thus go through a period of cell death and synaptic pruning. As a rule of thumb, the larger the species brain, the more pruning and cell death there is. Finally, myelin forms, a process that can be quite prolonged in large-brained animals such as primates.

The process of brain development in different species has important implications for experimental models of brain plasticity and behavior. Species such as laboratory rats and mice, as well as cats, are precocial and thus are born fairly early in the process of brain development. In contrast, the brains of altricial species such as guinea pigs are quite developed at birth. As a result, age relative to birth date is not a useful measure in developing models of early brain injury. Rather, it is the precise stage of neural development that is critical.

Table 1
Stages of brain development

1. Cell birth (neurogenesis, gliogenesis)
2. Cell migration
3. Cell differentiation
4. Cell maturation (dendrite and axon growth)
5. Synaptogenesis (formation of synapses)
6. Cell death and synaptic pruning
7. Myelogenesis (formation of myelin)

1.1 Early Brain Injury   The first systematic studies of early brain injury in laboratory animals were done on monkeys by Margaret Kennard in the early 1940s [1]. One difficulty with the Kennard studies is that monkeys are more developmentally advanced at birth than humans and laboratory rats, hamsters, and cats. Surprisingly, however, early injuries in other laboratory animals were not done in any systematic way until the early 1970s (e.g., 7–11). Patricia Goldman [6] made prenatal lesions in monkeys with the goal of more closely approximating the age of injury in rodents and cats. The general conclusion from studies of early cerebral injury in laboratory animals is that focal damage during the period of cerebral neurogenesis and later rapid dendritic expansion allows for extensive spontaneous neural plasticity whereas damage during the neural migration has especially bad behavioral and anatomical consequences (*see* 3 for a review). In contrast, studies of more global injuries, such as hemidecortication, decortication, and hypoxia-ischemia provide a different timeline. Specifically, it appears that earlier is indeed better with these types of injuries (e.g., 5, 12, 13, 14).

Although the effects of focal damage to different cerebral regions has somewhat different plastic responses, the general findings are remarkably consistent across sensory, motor, cingulate, and prefrontal cortex. The least plasticity appears to follow posterior parietal cortex lesions, although it is unclear why this is so.

1.2 Early Drug Exposure   It has long been known for at least 60 years that prenatal alcohol and barbiturate exposure have serious effects on brain and behavioral development (e.g., 15) but it has only been relatively recently that it has become clear that virtually every psychoactive drug, including prescription drugs, can influence later behavior. However, although these drugs are known to change behavior, far less is known about how early exposure to drugs alters brain plasticity. One clue has come from studies that gave adult rats psychomotor stimulants (amphetamine, cocaine, or nicotine) and then later placed the animals in complex environments, which on their own are known to produce large changes in synaptic organization. The surprising result was that the drug exposure appears to completely block the later effect of experience on synaptic plasticity [16]. Similar studies in developing animals has shown that prenatal or postnatal exposure to psychoactive drugs reduces later synaptic plasticity [17].

1.3 Perinatal Sensory Experiences   The simplest way to manipulate experience across ages is to compare brain structure in animals living in standard laboratory caging to animals placed either so-called enriched environments or in severely impoverished environments. Perhaps the oldest way to enhance sensory and motor functions is to place animals in complex environments in which there is an opportunity for animals to interact with a changing sensory and social environment and to engage in far more motor activity than regular caging. Such studies have identified a large range of neural changes associated with this form of "enrichment." These include increases in brain size, cortical thickness, neuron size, dendritic branching, spine density, synapses per neuron, glial numbers and complexity, and vascular arborization (e.g., 18, 19).

Raising animals in deprived environments such as in darkness, silence, or social isolation clearly retards brain development. For example, dog puppies raised alone show a wide range of behavioral abnormalities, including a virtual insensitivity to painful experiences [3]. Similarly, raising animals as diverse as monkeys, cats, and rodents in the dark severely interferes with development of the visual system. Perhaps the best-known deprivation studies are those of Weisel and Hubel [20] who sutured one eyelid of kittens closed and later showed that when the eye was opened there was an enduring loss of spatial vision (amblyopia) (e.g., 21). It has only been recently, however, that investigators considered the opposite phenomenon, namely giving animals enriched visual experiences to determine if vision could be enhanced. In one elegant study, Prusky et al. [22] used a novel form of visual stimulation in which rats were placed in a virtual optokinetic system in which vertical lines of differing spatial frequency moved past the animal. If the eyes are open and oriented towards the moving grating, it is impossible for animals, including humans, to avoid tracking the moving lines, if the spatial frequency is within the perceptual range. The authors placed animals in the apparatus for about 2 weeks following the day of eye opening (postnatal day 15). When tested for visual acuity in adulthood, the animals showed about a 25 % enhancement in visual acuity relative to animals without the early treatment.

Tactile experience can also be manipulated using a procedure first devised by Schanberg and Field [23]. In these studies infant rats were given tactile stimulation with a small brush for 15 min three times per day for 10–15 days beginning at birth. Kolb and Gibb [24] have shown that when the infants were studied in adulthood they showed both enhanced skilled motor performance and spatial learning as well changes in synaptic organization across the cerebral cortex.

Early stress can also have profound effects on later brain plasticity. The most commonly used procedure is a form of maternal separation whereby animals are removed from their nest (and mother) for anywhere from 1 to 3 h per day. Such treatments have chronic effects both on behavior and neuronal morphology in rodents, monkeys and humans [25, 26]. For example, enhanced alcohol preference and intake has been reported in maternally separated rats [27].

Finally, although they are less well studied, prenatal experiences can also profoundly affect later brain and behavioral plasticity.

The best studied example is prenatal stress. Prenatal stress has been studied in rats by employing a number of stress paradigms with variable frequency and duration. The two most common procedures involve either straining the pregnant dams with restraint stress or placement on a small platform in an empty room. The latter procedure has been shown to alter adult behavior and neuronal spine density (e.g., 28).

1.4 Gonadal Hormones

Gonadal hormones profoundly affect the development and function of neurons. Testosterone is released for a brief period in the course of prenatal development and acts to masculinize the brain. In rats there is also another period of testosterone release that occurs just after birth, further changing the brain. Early testosterone release normally only happens in males (i.e., animals with a Y chromosome) so if females (no Y chromosome) are exposed to testosterone their brain will be masculinized [29–31]. Similarly, if the gonads are removed at birth males will not get the second dose of testosterone and the brain is less masculinized. But gonadal hormones also act in adolescence to alter brain and behavior in both sexes, leading to sexual dimorphism across the cerebral hemispheres (e.g., 32). In addition, gonadal hormones influence recovery from early cortical lesions (e.g., 33, 34).

1.5 Measuring Brain Plasticity

Changes in the brain can be inferred from a wide range of techniques ranging from changes in behavior to changes in gene expression (*see* Table 2). None of these is the best, rather it depends upon the question being asked. For example, if one is interested in the effects of early experiences on language acquisition, it makes little sense to measure changes in gene expression because this measure does not easily map onto such complex behavior. On the other hand, if one is interested in later sensitivity to addictive drugs, measuring gene expression may provide strong insights into the underlying mechanisms.

Table 2
Types of analyses of brain plasticity

1. Behavior
2. Maps—noninvasive and invasive
3. Physiology (e.g., cell recording)
4. Neuronal morphology
5. Proteins and other molecules
6. Genetics and epigenetics

2   Methods

Our goal here is to outline the methods that have been used (or can be used) to study age-related effects on normal and abnormal brain development.

2.1   Studying the Effects of Brain Injury

The brain can be damaged selectively in a wide variety of ways, each of which requires slightly different equipment. In some procedures a craniotomy is performed and the cerebral cortex is removed or damaged by gentle aspiration, or by stripping away of the fine surface blood vessels to produce ischemia (pial stripping). The cortex can also be damaged by injecting drugs that produce vasoconstriction such as endothelin-1 (ET-1). A final procedure is to give an i.v. injection of a photosensitive dye, Rose Bengal. When the skull of rose-bengal treated rats is illuminated with a focused beam of light for 10 min, cortical thrombosis is induced in the underlying tissue, at least partly due to the induction of apoptosis. The advantage of Rose Bengal is that no craniotomy or small burr hole is necessary.

In those procedures using a craniotomy the skull is exposed and the size of the desired motor injury is marked on the skull. In animals younger than about 15 days, the bone is soft enough to be cut with fine scissors (e.g., iris scissors). In this case, a small incision can be made on the skull with a number 11 scalpel blade, which allows for the scissor to gain purchase on the skull. One difficulty with early craniotomies is that the brain has considerable pressure and will press out of the craniotomy. It is therefore not possible to have "sham" controls with a craniotomy. The only brain injury procedure with a craniotomy that can be used on animals younger than about 15 days is aspiration because the consistency of the brain material and the development of the cerebral vasculature precludes other techniques.

For older animals the bone is thinned with a high speed drill and then the remaining bone is removed with fine rongeurs. The size of the opening will vary with the region of interest. For aspiration and pial stripping the dura is then cut with a pointed (#11) scalpel blade and removed. We have found it best to also cut through the grey matter along the edges of the craniotomy in order to make it easier to delineate the size of the injury. The aspiration procedure involves using gentle suction and removing only the grey matter within the delineated region. For pial stripping, the cotton ball is dipped in sterile saline and then used to gently wipe off the blood vessels (e.g., 35).

There are several methods for the application of ET-1. These include the topical application across the surface of the cortex in the region of the craniotomy, multiple injections of the ET-1 into the grey and/or white matter, and injection of ET-1 adjacent to the middle cerebral artery. Dosages vary considerably across laboratories and require pilot work but in general, the higher the dose, the larger the injury.

For the Rose Bengal procedure the photochemical Rose Bengal is infused into the femoral vein via a microinjection pump within 2 min (20 mg/kg). The light is then turned on for 10 min. If the light is not cold (fiber optic), then the skull surface temperature must be monitored and the skull kept cool by cool air flow (e.g., 36).

A more global form of early injury is hypoxia-ischemia. This procedure is described in detail in the Chapter 1.

One problem for most infant surgery is anesthesia. The two preferred methods are cryoanesthesia, in which the animal's core temperature is lowered to about 20 °C, and inhalant anesthetics, most notably isoflurane. Cooling can be accomplished by placing pups on bedding in a cage in a cooling device (e.g., freezer) that is set to about –5 °C [37]. The problem with both types of anesthetic is that they are known to induce apoptosis in very young animals. As a result, control animals must undergo exactly the same anesthetic procedures as the experimental animals. Animals given hypoxia-ischemia do not have anesthesia because the brain injury is caused by lowering oxygen by placing animals in a separate enclosed environment.

One final challenge for inducing early focal injury is the fixation of the head during the surgical procedures. Aspiration of cerebral cortex in very young animals can be done by simply holding the head between the thumb and index finger of one hand. By the time the animals are about 2 weeks old it is possible to use a stereotaxic device in which the animals are placed upon a raised base. Fixing the skull for stereotaxic injections in very young animals is tricky because their ear canals are not large enough to allow penetration of the earbars. We have found that a plasticine mold can be constructed to fit animals at different ages. Anesthetized animals can then be placed into the mold and gently held in place with masking tape. It is best to rub the tape to reduce its stickiness before using it to hold the animals. The stereotaxic placement then can be done but pilot animals are needed to identify appropriate coordinates.

Each of the measures of brain plasticity summarized in Table 2 can be applied to animals at a variety of ages and recovery times.

2.2 Studying the Effects of Early Drug Exposure

Early drug exposure has been studied both by administering prescription or recreational drugs to pregnant animals or by administering the drugs directly to the developing infants. Given that most such studies are done in rodents, which are born developmentally younger than humans, it is often assumed that drug administration in the first week of rat or mouse is equivalent to the third trimester in humans. Early psychoactive drug exposure generally has profound effects on brain development, the most obvious example being alcohol but drugs such as antipsychotics and antidepressants seem to have effects that are nearly as large (e.g., 38). It is not wise to give i.p. injections to pregnant dams so drugs need to be injected subcutaneously or given in food or water. Doses will depend upon the question being asked and requires pilot studies. Even very small doses of some drugs (e.g., 0.3 mg/kg nicotine, which is equivalent to about one cigarette in human terms) can produce large chronic effects on both brain and behavior. There is good reason to believe that early drug exposure can interact with other perinatal experiences to produce profound neurodevelopmental disorders. For example, we have found that administering fluoxetine to pregnant rats leads to smaller brains and impaired motor and cognitive behaviors in otherwise normal rats but can completely block recovery from cortical injury in the second week of life (R. Gibb and B. Kolb, unpublished observations).

2.3 Studying the Effects of Experience

Over the past 20 years it has become clear that even fairly innocuous-looking experiences can profoundly affect brain development and that the range of experiences that can alter brain development is much larger than had been once believed and includes both prenatal and postnatal experiences (*see* Table 3). We highlight some of the most well-studied effects.

Sensory deprivation. Raising animals in deprived environments such as in darkness, silence, or social isolation clearly retards brain development. Perhaps the best-known deprivation studies are those of Weisel and Hubel [20] who sutured one eyelid of kittens closed and later showed that when the eye was opened there was an enduring loss of spatial vision (amblyopia) (e.g., 21). In these studies the animals are briefly anesthetized and the eyelids sewn shut prior to their opening, which is at 15 days in rats.

Table 3
Types of sensory-motor experiences during development

1. Complex housing
2. Visual stimulation
3. Tactile stimulation
4. Auditory stimulation
5. Infant handling
6. Sensory deprivation
7. Eyelid suturing
8. Maternal separation stress
9. Prenatal stress

Visual stimulation. Prusky et al. [22] used a novel form of visual stimulation in which rats were placed in a virtual optokinetic system in which vertical lines of differing spatial frequency moved past the animal. If the eyes are open and oriented towards the moving grating, it is impossible for animals, including humans, to avoid tracking the moving lines, if the spatial frequency is within the perceptual range. The authors placed animals in the apparatus for about 2 weeks following the day of eye opening (postnatal day 15). When tested for visual acuity in adulthood, the animals showed about a 25 % enhancement in visual acuity relative to animals without the early treatment. The beauty of the Prusky study is that improved visual function was not based upon specific training, such as in learning a problem, but occurred naturally in response to enhanced visual input. Another advantage of this procedure is that the treatment can begin at any age, which allows for investigation of the role of age in the observed behavioral and presumed neural changes. The software and hardware necessary for such stimulation can be purchased from Cerebral Mechanics (www. cerebralmechanics.com).

Tactile stimulation. The procedure was first devised by Schanberg and Field [23] who gave infant rats tactile stimulation with a small brush for 15 min three times per day for 10–15 days beginning at birth. This procedure not only leads to enhanced motor and cognitive abilities later in life but also facilitates recovery from early brain injury and is correlated with synaptic reorganization in the cerebral cortex [24]. The infants are removed from their nests and taken in a box to a cage that can be warmed with a heat lamp or heating pad to keep the pups' body temperature in the normal range. When they are young, the animals can be lined up and gently stroked with a blush brush. They typically go into REM sleep when the stimulation is applied. As they become more mobile, they need to be followed around the cage for stroking. Control animals have a similar procedure but no tactile stimulation. In order to distract the dams when collecting the pups it is handy to provide treats such as cheese twists or sweetened cereal. Although the role of age in the benefits of tactile stimulation has not been studied yet, they certainly could be related.

Tactile stimulation of pregnant rats using a child's hairbrush three times daily throughout the pregnancy has similar effects (R. Gibb and B. Kolb, unpublished). The pregnant animals are lightly held on an experimenters lap and gently brushed with the hairbrush. It helps to provide treats because the animals become quite motivated to sit on the investigator's lap. Control animals are removed from their cage and placed in a novel one where they get similar treats.

Auditory stimulation. Less is known about the effects of early auditory stimulation but one study is instructive. De Villers-Sidani et al. [39] exposed developing rat pups either to spectrally limited acoustic noise or broad noisy incoherent inputs. The continual presentation of 70 dB noise at 5–20 kHz (limited range) or random noise throughout the hearing spectrum of rats differentially changed the tonotopic maps. This type of method has not yet been applied to studying cognitive functions or interactions with other treatments but does provide a proof of principle using a very simple technique. As with visual and tactile stimulation, age can be manipulated to look at the role of specific auditory experiences at different developmental periods.

"Enriched" housing. Perhaps the most effective way to enhance sensory and motor functions is to place animals in complex environments in which there is an opportunity for animals to interact with a changing sensory and social environment and to engage in far more motor activity than regular caging. Such studies have identified a large range of behavioral and neural changes associated with this form of "enrichment." These include increases in brain size, cortical thickness, neuron size, dendritic branching, spine density, synapses per neuron, glial numbers and complexity, and vascular arborization (e.g., 18, 19). Typically 6–8 animals are housed in such environments for at least 30 days, although synaptic changes can be seen within 2 weeks (e.g., 40). The environments can take many forms, although most use a large enclosure such as livestock watering tank, which is usually about 4–8 ft in diameter, at least 2 ft high, and fitted with a screen top. Bedding such as sawdust is placed on the floor and normally there are many plastic or metal children's toys, PVC piping, and so on provided. These toys are changed at least weekly both to allow cleaning as well as Water bottles can be attached to the sides of the tank and food is simply scattered about the tank. A major advantage of the livestock tanks is that they are easy to clean. Another type of environment is a more vertically oriented one that can be made from stainless steel. Typically these are about 2 ft deep × 6 ft high × 4 ft wide. The side walls and front door are made of hardware cloth and the animals quickly learn to race up and down the walls. Platforms can be attached to the rear and side walls, as can water bottles. As with the stock tank environments, toys are scattered on the floor and changed regularly.

Developing animals can either be born and raised in the complex environments or they can be moved there at different times afterwards. The synaptic changes that are seen after housing from birth are different from those seen when animals are placed in the environments at weaning, which again is different from the effects later in life (e.g., 40). Complex housing is one of the most powerful treatments in reversing the effects of early brain injury, especially injury in the first week of life (e.g., 41, 42). There is less effect of the treatment on focal injury in the second week, which is reasonable given that the behavioral effects are smaller in the latter case.

Handling. The procedure known as handling involves separating infants from their mothers for brief periods of time (e.g., 43, 44). The earliest descriptions of the handling procedure detail a process in which infant rats were removed from the nest and placed individually in a small holding container for 3 min, after which the pup was returned to the nest [44]. Today the procedure can mean removal from the nest, either individually or as a group, for 3–20 min (e.g., 45). Rats given such treatment are reported to have higher body weights in adulthood [43], superior cognitive abilities [46], and a lack of cognitive decline in senescence [47]. Handling leads to a variety of physiological changes, including changes in plasma levels of norepinephrine [48], and increased hippocampal glucocorticoid expression [49] among others. Handling has been found to enhance recovery from perinatal hypoxia-ischemia [50] but to have no effect on recovery from focal cortical excisions [45].

Prenatal and postnatal stress. There are many methods for inducing prenatal stress. One simple procedure is to place pregnant animals on a clear Plexiglas platform ($\sim$20 cm $\times$ 20 cm) that is about a meter off the floor and placed in the center of an otherwise empty room with the lights on [51]. Room size is usually at least 2 m $\times$ 2 m. The animals are placed on the platform for 20–30 min once or twice per day depending upon the desired stress level. Animals will occasionally jump off the platform, in which case they are immediately placed back on it. The number of days and timing of the stressing procedure depends upon the question being asked but it is most common to stress the dams during part or all of the period of cerebral neurogenesis, which begins about embryonic day 12 and continues until birth. It is probably best to terminate the stress a couple of days before term so that the dam can prepare for the infants' arrival. Another procedure that is commonly used is to give 20 min of restraint stress alternated with 5 min of forced swim, twice daily from gestation days 12–18 [52]. In this procedure the mom is placed into a clear plastic tube about 20 cm long and $\sim$8 cm in diameter with both ends blocked. The dam cannot turn around but is otherwise able to move. The forced swim is typically done in a tank of 18–20 °C water in which the animal cannot touch bottom without going under the water. Stress can be reduced if the water temperature is raised to over 25 C.

Postnatal stress is usually induced by removing pups from the nest for 1–3 h per day, at approximately the same time each day (e.g., 53). The rat pups are transported in a box with new bedding to a separate room and placed on a warming pad with a temperature of $\sim$34 °C. Most studies begin the treatment at about postnatal day 3 and continue until weaning. This procedure leads to a variety of behavioral and neuroanatomical effects in later in life. Examples include feminization of play behavior in males and an altered response to psychomotor stimulants in both sexes (e.g., 54).

2.4 Studying the Effects of Gonadal Hormone Manipulations

In our experience, virtually every developmental manipulation produces sexually dimorphic effects. Many of these effects relate to how gonadally produced steroid hormones affect the organism. As noted above, steroid hormones have two main times of production: prenatal and perinatal periods of steroid hormone activity (so called organizational effects); and those that occur during puberty and afterwards (so called activational effects; 55). To quickly review, gonadal hormones are steroids that are derived from cholesterol and consist of three major classes: estrogens (of which the prototypical hormone is estradiol, E), progestins (of which the prototypical hormone is progesterone, P), and androgens (of which the prototypical hormone is testosterone, T). There are numerous areas of the brain that exhibit sexual dimorphisms following organizational exposure to gonadal hormones during critical periods, including the hypothalamus and preoptic areas, areas associated with endocrine and sexual function [56]. However, other areas of the brain, e.g., the amygdala and hippocampus, that are involved in higher cognitive or non-reproductive behaviors are susceptible to variation in concentrations of gonadal hormones during the critical period [56–60]. Further, these changes extend into adulthood (e.g., 58). Thus, this exposure to gonadal hormones has widespread and longlasting effects on brain and behavior.

For rats, the critical organizational period for gonadal hormones ranges from embryonic day 18 to approximately 10 days after birth [61]. Thus, studies that want to affect the organizational period often do so by gonadectomizing the neonate on postnatal day 1. Gonadectomizing the pup requires aseptic technique, analgesia, and anesthesia. Orchidectomy (removal of the testes) involves making an incision in the scrotum and the removal of the testes and the ligation of the vas deferens. Ovariectomy involves making an incision in the trunk of the pup, removing the ovaries and ligating the fallopian tubes. In the pup, incisions are often closed with surgical glue as the skin of the pups is too delicate for many of the more standard closure techniques. Another common manipulation is to implant a silastic implant or pellet containing a known concentration of a gonadal hormone at this time into a subset of pups. The purpose of this manipulation is to contrast how a known concentration of a gonadal hormone with the untreated but gonadectomized controls. Thus, in this way, the effects that androgens and estrogens have on brain and behavior during this critical period can be manipulated.

Although exposure to a specific gonadal hormone (e.g., T) results in the masculinization of the developing brain, it may be the metabolites of the hormone that are more important in producing the observed effects. For instance, T is metabolized to E and dihydrotestosterone (DHT). In a series of classic experiments, when E was given during the critical period to gonadectomized male rats, the rats behaved in a similar fashion to intact males [62].

E does not have this effect on female rats, as fetal ovaries do not produce T or E, and thus fetal female rats are not exposed to these gonadal hormones (for a review, *see* 60). However, androgens also have direct effects on brain via androgen receptors (ARs), particularly in primates [60] and humans, at least with respect to gender identity and sexual orientation [63].

Once individuals reach puberty, the gonads once again begin to produce steroid hormones and changes in behavior are observed. These so-called "activational effects" are result from the initial effects that E and T had on neural structures during the organizational period [55, 64]. Once mammals reach reproductive viability, endogenous concentrations of gonadal hormones fluctuate predictably and reflect cyclical endogenous variation in concentrations of E and T [65]. Many studies examine how gonadal hormones affect the behavior of adult rodents. As with pups, gonadectomy is the primary method for manipulating concentrations of gonadal hormones. The technique for gonadectomy in adults is somewhat easier as the size of the organism and anesthetic techniques make the surgery more straightforward. Again, once the rat is gonadectomized, exogenous supplementation of gonadal hormone often occurs (e.g., 66). This may be via the implantation of a silastic capsule containing the hormone of interest or by daily injection of the hormone of interest dissolve in an oil vehicle.

However, there are other ways of examining endogenous fluctuations of gonadal hormones that does not involve invasive techniques. In adult males, androgen concentrations fluctuate daily and seasonally and variations in brain and behavior have been noted (e.g., 67–69). In these studies, rats are typically studied at different times of the year. For instance, wild living Richardson's ground squirrels [68, 69] were trapped during either the breeding or non-breeding seasons. Notably, the concentrations of gonadal hormones were assessed indirectly via observation of gonadal weight (gonadal weight in this species is higher during breeding) and behavioral observation (squirrels were observed engaging in pre-mating behaviors). However, other studies that utilized these techniques have studied captive populations and thus were able to control access to breeding directly and assess hormone concentrations via blood samples or stool samples (e.g., 70, 71).

In adult female mammals, concentrations of estrogens and progestins vary endogenously across the estrus cycle, which has a unique profile of peaks and troughs that is species specific. In the rat, the estrous cycle is 4–5 days long and can be broken down into four phases, with E peaking at proestrus and at lowest levels at diestrus [72]. The stages of estrous in the rat can be assessed by sampling vaginal epithelium daily [73, 74]. When assessing stage, the cells should be collected at about the same time each day. Vaginal smears are collected with a sterile cotton swab, which is then transferred to a clear glass slide. The epithelia on the slides are examined under 10× magnification and the stage of estrous determined. Typically, collection occurs over two consecutive cycles to ensure accurate staging (e.g., 75). A number of studies have demonstrated that these techniques result in changes in both behavior and neural structures (e.g., 76, 77). On a related note, in the rat, circulating concentrations of estrogen peaks during the last 3–4 days of pregnancy [75, 78], with relatively low concentrations of E and P during the rest of the term [78, 79]. In the female rat, large cognitive and neural effects occur during late pregnancy (e.g., 78).

3   Summary

The rat is an excellent model for the study of the effect of early experiences on brain and behavioral development. There are several methods for perturbing the developing brain to examine the effects of early injury on brain development, each of which has its own advantages and disadvantages. Stress has powerful effects on brain development, including both gestational and postnatal stress and these early effects interact with later experiences to alter brain plasticity. One especially powerful early experience is that of gonadal hormones, which act both in prenatal and perinatal periods of steroid hormone activity (organizational effects) and those that occur during puberty and afterwards (activational) effects.

References

1. Kennard M (1942) Cortical reorganization of motor function. Arch Neurol 48:227–240

2. Hebb DO (1949) The organization of behavior. McGraw-Hill, New York

3. Kolb B (1995) Brain plasticity and behavior. Lawrence Erlbaum Associates, Philadelphia, PA

4. Schmanke TD, Villablanca JR (2001) A critical maturational period of reduced brain vulnerability to injury. A study of cerebral glucose metabolism in cats. Dev Brain Res 131:127–141

5. Villablanca JR, Hovda DA, Jackson GF, Infante C (1993) Neurological and behavioral effects of a unilateral frontal cortical lesion in fetal kittens: II. Visual system tests, and proposing a 'critical period' for lesion effects. Behav Brain Res 57:79–92

6. Goldman PS, Galkin TW (1978) Prenatal removal of frontal association cortex in the fetal rhesus monkey: anatomical and functional consequences in postnatal life. Brain Res 152: 451–485

7. Schneider J (1973) Early lesions of superior colliculus: factors affecting the formation of abnormal retinal projections. Brain Behav Evol 8:73–109

8. Castro AJ (1990) Plasticity in the motor system. In: Kolb B, Tees R (eds) The cerebral cortex of the rat. MIT, Cambridge, MA, pp 563–588

9. Cornwell P, Overman W, Ross C (1942) Extent of recovery from neonatal damage to the cortical visual system in rats. J Comp Physiol Psychol 92:255–270

10. Hicks S, D'Amato C (1975) Motor-sensory corticospinal system and developing locomotion placing in rats. Am J Anat 143:1–42

11. Kolb B, Nonneman AJ (1978) Sparing of function in rats with early prefrontal cortex lesions. Brain Res 151:135–148

12. Kolb B, Tomie J (1988) Recovery from early cortical damage in rats. IV. Effects of hemidecortication at 1, 5, or 10 days of age. Behav Brain Res 28:259–274

13. Kolb B, Whishaw IQ (1981) Decortication of rats in infancy or adulthood produced comparable functional losses on learned and species

typical behaviors. J Comp Physiol Psychol 95:468–483

14. Vannucci R, Vannucci SJ (1978) Cerebral carbohydrate metabolism during hypoglycemia and anoxia in newborn rats. Ann Neurol 4:73–79

15. Fishman RH, Yanai J (1983) Long-lasting effects of early barbiturates on central nervous system and behavior. Neurosci Biobehav Rev 7:19–28

16. Kolb B, Gorny G, Li Y, Samaha AN, Robinson TE (2003) Amphetamine or cocaine limits the ability of later experience to promote structural plasticity in the neocortex and nucleus accumbens. Proc Natl Acad Sci U S A 100:10523–10528

17. Kolb B, Mychasiuk R, Muhammad A, Li Y, Frost DO, Gibb R (2012) Experience and the developing prefrontal cortex. Proc Natl Acad Sci USA 109(Suppl 2):17186–17193

18. Greenough WT, Chang FF (1989) Plasticity of synapse structure and pattern in the cerebral cortex. In: Peters A, Jones EG (eds) Cerebral cortex, vol 7. Plenum, New York, pp 391–440

19. Sirevaag AM, Greenough WT (1987) A multivariate statistical summary of synaptic plasticity measures in rats exposed to complex, social and individual environments. Brain Res 441:386–392

20. Wiesel TN, Hubel DH (1963) Single-cell responses in striate cortex of kittens deprived of vision in one eye. J Neurophysiol 26:1003–1017

21. Giffin F, Mitchell DE (1978) The rate of recovery of vision after early monocular deprivation in kittens. J Physiol 274:511–537

22. Prusky GT, Silver BD, Tschetter WW, Alam NM, Douglas RM (2008) Experience-dependent plasticity from eye opening enables lasting, visual cortex-dependent enhancement of motion vision. J Neurosci 28:9817–9827

23. Schanberg SM, Field TM (1987) Sensory deprivation stress and supplemental stimulation in the rat pup and preterm human neonate. Child Dev 58:1431–1447

24. Kolb B, Gibb R (2010) Tactile stimulation facilitates functional recovery and dendritic change after neonatal medial frontal or posterior parietal lesions in rats. Behav Brain Res 214:115–120

25. Bock J, Gruss M, Becker S, Braun K (2005) Experience-induced changes of dendritic spine densities in the prefrontal and sensory cortex: correlation wit developmental time windows. Cereb Cortex 15:802–808

26. Higley JD, Hasert MF, Suomi SJ, Linnoila M (1991) Nonhuman primate modle of alcohol abuse: effects of early experience, personality, and stress on alcohol consumption. Proc Natl Acad Sci U S A 88:7261–7265

27. Huot RL, Thrivikraman KV, Meaney MJ, Plotsky PM (2001) Development of adult ethanol preference and anxiety as a consequence of neonatal maternal separation in Long Evans rats and reversal with antidepressant treatment. Psychopharmacology 158:366–373

28. Muhammad A, Kolb B (2011) Prenatal tactile stimulation attenuates drug-induced behavioral sensitization, modifies behavior, and alters brain architecture. Brain Res 1400:53–65

29. Goy RW (1966) Role of androgens in the establishment and regulation of behavioral sex differences in mammals. J Anim Sci 25(Suppl):21–35

30. Goy RW, Phoenix CH, Meidinger R (1967) Postnatal development of sensitivity to estrogen and androgen in male, female and psuedohermaphroditic guinea pigs. Anat Rec 157:87–96

31. Kolb B, Stewart J (1991) Sex-related differences in dendritic branching of cells in the prefrontal cortex of rats. J Neuroendocrinol 3:95–99

32. Goldstein JM, Seidman LJ, Horton NJ, Makris N, Kennedy DN, Caviness VS Jr, Faraone SV, Tsuang MT (2001) Normal sexual dimorphism of the adult human brain assessed by in vivo magnetic resonance imaging. Cereb Cortex 11:490–497

33. Saucier DM, Yager JY, Armstrong EA (2010) Housing environment and sex affect behavioral recovery from ischemic brain damage. Behav Brain Res 214:48–54

34. Kolb B, Stewart J (1995) Changes in neonatal gonadal hormonal environment prevent behavioral sparing and alter cortical morphogenesis after early frontal cortex lesions in male and female rats. Behav Neurosci 109:285–294

35. Kolb B, Cote S, Ribeiro-da-Silva A, Cuello AC (1997) NGF stimulates recovery of function and dendritic growth after unilateral motor cortex lesions in rats. Neuroscience 76:1139–1151

36. Labat-gest V, Tomasi S. Photothrombotic ischemia: a minimally invasive and reproducible photochemical cortical lesion model for mouse stroke studies. J Vis Exp 2013. doi:10.3791/50370

37. Kolb B, Cioe J (2001) Cryoanethesia on postnatal day 1, but not day 10, affects adult behavior and cortical morphology in rats. Dev Brain Res 130:9–14

38. Frost DO, Cerceo S, Carroll C, Kolb B (2009) Early exposure to haloperidol or olanzapine induces long-term alterations of dendritic form. Synapse 64:191–199

39. De Villers-Sidani E, Merzenich MM (2011) Lifelong plasticity in the rat auditory cortex: basic mechanisms and role of sensory experience. Prog Brain Res 191:119–131

40. Comeau W, McDonald R, Kolb B (2010) Learning-induced structural changes in the prefrontal cortex. Behav Brain Res 214:91–101

41. Kolb B, Gibb R, Gorny G (2003) Experience-dependent changes in dendritic arbor and spine density in neocortex vary with age and sex. Neurobiol Learn Mem 79:1–10

42. Kolb B, Elliott W (1987) Recovery from early cortical damage in rats. II. Effects of experience on anatomy and behavior following frontal lesions at 1 or 5 days of age. Behav Brain Res 26:47–56

43. Denenberg VH, Karas GG (1959) Interactive effects of infantile and adult experiences upon weight gain and mortality in the rat. J Comp Physiol Psychol 54:685–689

44. Levine S, Lewis GW (1959) The relative importance of experimenter contact in an effect produced by extra-stimulation in infancy. J Comp Physiol Psychol 52:368–369

45. Gibb R, Kolb B (2005) Neonatal handling alters brain organization but does not influence recovery from perinatal cortical injury. Behav Neurosci 19:1375–1383

46. Lehmann J, Pryce CR, Jongen-Rêlo AL, Stöhr T, Pothuizen HH, Feldon J (2002) Comparison of maternal separation and early handling in terms of their neurobehavioral effects in aged rats. Neurobiol Aging 23:457–466

47. Meaney MJ, Aitken DH, van Berkel C, Bhatnagar S, Sapolsky RM (1988) Effect of neonatal handling on age-related impairments associated with the hippocampus. Science 239:766–768

48. McCarty R, Horbaly WG, Brown MS, Baucom K (1981) Effects of handling during infancy on the sympathetic-adrenal medullary system of rats. Dev Psychobiol 14:533–539

49. Meaney MJ, Aitken DH, Viau V, Sharma S, Sarrieau A (1989) Neonatal handling alters adrenocortical negative feedback sensitivity and hippocampal type II glucocorticoid receptor binding in the rat. Neuroendocrinology 50:597–604

50. Chou IC, Trakht T, Signori C, Smith J, Felt BT, Vazquez DM, Barks JD (2001) Behavioral–environmental intervention improves learning after cerebral hypoxia-ischemia in rats. Stroke 32:2192–2197

51. Wong TP, Howland JG, Robillard JM, Ge Y, Yu W, Titterness AK, Brebner K, Liu L, Weinberg J, Christie BR, Phillips AG, Wang YT (2007) Hippocampal long-term depression mediates acute stress-induced spatial memory retrieval impairment. Proc Natl Acad Sci U S A 104:11471–11476

52. Metz GA, Jadavji NM, Smith LK (2005) Modulation of motor function by stress: a novel concept of the effects of stress and corticosterone on behavior. Eur J Neurosci 22:1190–1200

53. Plotsky PM, Meaney M (1993) Early, postnatal experience alters hypothalamic corticotropin-releasing factor (CRF) mRNA, median eminence CRF content and stress-induced release in adult rats. Mol Brain Res 18:195–200

54. Muhammad A, Kolb B (2011) Maternal separation during development alters dendritic morphology in the nucleus accumbens and prefrontal cortex. Neuroscience 216:103–109

55. McEwen BS (2001) Estrogens effects on the brain: Multiple sites and molecular mechanisms. J Appl Physiol 91:2785–2801

56. Leranth C, MacLusky NJ, Hajszan T (2008) Sex differences in neuroplasticity. In: Becker JB, Berkley KJ, Geary N, Hampson E, Herman JP, Young EA (eds) Sex differences in the brain From genes to behavior. Oxford, New York, pp 201–226

57. Breedlove SM (1997) Neonatal androgen and estrogen treatments masculinize the size of motor neurons in the rat spinal nucleus of the bulbocavernosus. Cell Mol Neurobiol 17:687–697

58. Jost A, Vigier B, Prepin J, Perchellet JP (1973) Studies on sex differentiation in mammals. Recent Prog Horm Res 29:1–41

59. Meaney MJ (1989) The sexual differentiation of social play. Psychiatr Dev 7:247–261

60. Zuloaga DG, Puts DA, Jordon CL, Breedlove SM (2008) The role of androgen receptors in the masculinization of brain and behavior: what we've learned from the testicular feminization mutation. Horm Behav 53:613–626

61. Arnold AP, Gorski RA (1984) Gonadal steroid induction of structural sex differences in the central nervous system. Annu Rev Neurosci 7:413–442

62. Gorski R (1971) Gonadal hormones and the perinatal development of neuroendocrine function. Oxford, New York

63. Swaab DF (2004) Sexual differentiation of the human brain: Relevance for gender identity, transsexualism and sexual orientation. Gynecol Endocrinol 19:301–312

64. Halbreich U, Lumley LA, Palter S, Manning C, Gengo F, Joe S (1995) Possible acceleration of age effects on cognition following menopause. J Psychiatr Res 29:153–163

65. Hausmann M, Slabbekoorn D, Van Goozen SH, Cohen-Kettenis PT, Güntürkün O (2000) Sex

hormones affect spatial abilities during the menstrual cycle. Behav Neurosci 114:1245–1250

66. Woolley CS, McEwen BS (1994) Estradiol mediates fluctuation in hippocampal synapse density during the estrous cycle in the adult rat. J Neurosci 12:2549–2554

67. Galea LAM (2007) Gonadal hormone modulation of neurogenesis in the dentate gyrus of adult male and female rodents. Brain Res Rev 57:332–341

68. Burger DK, Saucier JM, Iwaniuk AN, Saucier DM (2013) Seasonal and sex differences in the hippocampus of a wild rodent. Behav Brain Res 236:131–138

69. Burger DK, Gulbrandsen T, Saucier DM, Iwaniuk AN (2014) The effects of season and sex on detate gyrus size and neurogenesis in a wild rodent, Richardson's ground squirrel (Urocitellus rishadsonii). Neuroscience 9:240–251

70. Harper JM, Austad SN (2000) Fecal glucocorticoids: a non-invasive method of measure adrenal activity in wild and captive rodents. Physiol Biochem Zool 73:12–22

71. Kamel F, Wright WW, Mock EJ, Frankel AI (1977) The influence of mating and related stimuli on plasma levels of luteinizing hormone, follicle stimulating hormone, prolactin, and testosterone in the male rat. Endocrinology 101:421–429

72. Westwood FR (2008) The female rat reproductive cycle: a practical histological guide. Toxicol Pathol 36:375–384

73. Goldman JM, Murr AS, Cooper RL (2007) The rodent estrous cycle: characterization of vaginal cytology and its utility in toxicological studies. Birth Defects Res 80:84–97

74. Marcondes FK, Bianchi FJ, Tanno AP (2002) Determination of the estrous cycle phases of rats: some helpful considerations. Braz J Biol 62:609–614

75. Pawluski J, Brummelte S, Barha CK, Croizer TM, Galea LAM (2009) Effects of steroid hormones on neurogenesis in the hippocampus of the adult female rodent during the estrous cycle, pregnancy, lactation and aging. Front Neuroendocrinol 30:343–357

76. Becker JB, Robinson TE, Lorenz KA (1982) Sex difference and estrous cycle variations in amphetamine-elicited rotational behaviour. Eur J Pharmacol 80:65–72

77. Woolley CS, McEwen BS (1992) Roles of estradiol and progesterone n regulation of hippocampal dendritic spine density during the estrous cycle in the rat. J Comp Neurol 101:293–306

78. Galea LAM, Ormerod BK, Sampath S, Kostaras X, Wilkie DM, Phelps MT (2000) Spatial working memory and hippocampal size across pregnancy in rats. Horm Behav 37:86–95

79. Rosenblatt SJ, Mayer AD, Giordano AL (1988) Hormonal basis during pregnancy for the onset of maternal behaviour in the rat. Psychoneuroendocrinology 13:29–46

Jerome Y. Yager (ed.), *Animal Models of Neurodevelopmental Disorders*, Neuromethods, vol. 104,
DOI 10.1007/978-1-4939-2709-8, © Springer Science+Business Media New York 2015